Charles Barney Cory

The birds of the West Indies

Including all speciesknown to occur in the Bahama Islands

Charles Barney Cory

The birds of the West Indies
Including all speciesknown to occur in the Bahama Islands

ISBN/EAN: 9783337221867

Printed in Europe, USA, Canada, Australia, Japan

Cover: Foto ©berggeist007 / pixelio.de

More available books at **www.hansebooks.com**

THE

BIRDS OF THE WEST INDIES.

INCLUDING

ALL SPECIES KNOWN TO OCCUR IN THE BAHAMA ISLANDS, THE GREATER
ANTILLES, THE CAYMANS, AND THE LESSER ANTILLES, EXCEPTING
THE ISLANDS OF TOBAGO AND TRINIDAD.

BY

CHARLES B. CORY,

*Curator of Birds in the Boston Society of Natural History, Fellow of the
Linnæan and Zoölogical Societies of London, Member of the American Orni-
thologists' Union, of the British Ornithologists' Union, of the Société
Zoölogique de France, etc., Honorary Member of the California
Academy of Sciences, Cor. Member of the New York Academy
of Sciences, of the Chicago Academy of Sciences, etc., etc.*

AUTHOR OF

"THE BEAUTIFUL AND CURIOUS BIRDS OF THE WORLD," "THE BIRDS OF
THE BAHAMA ISLANDS," "THE BIRDS OF HAITI AND SAN DOMINGO,"
"A NATURALIST IN THE MAGDALEN ISLANDS," "A LIST
OF THE BIRDS OF THE WEST INDIES," ETC., ETC.

Illustrated.

ESTES & LAURIAT,
BOSTON, U. S. A.
1889.

Alfred Mudge & Son, Printers,
24 Franklin Street,
Boston.

TO MY MOTHER.

WHOSE INTEREST IN THE STUDY OF NATURAL
HISTORY TENDED TO ENCOURAGE AND STIMULATE THE
INVESTIGATIONS OF HER SON,

This book is affectionately dedicated.

INTRODUCTION.

In preparing the present work, the writer has examined a large series of birds from nearly all of the islands of the West Indies, the combined collections representing many thousands of specimens. The writer has personally made five trips to different parts of the West Indies, and, aside from such collections as were made under his personal supervision, a number of collectors were sent to the more important islands for the purpose of obtaining as complete a series as possible.

Several of the collectors were absent from six to eighteen months, and it is fair to assume that their collections contained nearly all of the resident species of the islands which they visited. The collections made by Mr. W. B. Richardson in Grand Cayman and some of the Lesser Antilles proved especially interesting, being very rich in novelties, his combined collections containing no less than seventeen species new to science.

Most of the matter contained in the present work appeared in the " Auk " for 1886, 1887, and 1888; but since that time a large number of species have been added to the West Indian avi-fauna, which were either new to science or had not previously been recorded from that locality; descriptions of these are given in the Appendix, unless included in their proper order in the body of the work. A number of alterations and corrections have been made in the original plates, and several new illustrations have been added.

The following species have been recorded since the " Birds of the West Indies " was first published in the " Auk " : —

Mimocichla ravida Cory. Grand Cayman.
Margarops montanus rufus Cory. Dominico.
Margarops montanus albiventris (Lawr.). Grenada.
Rhamphocinclus sanctæ-luciæ Cory. St. Lucia.
Polioptila cærulea cæsiogaster Ridgw. Bahamas.
Thryothorus guadeloupensis Cory. Grand Terre, Guadeloupe.
Dendroica vittellina Cory. Grand Cayman.
Dendroica aurocapilla Ridgw. Grand Cayman.
Geothlypis coryi Ridgw. Eleuthera I., Bahamas.
Geothlypis tanneri Ridgw. Abaco I., Bahamas.
Certhiola sharpei Cory. Grand Cayman.

Vireo alleni Cory.   Grand Cayman.
Vireo crassirostris flavescens Ridgw.   Bahamas.
Vireo caymanensis Cory.   Grand Cayman.
Calliste cuccullata (Swains.).   Grenada.
Spindalis salvini Cory.   Grand Cayman.
Spindalis zena townsendi Ridgw.   Abaco I., Bahamas.
Lexigilla barbadensis Cory.   Barbadoes.
Lexigilla richardsoni Cory.   St. Lucia.
Volatinia jacarina (Linn.).   Grenada.
Spermophila gutturalis (Licht.).   Grenada.
Icterus bairdi Cory.   Grand Cayman.
Quiscalus caymanensis Cory.   Grand Cayman.
Elainea barbadensis Cory.   Barbadoes.
Elainea pagana Licht.   Grenada.
Myiarchus denigratus Cory.   Grand Cayman.
Myiarchus berlepschii Cory.   St. Kitts.
Blacicus martinicensis Cory.   Martinique.
Blacicus flaviventris Lawr.   Grenada.
Milvulus tyrannus (Linn.).   Grenada.
Chætura cinereiventris Sclater.   Grenada.
Chætura brachyura (Jardine).   St. Vincent and Grenada.
Centurus caymanensis Cory.   Grand Cayman.
Centurus nyeanus Ridgw.   Watlings I., Bahamas.
Centurus blakei Ridgw.   Abaco I., Bahamas.
Colaptes gundlachi Cory.   Grand Cayman.
Coccygus maynardi Ridgw.   Bahama Islands.
Chrysotis caymanensis Cory.   Grand Cayman.
Vanellus vanellus (Linn.).   Barbadoes.
Rallus coryi Maynard.   Andros I., Bahamas.
Ardea bahamensis Brewster.   Bahamas.
Hydrochelidon leucoptera (Temm.).   West Indies.
Stercorarius parasiticus (Linn.).   Barbadoes.

No descriptions are given of well-known North American birds, and the references to such are mainly restricted to the citation of works and papers on West Indian ornithology.

In concluding, I wish to express my thanks to my ornithological brethren for loans of specimens and much valuable information and assistance. In this connection, I am especially indebted to my friends, Dr. Jean Gundlach, Mr. J. A. Allen, Mr. Robert Ridgway, and Mr. Geo. N. Lawrence.

CHARLES B. CORY.

A LIST

OF THE

PRINCIPAL WORKS AND PAPERS

RELATING TO

WEST INDIAN ORNITHOLOGY,

REFERRED TO IN THIS WORK.

The Natural History of Carolina, Florida, and the Bahama Islands,
by Mark Catesby.   *London*, 1754.

Letter on Various Zoölogical Subjects, relating to the Island of
Haiti, by J. Hearne. P. Z. S., 1834, pp. 25, 110.
London, 1834.

Notice of a Collection of Bird Skins, formed by J. Hearne, in Haiti.
P. Z. S., 1835, p. 105.   *London*, 1835.

Notes contained in a letter, relating to a bird called, in Cuba, the
Musician, by C. Clarke. P. Z. S., 1840, p. 153.
London, 1840.

Historie Physique, Politique, et Naturelle de L'Ile de Cuba, par
M. Ramon De La Sagra. Ornithologie, par Alcide D'Orbigny.
Paris, 1840.

Observations of the Nests of the Birds of Jamaica, by R. Hill.
P. Z. S., 1841, p. 69.   *London*, 1841.

Donation of Birds' Skins from Jamaica, by R. Hill. P. Z. S.,
1844, p. 1.   *London*, 1844.

The Birds of Jamaica, by P. H. Gosse.   *London*, 1847.

A few Remarks on the Geographical Distribution of Birds in the
West Indies, by Wm. Denny. P. Z. S., 1847, p. 36.
London, 1847.

Description of a new rapacious Bird, in the Museum of the Acad-
emy of Natural Sciences, of Philadelphia, by John Cassin.
Pr. Acad. Nat. Sci., Phila., III, p. 199 (1847) (Cuba).
Philadelphia, 1847.

Aves de la Isla de Cuba, par Juan Lembeye.   *Habana*, 1850.

Dr. J. Gundlachs. Beiträge zur Ornithologie Cubas.   J. f. O., 1855,
    p. 465; *ib.* 1856, pp. 1, 97.   (Cassel) *Leipzig*, 1855–56–57.
Ueber die von Dr. Gundlach eingesendeten Eier und Nester cuban-
    ischer Vögel Gegeben von Dr. F. A. Thienemann.   J. f. O.,
    1857, p. 145.   (Cassel) *Leipzig*, 1857.
Liste des Oiseaux rapportes et observes dans la Republique Domin-
    icaine.   Ancienne partie Espagnole de L'Ile St. Domingue ou
    D'Haiti, par M. C. Sallé, pendant son voyage de 1849 à 1851.
    (Communicated by P. L. Sclater.)   P. Z. S., 1857, p. 230.
                                                            London, 1857.
Notes on an unnamed Parrot, from the Island of San Domingo,
    now living in the Societies Gardens, and on some other Species
    of the same Family, by P. L. Sclater.   P. Z. S., 1857, p. 224.
                                                            London, 1857.
Description of five new Species of Birds and other Ornithological
    Notes of Cuban Species, by Dr. John Gundlach.   Journ. Bost.
    Soc. Nat. Hist., VI, p. 313 (1857).   *Boston*, 1857.
Description of a new Species of Bird of the Genus Sylvicola,
    Swainson, by J. Gundlach.   Ann. N. Y. Lyc. Nat. Hist., VI,
    p. 160 (1858) (Cuba).   *New York*, 1858.
Notes on some Cuban Birds, with Descriptions of three new Species,
    by J. Gundlach.   Ann. N. Y. Lyc. Nat. Hist., VI, p. 267 (1858),
    and observations on the above by Geo. N. Lawrence, p. 275.
                                                            New York, 1858.
Description of new Species of Birds of the Genera Chordeiles,
    Swainson, and Polioptila, Sclater, by Geo. N. Lawrence.   Ann.
    N. Y. Lyc. Nat. Hist., VI, p. 165 (1858) (Cuba).
                                                            New York, 1858.
Observations on the Birds of St. Croix, made between Feb. 20 and
    Aug. 6, by Alfred Newton, and between March 4 and Sept. 28,
    1858, by Edward Newton.   Ibis, 1859, p. 150.   *London*, 1859.
A List of the Birds seen at the Bahamas, from Jan. 20 to May 14,
    1859, with Descriptions of new or little-known Species, by Dr.
    H. Bryant.   Pr. Bost. Soc. Nat. Hist., VII, p. 102, 1859.
                                                            Boston, 1859.
Ornithologisches aus Briefen von Cuba verfasst, von Dr. J. Gund-
    lach.   J. f. O., 1859, pp. 294, 317.   *Leipzig*, 1859.
List of the Birds of Cuba, compiled from two lists furnished by Dr.
    John Gundlach, by Dr. T. M. Brewer.   Pr. Bost. Soc. Nat. Hist.,
    VII, p. 306 (1860).   *Boston*, 1860.

Description of a new Tyrant-bird of the Genus Elainea, from the Island of St. Thomas, West Indies, by P. L. Sclater. P. Z. S., 1860, p. 313. *London, 1860.*

List of a Collection of Birds made by the late Mr. W. Osborn, in Jamaica, with notes by P. L. Sclater. P. Z. S., 1861, p. 69. *London, 1861.*

Zur Ornithologie Cubas, nach Geo. N. Lawrence und J. Gundlach, mitgetheilt von R. Albrecht. J. f. O., 1861, p. 198. *Leipzig, 1861.*

Catalogue of Birds from the Island of St. Thomas, West Indies, collected and presented to the Academy of Natural Sciences, by Mr. Robert Swift, with notes by John Cassin. Pr. Acad. Nat. Sci., Phila., 1860, p. 374. *Phila., 1861.*

Tabellarische Uebersicht aller bisher auf Cuba beobachteten Vogel, von Dr. J. Gundlach. J. f. O., 1861, p. 321. *Leipzig, 1861.*

Zusatze und Berichtigungen zu den Beitrragen zur Ornithologie Cubas, von Dr. J. Gundlach. J. f. O., 1861, p. 401; *ib.* 1862, pp. 81, 177. *Leipzig, 1861–62.*

Zur Ornithologie von Jamaica, nach Osborn, Sclater, und Gosse, zusammengestellt von R. Albrecht. J. f. O., 1862, p. 192. *Leipzig, 1862.*

Notes on some Cuban Birds, with Descriptions of several new Species, by Geo. N. Lawrence. Ann. N. Y. Lyc. Nat. Hist., VII, p. 247 (1862). *New York, 1862.*

Descriptions of new Species of Birds of the Families Vireonidæ and Rallidæ, by Geo. N. Lawrence (Sombrero). Pr. Acad. Nat. Sci., Phila., 1863, p. 106. *Phila., 1863.*

Descriptions of two Birds from the Bahama Islands, hitherto undescribed, by Dr. H. Bryant. Pr. Bost. Soc. Nat. Hist., IX, p. 279 (1864). *Boston, 1864.*

Five Months in the West Indies, by E. C. Taylor (including notes on the avi fauna of Porto Rico, Dominica, and Martinique). Ibis, 1864, p. 157. *London, 1864.*

Notes on the Birds of Jamaica, by W. T. March, with remarks by S. F. Baird. Pr. Acad. Nat. Sci., Phila., 1863, pp. 150, 283; *ib.* 1864, p. 62. *Phila., 1863–64.*

Remarks on a rare Parrot from Dominica, by P. L. Sclater. P. Z. S., 1865, p. 437. *London, 1865.*

Remarks on the Genus Galeoscoptes, Cabanis, with the Characters of two new Genera, and a Description of Turdus plumbeus

Linn, by Dr. H. Bryant. Pr. Bost. Soc. Nat. Hist., IX, p. 369
  (1865) (Bahamas).   *Boston,* 1865.
Revista y Catalogo de las Aves Cubana's, por J. Gundlach. Repert
  Fisico, Nat. Cuba I, pp. 165, 221, 281, 347, 386 (1865-66).
                                       Habana, 1865-1866.
On an undescribed Species of Petrel from the Blue Mountains of
  Jamaica, by Alexander Carte. P. Z. S., 1866. p. 93.
                                       London, 1866.
List of Birds from Porto Rico, presented to the Smithsonian Insti-
  tution, by Messrs. Robert Swift and George Latimer, with
  Descriptions of new Species or Varieties by Dr. H. Bryant.
  Pr. Bost. Soc. Nat. Hist., X, p. 248 (1866).   *Boston,* 1866.
Additions to a List of Birds seen at the Bahamas, by Dr. H. Bryant.
  Pr. Bost. Soc. Nat. Hist., XI. p. 63 (1866).   *Boston,* 1866.
A List of the Birds of San Domingo, with Descriptions of some new
  Species by Dr. H. Bryant. Pr. Bost. Soc. Nat. Hist., XI,
  p. 89 (1866).   *Boston,* 1866.
Briefliches von Cuba, von Dr. Gundlach. J. f. O., 1866, p. 352.
                                       Leipzig, 1866.
Descriptions of new Species of Birds of the Families Tanagridæ,
  Cuculidæ, and Trochilidæ, by Geo. N. Lawrence. Ann. N. Y.
  Lyc. Nat. Hist., VIII, p. 41 (1867) (W. I.). *New York,* 1867.
Catalogue of Birds collected at the Island of Sombrero, West Indies,
  with observations by A. A. Julien, by Geo. N. Lawrence. Ann.
  N. Y. Lyc. Nat. Hist., VIII, p. 92 (1867). *New York,* 1867.
Note on Geotrygon sylvatica, by Richard Hill, Pr. Acad. Nat. Sci.,
  Phila., 1867, p. 130 (Jamaica).   *Phila.,* 1867.
Description of seven new Species of American Birds from various
  Localities, by Geo. N. Lawrence, Pr. Acad. Nat. Sci., Phila.
  1868, p. 360 (Barbadoes).   *Phila.,* 1868.
Foglarnepa on Portorico, efter Hr Hjalmarsons insambingar frams-
  tällda af Carl J. Sundevall. Ofv. Af. K. Vet. Akad. Forh,
  1869, p. 593.   *Stockholm,* 1870.
St. Barthelemy efter de af Dr. A. von Goes, penisanda samlingarna
  bestamde af Carl J. Sundevall. Ofv. Af. K. Vet. Akad. Forh,
  1869, p. 579.   *Stockholm,* 1870.
On the Birds of the Island of Santa Lucia, West Indies, by P. L.
  Sclater. P. Z. S., 1871, p. 263.   *London,* 1871.
Observations of the Birds of Santa Lucia, by J. E. Semper, with
  notes by P. L. Sclater. P. Z. S., 1872, p. 647. *London,* 1872.

The Humming-birds of the West Indies, by D. G. Elliot. Ibis,
1872, p. 345.   *London, 1872.*
On a small Collection of Birds from Barbadoes, West Indies, by
P. L. Sclater. P. Z. S., 1874, p. 174.   *London, 1874.*
Beitrage zur Ornithologie der Insel Portorico, von Dr. Juan Gund-
lach. J. f. O., 1874, p. 304.   *Leipzig, 1874.*
On a new Genus and Species of Bird, by R. B. Sharpe (Jamaica).
P. Z. S., 1874, p. 427.   *London, 1874.*
Descriptions of five new Species of American Birds, by George N.
Lawrence (Porto Rico). Ibis, 1875, p. 383.   *London, 1875.*
Notes on several West Indian Birds, by J. Cabanis. J. f. O., 1875,
p. 222.   *Leipzig, 1875.*
Neue Beitrage zur Ornithologie Cuba's, von Dr. Juan Gundlach.
J. f. O, 1875, pp. 293, 353.   *Leipzig, 1875.*
Contribucion a la Ornitologia Cubana, por el Dr. Juan Gundlach.
                                                Habana, 1876.
Description of a new Species of Bird of the Genus Pitangus, by
George N. Lawrence (San Domingo). Ann. N. Y. Lyc. Nat.
Hist., XI, p. 288 (1876).   *New York, 1876.*
On some additional Species of Birds from Santa Lucia, West Indies,
by P. L. Sclater. P. Z. S., 1876, p. 13.   *London, 1876.*
List of Birds, chiefly Visitors from North America, seen and killed
in the Bahamas in July, August, October, November, and De-
cember, 1876, by N. B. Moore, Pr. Bost. Soc. Nat. Hist., XIX,
p. 241 (1877).   *Boston, 1877.*
Notes of a few Birds observed at New Providence, Bahama Islands,
not included in Dr. Bryant's List of 1859, by L. J. K. Brace.
Pr. Bost. Soc. Nat. Hist., XIX, p. 240 (1877). *Boston, 1877.*
Catalogue of a Collection of Birds obtained in Guadeloupe for the
Smithsonian Institution by Mr. Fred. A. Ober, by Geo. N.
Lawrence. Pr. U. S. Nat. Mus., I, p. 449 (1878).
                                                Washington, 1878.
Descriptions of new Species of Birds from the Island of Dominica,
by Geo. N. Lawrence. Ann. N. Y. Acad. Sci., I, p. 46 (1878).
                                                New York, 1878.
Catalogue of the Birds of Dominica, from Collections made for the
Smithsonian Institution by Frederick A. Ober, together with
his notes and observations, by Geo. N. Lawrence. Pr. U. S.
Nat. Mus., I, p. 48 (1878).   *Washington, 1878.*
Catalogue of the Birds of Grenada, from a Collection made by Mr.

Fred. A. Ober for the Smithsonian Institution, including others seen by him but not obtained, by Geo. N. Lawrence. Pr. U. S. Nat. Mus., I, p. 265 (1878).   *Washington, 1878.*

Catalogue of the Birds collected in Martinique by Mr. F. A. Ober for the Smithsonian Institution, by Geo. N. Lawrence. Pr. U. S. Nat. Mus., I, p. 349.   *Washington, 1878.*

Descriptions of seven new Species of Birds from the Island of St. Vincent, West Indies, by Geo. N. Lawrence. Ann. N. Y. Acad. Sci., I. p. 146 (1878).   *New York, 1878.*

Catalogue of the Birds of St. Vincent, from Collections made by Mr. Fred. A. Ober, under the directions of the Smithsonian Institution, with his notes thereon, by Geo. N. Lawrence. Pr. U. S. Nat. Mus., I, p. 185 (1878).   *Washington, 1878.*

Catalogue of the Birds of Antigua and Barbuda, from Collections made for the Smithsonian Institution by F. A. Ober, with his observations, by Geo. N. Lawrence. Pr. U. S. Nat. Mus., I, p. 232 (1878).   *Washington, 1878.*

Apuntes para la Fauna Puerto Riquena, por Don Juan Gundlach. Anal. Soc. Esp. Hist. Nat, VII, p. 135 (1878). *Habana, 1878.*

Brietliches uber eine neue Dysporus-Art auf Cuba, von Dr. Jean Gundlach. J. f. O , 1878, p. 298.   *Leipzig, 1878.*

Neue Beitrage zur Ornithologie der Insel Portorico, mitgetheilt von Dr. Jean Gundlach. J. f. O., 1878, p. 157.   *Leipzig, 1878.*

Observations on some Birds seen near Nassau, New Providence, in the Bahama Islands, by N. B. Moore, Pr. Bost. Soc. Nat. Hist., XIX, p. 243 (1877).   *Boston, 1878.*

Descriptions of new Species of Birds of the Families Trochilidæ and Tetraonidæ, by Geo. N. Lawrence (New Providence). Ann. N. Y. Acad. Sci., I. p. 50 (1879).   *New York, 1879.*

Descriptions of supposed new Species of Birds from the Islands of Grenada and Dominica, West Indies, by Geo. N. Lawrence. Ann. N. Y. Acad. Sci., I, p. 160 (1879).   *New York, 1879.*

Exhibition of and Remarks upon a small Collection of Birds from the Island of Montserrat, by P. L. Sclater. P. Z. S., 1879, p. 764.   *London, 1879.*

Notes upon some West Indian Birds, by P. L. Sclater. Ibis, 1880, p. 71.   *London, 1880.*

Description of a new Species of Icterus from the West Indies, by Geo. N. Lawrence. Pr. U. S. Nat. Mus., III, p. 351 (1880). *Washington, 1880.*

Description of a new Species of Bird of the Family Turdidæ, from the Island of Dominica, West Indies, by Geo. N. Lawrence. Pr. U. S. Nat. Mus., III, p. 16 (1880).
Washington, 1880.

Description of a new Species of Parrot, of the Genus Chrysotis, from the Island of Dominica, by Geo. N. Lawrence. Pr. U. S. Nat. Mus., III, p. 254 (1880). *Washington,* 1880.

List of the Birds of the Island of Santa Lucia, West Indies, by J. A. Allen. Bull. Nutt. Orn. Club, V, p. 163 (1880).
Cambridge, 1880.

Field Notes on the Birds of St. Vincent, West Indies, by C. E. Lister. Ibis, 1880, p. 381. *London,* 1880.

The Birds of the Bahama Islands, by C. B. Cory. *Boston,* 1880.

Descriptions of four new Species of Haitien Birds, by C. B. Cory. Bull. Nutt. Orn. Club, VI, p. 129 (1881). *Cambridge,* 1881.

A list of the Birds of Haiti, taken in different parts of the island, between Jan. 1 and March 12, 1881, by C. B. Cory. Bull. Nutt. Orn. Club, VI, p. 151 (1881). *Cambridge,* 1881.

Description of a new Owl from Porto Rico, by Robert Ridgway. Pr. U. S. Nat. Mus., IV, p. 366 (1881). *Washington,* 1881.

Remarks on a Skin of a Chrysotis from Santa Lucia, by P. L. Sclater. P. Z. S., 1881, p. 627. *London,* 1881.

Description of a new Sub-Species of Loxigilla from the Island of St. Christopher, by Geo. N. Lawrence. Pr. U. S. Nat. Mus., IV, p. 204 (1881). *Washington,* 1881.

Handbook of Jamaica, by A. and E. Newton. *Kingston,* 1881.

Nachtrage zur Ornithologie Cuba's, von Dr. Jean Gundlach. J. f. O., 1881, p. 400. *Leipzig,* 1881.

Nachtrage zur Ornithologie Portoricos, von D. Jean Gundlach. J. f. O., 1881, p. 401. *Leipzig,* 1881.

Briefliches zur Fortpflanzungsgeschichte des Chlorospingus speculiferus. von Dr. Jean Gundlach. J. f. O.. 1882, p. 161 (Cuba).
Leipzig, 1882.

Synopsis of the West Indian Myiadestes, by Leonhard Stejneger. Pr. U. S. Nat. Mus. V. p. 15 (1882). *Washington,* 1882.

On the Birds of Montserrat, by T. Grisdale. Ibis, 1882, p. 485.
London, 1882.

Description of a new Warbler from the Island of Santa Lucia, by Robert Ridgway. Pr. U. S. Nat. Mus. V, p. 525 (1882).
Washington, 1882.

Descriptions of three new Species of Birds from San Domingo, by
C. B. Cory.   Journ. Boston Zoölogical Society, II, p. 451 (1883).
                                                Boston, 1883.
Notes upon some rare Species of Neotropical Birds, by Robert
Ridgway.   Ibis, 1883, p. 399.   *London*, 1883.
Characters of a new Species of Pigeon, of the Genus Engyptila,
from the Island of Grenada, by Geo. N. Lawrence.   Auk, I,
p. 180 (1884).   *Boston*, 1884.
Descriptions of several new Birds from San Domingo, by C. B.
Cory.   Auk, I, p. 1 (1884).   *Boston*, 1884.
On a Collection of San Domingo Birds, by H. B. Tristram.   Ibis,
1884, p. 167.   *London*, 1884.
A List of the Birds of the West Indies, by C. B. Cory.
                                                Boston, 1885.
On a Collection of Birds made by Messrs. J. E. Benedict and W.
Nye, of the United States Fish Commission steamer Albatross,
by Robert Ridgway (St. Thomas).   Pr. U. S. Nat. Mus., VII,
p. 172 (1884).   *Washington*, 1885.
Description of a new Species of Coot from the W. Indies, by
Robert Ridgway (Guadeloupe and St. John's).   Pr. U. S.
Nat. Mus., VII, p. 358 (1884).   *Washington*, 1885.
Descriptions of new Species of Birds of the Family Columbidæ, by
Geo. N. Lawrence (Grenada).   Auk, II, p. 357 (1885).
                                                Boston, 1885.
The Birds of Haiti and San Domingo, by C. B. Cory.
                                                Boston, 1885.
Helinaia Swainsoni in Jamaica, by C. H. Merriam.   Auk, II,
p. 377 (1885).   *New York*, 1885.
A List of the Birds of the West Indies, by C. B. Cory.   (Revised
edition.)   *Boston*, 1886.
Description of four new Species of Birds from the Bahama Islands,
by. R. Ridgway.   Auk, III, p. 334 (1886).   *New York*, 1886.
Description of a new Genus of Tyrannidæ from San Domingo, by
R. Ridgway.   Auk, III, p. 382 (1886).   *New York*, 1886.
The Birds of the West Indies, including the Bahama Islands, the
Greater and the Lesser Antilles, excepting the Islands of
Tobago and Trinidad, by C. B. Cory.   Auk, III, pp. 1, 187,
337, 454 (1886).   *New York*, 1886.
Descriptions of new Species of Birds from the West Indies, by C.
B. Cory.   Auk, III, p. 381 (1886).   *New York*, 1886.

On a Collection of Birds from several little-known Islands of the
West Indies, by C. B. Cory. Ibis, 1886, p. 471.

London, 1886.

List of a few Species of Birds new to the fauna of Guadeloupe, with
a Description of a new Species of Ceryle, by G. N. Lawrence.
Pr. U. S. Nat. Mus., VIII, p. 621 (1885). *Washington,* 1886.

A List of the Birds of Grenada, by J. G. Wells. *Grenada,* 1886·

List of the Birds of Grenada, by J. G. Wells and G. N. Lawrence.
Pr. U. S. Nat. Mus., IX, p. 609 (1886). *Washington,* 1886.

Descriptions of thirteen new Species of Birds from the Island of
Grand Cayman, West Indies, by C. B. Cory. Auk, III, p. 497
(1886). *New York,* 1886.

A List of the Birds collected in the Island of Grand Cayman, West
Indies, by W. B. Richardson, during the summer of 1886, by
C. B. Cory. Auk, III, p. 501 (1886). *New York,* 1886.

Description of a new Species of Thrush from the Island of Grenada,
West Indies, by G. N. Lawrence. Ann. N. Y. Acad. Sci., IV,
p. 23 (1887). *New York,* 1887.

A new Vireo from Grand Cayman, W. I., by C. B. Cory. Auk, IV,
p. 6, 1887. *New York,* 1887.

The Birds of the W. I., including the Bahama Islands, the Greater
and the Lesser Antilles, excepting the Islands of Tobago and
Trinidad, by C. B. Cory. Auk, IV, pp. 37, 108, 223, and 311
(1887). *New York,* 1887.

Description of a new Species of Rhamphocinclus from Santa Lucia,
by C. B. Cory. Auk, IV, p. 94, 1887. *New York,* 1887.

List of the Birds collected by W. B. Richardson in Martinique,
W. I., by C. B. Cory. Auk, IV, p. 95, 1887.

New York, 1887.

Five new Species of Birds from the Bahamas, by C. J. Maynard.
The American Exchange and Mart and Household Journal,
Vol. III, No. 6, p. 69, Feb. 5. *Boston and New York,* 1887.

Description of a supposed new form of Margarops from Dominica,
by C. B. Cory. Auk, V, p. 47 (1888). *New York,* 1888.

An apparently new Elainea from Barbadoes, by C. B. Cory. Auk,
V, p. 47 (1888). *New York,* 1888.

The Birds of the West Indies, including the Bahama Islands, the
Greater and the Lesser Antilles, excepting the Islands of
Tobago and Trinidad, by C. B. Cory. Auk, V. pp. 48, 155
(1888, January and April). *New York,* 1888.

Description of supposed new Birds from Lower California, Sonora,
   and Chihuahua, Mexico, and the Bahamas, by William Brewster. Auk, V, p. 82 (1888).   *New York*, 1888.
Description of a new Myiarchus from the West Indies (St. Kitts),
   by C. B. Cory. Auk, V, p. 266 (1888).   *New York*, 1888.
The Jocanidæ, by D. G. Elliot. Auk, p. 288 (1888).
                                             New York, 1888.
Catalogue of a Collection of Birds made by Mr. Chas. H. Towns-
   end, on Islands in the Caribbean Sea and in Honduras (Grand
   Cayman), by Robert Ridgway. Pr. U. S. Nat. Mus., p. 572
   (1887).   *Washington*, 1888.
The European Lapwing in the Island of Barbados, by H. W.
   Feilden, Zoölogist, p. 301 (1888, August).   *London*, 1888.
Richardson's Skua in the Island of Barbados, by H. W. Feilden,
   Zoölogist, p. 350 (1888, September).   *London*, 1888.

THE BIRDS

THE WEST INDIES.

. ────

FAMILY TURDIDÆ.

GENUS Turdus LINN.

Turdus LINN. Syst. Nat. I, p. 291 (1766).

Turdus mustelinus GMEL.

Turdus mustelinus GMEL. Syst. Nat. I, p. 817 (1788). — D'ORB. in La Sagra's Hist. Nat. Cuba, Ois. p. 49 (1840). — GOSSE, Bds. Jam. p. 144 (1847) (Jamaica)? — GUNDL. J. f. O. 1855, p. 469 (Cuba); *ib.* 1872, p. 405 (Cuba). — BREWER, Pr. Bost. Soc. Nat. Hist. VII, p. 307 (1860) (Cuba). — ALBRECHT, J. f. O. 1862, p. 201 (Jamaica). — COUES, Bds. Colo. Vall. p. 28 (1878). — SEEBOHM, Cat. Bds. Brit. Mus. V, p. 196 (1881). — CORY, List Bds. W. I. p. 5 (1885).

This species is occasionally found in Cuba, and has been recorded from Jamaica, but its occurrence in the latter island is questioned.

Turdus fuscescens Steph.

Turdus fuscescens Steph. Shaw's Gen. Zóol. Bds. 1817, p. 182.—Gundl.
    J. f. O. 1861, p. 324; *ib.* Repert. Fisico-Nat. Cuba, I. p 288
    (1865) (Cuba).—Seebohm, Cat. Bds. Brit. Mus. V, p. 203 (1881).
    —Cory, List Bds. W. I. p. 5 (1885).
Turdus minor Less. D'Orb. in La Sagra's Hist. Nat. Cuba, Ois p. 47
    (1840).

Common in Cuba.

Turdus swainsoni Caban.

Turdus swainsoni Cab. Tschudi's Fauna Peruana, 1844; *ib.* J. f. O.
    1857, p. 241 (Cuba).—Brewer, Pr. Bost. Soc. Nat. Hist. VII, p.
    307 (1860).—Gundl. J. f. O. 1861, p. 324; *ib.* 1872, p. 405
    (Cuba).—Coues, Bds. Colo. Vall. p. 34 (1878).—Cory, List Bds.
    W. I. p. 5 (1885).

Accidental in Cuba.

Turdus aliciæ Baird.

Turdus aliciæ Bd. Cass. & Lawr. Bds. N. Am. p. 217 (1858).—
    Gundl. Repert. Fisico-Nat. Cuba, I, p. 229 (1865).—Bd. Bwr.
    & Ridgw. Hist. N. Am. Bds. I, p. 11 (1874).—Seebohm, Cat. Bds.
    Brit. Mus. V, p. 202 (1881).—Cory, Bds. Haiti & San Domingo, p.
    17 (1885); *ib.* List Bds. W. I. p. 51 (1885).

Cuba and San Domingo; not common.

Genus Merula Leach.

Merula Leach, Cat. Brit. Mus. p. 20 (1816).

Merula jamaicensis (Gmel.).

Turdus jamaicensis Gmel. Syst. Nat. I, p. 809 (1788).—Br. Consp. I,
    p. 271 (1850).—Scl. P. Z. S. 1859, p. 327; *ib.* 1861, p. 70.—Albrecht,
    J. f. O. 1862, p. 191.—March, Pr. Acad. Nat. Sci. Phila. 1863,
    p. 292.—Gray, Handl. Bds. I, p. 257 (1869).—Scl. & Salv. Nom.
    Avium Neotr. p. 1 (1873).—Seebohm, Cat. Bds. Brit. Mus. V,
    p. 208 (1881).—A. & E. Newton, Handb. Jamaica. p. 105 (1881).
Merula jamaicensis Gosse, Bds. Jam. p. 142 (1847).—Denny, P. Z. S.
    (1847), p. 38.—Cory, List Bds. W. I. p. 5 (1885).
Turdus capucinus "Hartl." fide Br. Consp. I, p. 271 (1850).
Turdus leucophthalmus "Hill," fide Br. Consp. I, p. 271 (1850).
Turdus leireboulleti Br. Compt. Rend. XXXVIII, p. 3 (1854).

SP. CHAR. *Male.*—Chin, and a band on the lower part of the throat showing white; the rest of entire head and throat reddish brown; back brown, with a faint tinge of olive, becoming grayish on the rump; underparts grayish brown, becoming dull white on the abdomen; wings and tail dark brown.

The sexes are similar.

Length (skin), 8.50; wing, 4.50; tail, 3.50; tarsus, 1.25; bill, .70.

HABITAT. Jamaica.

Merula migratoria (LINN.).

Turdus migratorius LINN. Syst. Nat. I, p. 292 (1766).—BD. BWR. & RIDGW. Hist. N. Am. Bds. I, p. 25 (1874).—SEEBOHM, Cat. Bds. Brit. Mus. V, p. 220 (1881).

Planesticus migratorius GUNDL. J. f. O. 1872. p. 405.

Merula migratoria CORY, List Bds. W. I. p. 5 (1885).

Accidental in Cuba.

Merula aurantia (GMEL.).

Turdus aurantius GMEL. Syst. Nat. I, p. 832 (1788).—BP. Consp. I, p. 275 (1850).—SCL. P. Z. S. 1861, p. 70; *ib.* Cat. Am. Bds. p. 6 (1862).—ALBRECHT, J. f. O. 1862, p. 192.—MARCH, Pr. Acad. Nat. Sci. Phila. 1863, p. 292.

Turdus leucogenus LATH. Ind. Orn. I, p. 341 (1790).—VIEILL. Nouv. Dict. XX. p. 254 (1818).

Merula saltator "HILL. Comp. Jam. Alm. 1842."—GOSSE, Bds. Jam. p. 140 (1847).

Merula leucogenys GOSSE, Bds. Jam. p. 136 (1847).

Catharus aurantius BP. Compt. Rend. XXXVIII, p. 3 (1854).

Semimerula aurantia BAIRD, Rev. Am. Bds. p. 84 (1864).—A. & E. NEWTON, Handb. Jamaica, p. 105 (1881).

Mimocichla aurantia SCL. & SALV. Nom. Avium Neotr. p. 2 (1873).

Merula aurantia SEEBOHM. Cat. Bds. Brit. Mus. V, p. 247 (1881).—CORY. List Bds. W. I. p. 5 (1885).

SP. CHAR. *Male:*—Top of head dark brown; chin white; abdomen dull white; the rest of plumage slaty brown; wings and tail dark brown; two of the greater wing-coverts next to the inner secondaries broadly edged with white, giving a noticeable white marking to the wing.

Female:—Appears to be similar to the male, but is perhaps somewhat paler. Some specimens do not seem to differ at all in coloration.

Length (skin), 9.50; wing, 5; tail, 4; tarsus, 1.80; bill, .85.

HABITAT. Jamaica.

Merula gymnophthalma (Caban.).

Turdus gymnophthalmus Cab. Schomb. Reis. Guian. III, p. 665 (1848).—Gray. Handl. Bds. I, p. 257 (1869).—Scl. & Salv. Nom. Avium Neotr. p. 1 (1873).—Seebohm, Cat. Bds. Brit. Mus. V, p. 212 (1881).

Turdus gymnopsis "Temm." fide Br. Consp. I, p. 272 (1850).
Turdus nudigenis Lafr. Rev. Zool. 1848, p. 4.—Leot. Ois. Trinid. p. 20 (1866).
Turdus caribbæus Lawr. Ann. N. Y. Acad. Sci. I, p. 160 (1878); *ib.* Pr. U. S. Nat. Mus. I, p. 486 (1878).
Turdus gymnogenys Scl. & Salv. Ibis, 1879, p. 357.
Merula gymnophthalma Cory, List Bds. W. I. p. 5 (1885).

Sp. Char. *Male:*—Above dull olive brown; underparts pale brown; throat pale, mottled with dull brown; belly pale, showing markings of dull white on the crissum; under wing-coverts pale rufous.

The sexes are similar.

Length (skin), 8.50; wing, 4.75; tail, 4; tarsus, 1.15.

Habitat. Grenada, Trinidad, and Tobago.

Specimens taken in Grenada vary slightly in size and coloration from South American examples, but are apparently the same.

Merula nigrirostris (Lawr.).

Turdus nigrirostris Lawr. Ann. N. Y. Acad. Sci. I, p. 147 (1878).—Lister, Ibis, 1880, p. 39.—Seebohm, Cat. Bds. Brit. Mus. V, p. 218 (1881).
Merula nigrirostris Cory, List Bds. W. I. p. 5 (1885).

"*Female :*—Front, crown, and occiput dark warm brown, each feather of the crown and occiput with a shaft-stripe of dull pale rufous; upper plumage reddish olivaceous brown, deeper in color on the upper part of the back and on the wing-coverts; the latter have their ends marked with small spots of bright rufous, which possibly may be an evidence of the example not being fully mature; the tail is of a dark warm brown, the shafts black; inner webs of quills blackish brown; the outer webs reddish brown, of the same color as the tail feathers; the shafts are glossy black; under lining of wings clear cinnamon red; under plumage light brownish ash, with the middle of the abdomen and the crissum white; on the upper part of the breast a few feathers end with dark reddish brown, forming an irregular narrow band; the throat unfortunately is soiled with blood, but as well as I can judge, it has stripes colored like the breast, and the feathers edged with whitish; the thighs are dull rufous; the bill is large and strong, the upper mandible is black, the under also, but showing a brownish tinge; tarsi and toes dark brown."

The sexes are similar.

"Length (fresh), 9¼ in.; wing, 4½; tail, 3½; tarsus, 1¼; bill from front, ⅜." (LAWR. orig. descr.)

HABITAT. St. Vincent.

This species is allied to *M. fumigatus*, but is perfectly distinct. It has thus far only been taken in the island of St. Vincent.

GENUS **Mimocichla** SCL.

Mimocichla SCLATER, P. Z. S. 1859, p. 336.

Mimocichla rubripes (TEMM.).

Turdus rubripes TEMM. Pl. Col. II, p. 409 (1826).—VIG. Zool. Journ. III, p. 439 (1827).—D'ORB. in La Sagra's Hist. Nat. Cuba. Ois. p. 46 (1840).—GUNDL. Bost. Journ. Nat. Hist. VI, p. 318 (1852).
Mimus rubripes BP. Consp. I, p. 276 (1850).
Galeoscoptes rubripes CAB. Mus. Hein. I, p. 82 (1850).—BREWER, Pr. Bost. Soc. Nat. Hist. VII, p. 307 (1860).
Mimocichla rubripes SCL. Cat. Am. Bds. p. 6 (1862).—BAIRD, Rev. Am. Bds. p. 38 (1864).—GRAY, Handl. Bds. I, p. 263 (1869).—GUNDL. J. f. O. 1872, p. 406.—SCL. & SALV. Nom. Avium Neotr. p. 2 (1873).—SEEBOHM, Cat. Bds. Brit. Mus. V, p. 283 (1881).—CORY, List Bds. W. I. p. 5 (1885).

SP. CHAR. *Male :*—Upper plumage dark slaty gray; feathers on the head darker in the centre; lores and ear-coverts very dark brown; chin and lower sides of the cheeks white; throat black, the lower portion having the feathers margined with gray; breast and upper part of the belly and sides slate gray; lower part of the belly and thighs chestnut; under tail-coverts white; quills, secondaries, and wing-coverts black, edged with slate color; tail brownish black, the four outer feathers on each side tipped with white, some of the feathers showing gray at the base of the outer webs; bill brownish black.

The sexes are similar.

Length (skin), 10; wing, 4.40; tail, 4.20; tarsus, 1.45; bill, .90.

HABITAT. Cuba.

Mimocichla schistacea BAIRD.

Mimocichla schistacea BAIRD, Rev. Am. Bds. p. 37 (1864). — GRAY, Handl. Bds. I, p. 263 (1869).—GUNDL. J. f. O. 1872, p. 407.—CORY, List Bds. W. I. p. 5 (1885).

SP. CHAR.—General appearance of *M. rubripes*, but lacking the reddish
on the belly, which is replaced by white; the crissum is also white;
bill heavier than in *rubripes;* otherwise the two forms are alike.
    Length, 10.50; wing, 5; tail, 5.10; tarsus, 1.50; bill, 1.20.

HABITAT.   Eastern part of Cuba.

Dr. Gundlach (J. f. O., l. c.) considers this a good species,
and says the eggs are smaller and more finely spotted than those
of *M. rubripes.* Seebohm (Cat. Bds. Brit. Mus. V, p. 283),
gives *M. schistacea* as a synonym of *M. rubripes,* but gives no
reasons for so doing. Although it would be strange if two species
of *Mimocichla* should be found to inhabit Cuba, yet, with our
present knowledge of the two forms, *M. schistacea* must be
considered distinct.

Mimocichla plumbea (LINN.).

Turdus plumbeus LINN. Syst. Nat. I, p. 294 (1766).—VIEILL. Ois. Am.
    Sept. II, p. 2, pl. 58.
Turdus ardosiaceus VIEILL. Ency. Méth. p. 646 (1823).
Galeoscoptes plumbea CAB. Mus. Hein. I, p. 82 (1850).—SALLÉ, P. Z. S.
    1857, p 231.—SCL. P. Z. S. 1859, p. 337.
Mimocichla plumbeus BAIRD, Rev. Am. Bds. p. 36 (1864).—SCL. & SALV.
    Nom. Avium Neotr. p. 2 (1873).—CORY, Bds. Bahama I. p. 45, pl.
    11 (1880); *ib.* List Bds. W. I. p. 5 (1885).
Turdus (Mimokitta) plumbeus BRYANT, Pr. Bost. Soc. Nat. Hist. XI, p.
    68 (1866).
Mimocitta plumbea NEWTON, Ibis, 1866, p. 121.
Mimokitta plumbeus GRAY, Handl. Bds. I, p. 263 (1869).
Mimocichla bryanti SEEBOHM, Cat. Bds. Brit. Mus. V, p. 280 (1881).

SP. CHAR. *Male:*— General plumage plumbeous; chin and small
    patch at base of lower mandible white; throat black; primaries
    and secondaries dark brown, except the first two, edged with
    slaty grey; tail very dark brown, almost black; the terminal third
    of the inner webs of the first two, and tips of first four feathers
    white; crissum plumbeous; legs and eyelids vermilion red; iris red-
    dish brown.
     Female:—Similar to the male, but appears to be slightly smaller.
    Cannot be distinguished otherwise than by dissection.
     Length, 10.25; wing, 5; tail, 5; tarsus, 1; bill, 90.

HABITAT.   Bahama Islands. Common at New Providence,
Andros, and Abbacco.

Mimocichla ardesiaca (VIEILL.).

Turdus plumbeus LINN. Syst. Nat. I, p. 294 (1866).—VIEILL. Ois. Am.
    Sept. II, p. 2 (1807); *ib.* Nouv. Dict. Hist. Nat. XX, p. 242
    (1818).

Turdus ardosiaceus VIEILL. Ency Méth. p. 646 (1823).—BRYANT, Pr. Bost.
Soc. Nat. Hist. XI, p. 92 (1866) ; *ib.* X, p. 25 (1866).
Mimus plumbeus GRAY, Gen. Bds. I, p. 221 (1844).—Br. Consp. I, p. 276
(1850).
Galeoscoptes plumbeus CAB. Mus. Hein. I, p. 82 (1850).—SALLÉ, P. Z. S.
1857, p. 231.
Mimocichla ardosiacea BAIRD, Rev. Am. Bds. p. 39 (1864).—GUNDL. J. f. O.
1878, p. 165; *ib.* Anal. Soc. Esp. Hist. Nat. VII, p. 171 (1878).
Turdus ardosiaceus var. *potoricensis* BRYANT, Pr. Bost. Soc. Nat. Hist.
XI, p. 93 (1866).
Mimokitta ardosiacea Gray, Handl. Bds. I, p. 263 (1869).
Mimokitta ardosiacea var. *portoricensis* GRAY, Handl. Bds. I, p. 263 (1869).
Mimocichla ardesiata SCL. & SALV. Nom. Avium Neotr. p. 2 (1873). —
SEEBOHM, Cat. Bds. Brit. Mus. V, p. 282 (1881).—CORY, Bull. Nutt.
Orn. Club, VI, p. 151 (1881) ; *ib.* Bds. Haiti & San Domingo, p. 18
(1885) ; *ib.* List Bds. W. I. p. 5 (1885).—TRISTRAM, Ibis, 1884, p. 168.

SP. CHAR. *Male:*—General plumage plumbeous ; a patch of black extending from below and in front of the eye to the base of the upper
mandible ; throat white, streaked heavily with black ; top of head
somewhat dotted with brown ; underparts pale plumbeous, becoming
white on the abdomen and crissum ; primaries dark brown, the outer
webs edged with plumbeous gray ; same marking, but much broader,
edging the secondaries ; tail dark brown, the outer feathers broadly
tipped with white, the white becoming less and less to the fourth,
which is only narrowly touched ; but the tail-marking varies in different specimens and seasons ; bill, eyelids, and legs vermilion
orange ; iris reddish brown.

The sexes are similar.

Length, 10 ; wing, 5.20 ; tail, 4.70 ; tarsus, 1.40 ; bill, .75.

HABITAT. San Domingo and Porto Rico.

GENUS **Cichlherminia** BONAPARTE.

Cichlherminia Bp. Comptes Rendus, XXXVIII, p. 2 (1854).

Cichlherminia herminieri (LAFR.).

Turdus herminieri LAFR. Rev. Zool. 1844, p. 167.—GRAY, Gen. Bds. I, p.
219 (1844).
Cichlherminia herminieri BP. Compt. Rend. XXXVIII, p. 2 (1854).—
SHARPE, Cat. Bds. Brit. Mus. VI, p. 327 (1881).—CORY, List Bds.
W. I. p. 5 (1885).
Cichlherminia bonapartii SCL. P. Z. S. 1859, p. 335.
Cichlherminia l'herminierii GRAY, Handl. Bds. I, p. 259 (1869).
Margarops herminieri SCL. & SALV. Nom. Avium Neotr. p. 2 (1873).—
LAWR. Pr. U. S. Nat. Mus. I, p. 52 (1878).—SCL. P. Z. S. 1880, p. 72.

SP. CHAR. *Male:*—Above brown; intermediate between *C. dominicensis*
and *C. sanctæ-luciæ;* the feathers on the crown showing faint dusky

margins; ear-coverts brown, showing pale shaft-lines; throat rufous brown, palest on the upper portion, the centre of the feathers
showing dull white; rest of underparts having the feathers white,
edged with brown, giving the feathers a clean-cut, pointed appearance, the white portion somewhat resembling a broad arrow head;
quills and tail brown; upper surface of tail-feathers showing a rufous
tinge; under mandible and tarsus pale.

The sexes are similar.

Length (skin), 9.40; wing, 5.25; tail. 3.50; tarsus, 1.70; bill, 1.

HABITAT.   Guadeloupe and Martinique.

Cichlherminia sanctæ-luciæ (SCL.).

Margarops herminieri SCL. P. Z. S. 1871, p. 268.—SCL. & SALV. Nom.
Avium Neotr. p. 2 (1873).
Margarops sanctæ-luciæ SCL. Ibis, 1880, p. 73.—ALLEN, Bull. Nutt. Orn.
Club, V, p. 165 (1880).
Margarops herminieri var. *semperi* LAWR. MS. Bull. Nutt. Orn. Club,
V, p. 165 (1880).
Cichlherminia sanctæ-luciæ SHARPE, Cat. Bds. Brit. Mus. VI, p. 328
(1881).—CORY, List Bds. W. I. p. 5 (1885).

SP. CHAR. *Male*:—Above light brown, showing a faint olive tinge, the
color paler than in *C. herminieri*; throat dull white, showing
brown shaft-markings; feathers of the breast brownish white,
edged with olive brown; abdomen white, showing the brown marking on the sides; quills and tail light brown; under surface of tail
ashy brown; under tail-coverts showing reddish brown at the base;
under mandible and tarsus dull yellow.

The sexes are similar.

Length (skin), 10; wing, 5.10; tail, 3.70; tarsus, 1.55; bill, .90.

HABITAT.   Santa Lucia.

Cichlherminia dominicensis (LAWR.).

*Margarops hermin-
ieri* LAWR. Pr. U.
S. Nat. Mus. I, p.
52 (1878).
*Margarops domini-
censis* LAWR. Pr. U.
S. Nat. Mus. III, p.
16, (1880).
*Cichlherminia do-
minicensis* SHARPE,
Cat. Bds. Brit. Mus.
VI, p. 328 (1881).
—CORY, List Bds.
W. I. p. 5 (1885).

" *Male :*—The entire upper plumage is of a rich dark brown, the crown is darker and has the edges of the feathers of a lighter shade; tail and quill-feathers of a darker brown than the back; axillars and under wing-coverts white; the lores are blackish brown; the feathers back of the eyes and the ear-coverts have narrow shaft-streaks of pale rufous; the feathers of the neck and upper part of the breast are of a warm dark brown, those of the chin and middle of the throat with light rufous centres, those of the lower part of the neck and the upper part of the breast have also light rufous centres, but in addition each feather has a light terminal spot; on the lower part of the breast and on the sides the feathers have white centres, bordered strikingly with brown; the markings of the breast-feathers are squamiform in shape, those of the sides lanceolate; the abdomen is white, a few feathers on the upper part are very narrowly margined with brown; under tail-coverts brown, terminating with white; outer feathers of thighs brown, the inner whitish; 'iris ten-color'; there is a naked space around the eye; bill yellow. with the basal half of the upper mandible dusky; tarsi and toes pale yellow."

The sexes are similar.

"Length (fresh), 9 inches; wing. 5; tail, 3½; tarsus, 1¾; bill from front, 15-16, from gape 1¼." (Lawr. orig. descr.)

Habitat. Dominica.

Genus Sialia Swains.

Sialia Swainson, Zool. Journ. III, p. 173 (1827).

Sialia sialis (Linn.).

Motacilla sialis Linn. Syst. Nat. I, p. 187 (1758); *ib.* I, p. 336 (1766).
Sialia sialis Gundl. J. f. O. 1861, p. 324; *ib.* 1862, p. 177; *ib.* 1872, p. 409; *ib.* Repert. Fisico-Nat. Cuba, I. p. 230 (1865) (Cuba).—Baird, Rev. Am. Bds. p. 62 (1864).—Cory, List Bds. W. I. p. 5 (1885).

Cuba; no other West Indian record.

Genus Myiadestes Swains.

Myiadestes Swainson, Nat. Libr. Ornith. p. 132 (1838).

Myiadestes sibilans Lawr.

Myadestes sibilans Lawr. Ann. N. Y. Acad. Sci. I, p. 148 (1878); *ib.* Pr. U. S. Nat. Mus. I, p. 188 (1878).—Lister, Ibis, 1880, p. 39.—Cory. List Bds. W. I. p. 5 (1885).
Myiadectes sibilans Sharpe, Cat. Bds. Brit. Mus. VI. p. 371 (1881).
Myadestes sibilans Stejn. Pr. U. S. Nat. Mus. V, p. 17 (1882).

SP. CHAR.—Tail shorter than wing; upper surface very dark-brown, almost black; a tinge of olive brown on the lower back and rump; chin and portion of malar stripe joining base of lower mandible white, the rest the color of throat; shafts of ear-coverts showing delicate lines of white; the lower eyelid is also white; throat bright rufous, tinged with orange, separated from the malar stripe by a narrow black line; breast and upper abdomen ashy gray, some of the feathers often tipped with orange rufous; rest of underparts like the throat; wings black; a white patch at base of inner webs of first six primaries reaching and extending to the base of outer web on the seventh, eighth, and ninth; central tail-feathers black, becoming grayish at base; outer tail-feather showing a wedge-shaped white mark on inner web, nearly reaching the base, which is brownish black; outer web showing brownish black on terminal half, next feather marked like outer feather, but having much less white, third narrowly tipped with white, rest black except the two central feathers, as above described; bill black; legs pale yellow; "iris hazel." Some specimens seem to ack the white spot at tip of of third outer tail feather.

Length about 7.20; wing, 3.30; tail, 2.75: tarsus, .95.

HABITAT. St. Vincent.

Myiadestes genibarbis SWAINS.

Myiadestes genibarbis SWAINS. Nat. Libr XIII, p 134 (1838).— BAIRD, Rev. Am. Bds. p. 423 (1864).— GRAY, Handl. Bds. I, p. 366 (1869). — LAWR. Pr. U. S. Nat. Mus. I, p. 352 (1878).— CORY, List Bds. W. I. p. 5 (1885).
Myiadectes genibarbis SHARPE, Cat. Bds. Brit. Mus. VI. p. 370 (1881).
Myiadestes genibarbis STEJN. Pr. U. S. Nat. Mus. V, p. 18 (1882).

"Upper surface pure slaty-plumbeous, forehead slightly washed with olivaceous; lores black; also a stripe below the white patch on the under eyelid, assuming the color of the back on the ear-coverts, each feather of which and the above-mentioned stripe having a narrow well-defined white central streak behind, very faintly washed with brownish. From the base of lower mandible a well-defined malar stripe runs backwards, the anterior third of which is white, while the lower two-thirds have the color of the throat, from which the malar stripe is separated by a narrow, but distinct, black stripe, reaching close to the lower edge of the mandible. Throat and chin chestnut rufous, the white bases of the feathers on the latter showing somewhat through. Breast and upper sides of abdomen lighter than the back, almost clear ash-gray, becoming gradually lighter towards the abdomen; remaining underparts of the same color as the throat, only somewhat paler, and assuming a faint olivaceous shade on the upper abdomen; tibia like the back, a few feathers being tipped with rufous. Wings blackish, with pale edges on the primaries and two ash-gray bars across the secondaries, leaving between them a deep black patch; wing-coverts, except the primary coverts, broadly edged with gray like the back; innermost secondaries almost entirely so; inner web of the quills white at the base, forming a broad bar on the under surface of the wing; edge of wing grayish white. Middle tail-feathers uniform slate-gray; the following pairs black, the outermost with a wedge-shape white spot on the inner web at the end, making on the innermost only one-fifth of the length of the quill, on the middle one about one-half, and on the outermost about two-thirds, the outer webs being light slate-gray for the same extent from the tip. Bill black, legs pale brownish yellow. The female seems to differ from the male in having the gray color of the breast less pure, this part being somewhat suffused with rufous-olive." (Stejn. l. c.)

Length, 7.30; wing, 3.40; tail, 3.25; tarsus, .82.

Habitat. Martinique.

Myiadestes sanctæ-luciæ Stejn.

Myiadestes genibarbis Scl. P. Z. S. 1871, p. 269.—Semper, P. Z. S. 1872. p. 649.—Scl. & Salv. Nom. Avium Neotr. p. 4 (1873).—Allen, Bull. Nutt. Orn. Club, V, p. 166 (1880).—Cory, List Bds. W. I. p. 5 (1885).
Myiadestes sanctæ-luciæ Stejn. Pr. U. S. Nat. Mus. V, p. 20 (1882).—Cory, Auk, p. 11, 1886; *ib.* Ibis, p. 475 (1886).

"Whole upper parts slaty plumbeous with a conspicuous olivaceous wash, becoming more intense on the lower back, but lacking on the rump and upper tail-coverts. The pattern of the head that of *M. genibarbis*, except that the black stripe below the eye extends further back on the auriculars, and that the white part of the malar stripe occupies the forward half. Chin pure white, this color abruptly defined against the throat, which is rufous chestnut. The remaining underparts like those of the Martinque bird, except that the gray of the breast extends more back on the abdomen. Wings and tail also have the same general appearance as in the above-mentioned-species; on the wing, however, the black speculum of the secondaries is more reduced, the adjacent gray cross-bands being broader, and on the tail the white is more extended, especially on the outer pair, in which the middle third of the outer web is white; besides, the outer webs of the three outermost rectrices are broadly tipped with white, and the following two pairs have also very distinct white tips. Bill black, feet pale yellow. In none of the seven specimens before me is the sex indicated; but as they show no differences the specimen described above, I presume there is no difference between the male and female." (Stejn. orig. descr.)

Length, 7.25; wing, 3.45; tail, 3.30; tarsus, .86.

Habitat. Santa Lucia.

Myiadestes dominicanus Stejn.

Myiadestes genibarbis Lawr. Pr. U. S. Nat. Mus. I, p. 53 (1878).—Cory, List Bds. W. I. p. 5 (1885).

Myadestes dominicanus Stejn. Pr. U. S. Nat. Mus. V. p. 22 (1882).

"Above slaty plumbeous, with a very faint tinge of olivaceous on head and back; lores and a narrow stripe above the eyes conspicuously suffused with olivaceous; almost the whole malar stripe whitish, the feathers the lower end tipped with chestnut, chin white, throat pure chestnut; breast, flanks, and abdomen. except the lower middle part of the latter, ash-gray, duller on the breast, more whitish on the abdomen, and very faintly washed with olivaceous, especially on the flanks, where more tinged with rufous; lower middle of abdomen, crissum, and under tail-coverts chestnut-rufous; wings and tail as in *M. sancta-inca*, the light basal spot on the outer web of the innermost primaries being very conspicuous and well defined; the black speculum on the secondaries larger and the amount of white on the outer tail feathers rather less than in that bird: bill black, feet pale yellow. The female differs only in having a stronger wash of olive on the back." (Stejn. orig. descr. l. c.)

Length, 7.20; wing, 3.40; tail, 3.25; tarsus, .85.

Habitat.   Dominica.

Myiadestes montanus Cory.

Myiadestes montanus Cory, Bull. Nutt. Orn. Club, VI, pp. 130, 151 (1881);
    ib. Bds. Haiti & San Domingo, p. 52 (1885); *ib.* List Bds. W. I. p.
    5 (1885).
Myiadectes montanus Sharpe, Cat. Bds. Brit. Mus. VI, p. 370 (1881).
Myiadestes montanus Stejn. Pr. U. S. Mus. V, p. 23 (1882).

Sp. Char. *Female:*—Upper parts and two central tail-feathers slaty gray; primaries and secondaries brownish black, showing white near the base of the inner webs; outer webs of primaries and terminal portion of the outer webs of secondaries edged with gray; no white spot on the chin; a spot of chestnut at the malar apex; lower eyelid whitish; throat, crissum, and belly, near the vent, reddish brown, intermediate between that of *M. solitarius* and *M. sibilans*, but approaching nearer the color of the former; rest of underparts pale gray; outer tail-feather white, with black shafts, showing a dark tinge near the extremity of the outer web; second feather black, with the central portion of the terminal half white, the black narrowing to the extremity, leaving the tip white; third feather showing a triangular patch of white at the tip; rest of tail-feathers, except the two central ones, black; bill black; legs and feet pale; iris brown.

Length, 7; wing, 3.35; tail, 3.38; tarsus, 1; bill, .38.

Habitat.   Haiti.   Inhabits the mountains.   The type, in my collection, is unique, although the bird is probably not uncommon in some of the mountains in the interior.

Myiadestes solitarius BAIRD.

Ptilogonys armillatus GRAY, Gen. Bds. I. p. 281 (1844).—GOSSE, Bds.
Jam. p. 198 (1847).—SCL. P. Z. S. 1861, p. 73.—ALBRECHT, J. f. O.
1862, p. 196.
Myiadestes armillatus BR. Consp. I, p. 335 (1850).—SCL. Cat. Am. Bds.
p. 47 (1862).
Myiadestes solitarius BAIRD, Rev. Am. Bds. p. 421 (1864).—GRAY. Handl.
Bds. I, p. 366 (1869).—SCL. & SALV. Nom. Avium Neotr. p. 4
(1873).—CORY, List Bds. W. I. p. 5 (1885).
Myiadestes solitarius SHARPE, Cat. Bds. Brit. Mus. VI, p. 369 (1881).—A.
& E. NEWTON, Handb. Jamaica, p. 107 (1881).
Myiadestes solitarius STEJN. Pr. U. S. Nat. Mus. V, p. 24 (1882).
Myiadestes armillatus MARCH, Pr. Acad. Nat. Sci. Phila. 1863, p. 294.

Sp. CHAR.—Upper surface slaty-plumbeous; faint tinge of olivaceous on
the forehead; cheeks dull black; lower eyelid and a small spot at
the malar apex and extremity of chin white, rest of throat chestnut;
underparts slaty-plumbeous, becoming lighter on the belly and
crissum; under tail-coverts chestnut; wings and tail as in other
species in character of marking.

Length, 7.45; wing, 3.6; tail, 3.6; tarsus, .80.

HABITAT. Jamaica.

Myiadestes elizabeth (LEMB.).

Muscicapa elizabeth LEMB. Aves Cuba, p. 39 (1850).
Myiadestes elizabeth CAB. J. f. O. 1856, p. 2.—BREWER, Pr. Bost. Soc.
Nat. Hist. VII, p. 307 (1860).—BAIRD, Rev. Am. Bds. p. 425 (1864).
—GRAY, Handl. Bds. I, p. 366 (1869).—GUNDL. J. f. O. 1872, p.
428; *ib.* Orn. Cuban Anales. 1873, p. 79.—CORY, List Bds. W. I. p.
5 (1885).
Myiadestes elizabethæ NEWTON, Ibis, 1859, p. 110.—ALBRECHT, J. f. O.
1861, p. 209.—SCL. & SALV. Exot. Orn. 1867, p. 55, pl. 28; *ib.* Nom.
Avium Neotr. p. 4 (1873).
Myiadestes elizabethæ SHARPE, Cat. Bds. Brit. Mus. VI, p. 372 (1881).
Myadestes elizabeth STEJN. Pr. U. S. Nat. Mus. V, p. 26 (1882).

Sp. CHAR.—Upper surface pale brownish olive, ashy on the head and
rump; wings dull brown margined with pale ashy olive; tail brown
margined with olive brown; central feathers dull brown, outer
feathers tipped with white; throat and abdomen dull white; breast
and sides shading into ashy; a faint tinge of white at the base of the
forehead; lores and feathers at the eye showing pale buff; ear-
coverts dull olive brown, with narrow white shaft-lines; flanks
showing a tinge of olive brown; axillaries ash colored, showing a
buff tinge; under wing-coverts pale buff.

Length, 7.90; wing, 3.45; tail, 3.35; tarsus, .88.

HABITAT. Cuba.

Myiadestes armillatus (VIEILL.).

Muscicapa armillata VIEILL. Ois. Am. Sept. p. 69, pl. 42 (1802); *ib.*
    Nouv. Dict. XXI. p. 448 (1818).
Ptilogonys armillatus GRAY. Gen. Bds. I. p. 281 (1844); *ib.* Handl. Bds.
    I, p. 366 (1869).
Myiadestes armillatus BAIRD, Rev. Am. Bds. p. 422 (1864).—SCL. P. Z.
    S. 1871, p. 270.—LAWR. Ann. N. Y. Acad. Sci. 1878, p. 149.—CORY,
    List Bds. W. I. p. 5 (1885).
Myiadectes armillatus SHARPE. Cat. Bds. Brit. Mus. VI, p. 370 (1881).
Myiadestes armillatus STEJN. Pr. U. S. Nat. Mus. V, p. 25 (1882).

If this bird is not one of the known species poorly described,
its true habitat yet remains to be discovered.   Professor Baird
gives the following translation (l. c.) of Vieillot's original
description.

"Bill blackish; a white spot on the sides of the throat, and at its origin
(the chin) immediately below the lower mandible (the two continuous);
the eye surrounded by the same color.   Head, back, rump, two intermediate tail-feathers, and the breast of a grayish-slate, paler below.   Wing
and tail feathers blackish; bordered externally by gray, the three lateral
on each side of the tail more or less white.   Belly and hinder parts
brownish rufous; a beautiful yellow in form of a bracelet on the feathers
of lower part of leg; feet brown; length 6 inches, 3 lines." (VIEILL. l. c.)
    Vieillot gives the habitat as "Martinique."

FAMILY MIMIDÆ.

GENUS Margarops SCLATER.

Margarops SCLATER, P. Z. S. 1859, p. 335.

Margarops fuscatus (VIEILL.).

Turdus fuscatus VIEILL. Ois. Am. Sept. II, p. 1 (1807).—BP. Consp. I,
    p. 276 (1850).
Colluricincla fusca GOULD, P. Z. S. 1836, p. 6.
Mimus fuscatus BP. Compt. Rend. XXXVIII, p. 2 (1854).
Cichlalopia fuscatus BP. Rev. Zool. 1857, p. 204.
Margarops fuscatus SCL. P. Z. S. 1859. p. 335.—BAIRD, Rev. Am. Bds.
    p. 42 (1864).—GRAY. Handl. Bds. I, p. 259 (1869).—SCL. & SALV.
    Nom. Avium Neotr. p. 2 (1873).—GUNDL. J. f. O. 1874. p. 310;
    ib. Anal. Soc. Esp. Hist. Nat. VII. p. 172 (1878).—CORY, Bds. Bahama
    I. p. 47 (1880;) *ib.* Bds. Haiti & San Domingo, p. 22 (1885); *ib.*
    List Bds. W. I. p. 6 (1885).
Cichlherminia fuscata A. & E. NEWTON, Ibis, 1859, p. 141.—SHARPE, Cat.
    Bds. Brit. Mus. VI, p. 329 (1881).
Merula fuscata CASSIN, Pr. Acad. Nat. Sci. Phila. 1860, p. 376.

Margarops fusca GRAY, Handl. Bds. I, p. 259 (1869).

SP. CHAR. *Male :*—Above brown, the feathers slightly edged with ash; throat and breast brown, feathers heavily edged with white, giving a mottled appearance which shows faintly on the belly and almost disappears at the vent; primaries brown, pale edged; upper tail-coverts tipped with white; tail brown, tipped with white; bill yellowish, with an olive tinge; upper mandible shading into brown at the base; legs pale olive; iris pale yellow.

The sexes are apparently similar.

Length, 10.25; wing, 5.20; tail, 4.50; tarsus, 1.40; bill, .76.

HABITAT. Inagua, Bahamas; Porto Rico, San Domingo? St. Thomas, St. Croix.

Margarops densirostris (VIEILL.).

Turdus densirostris VIEILL. Nouv. Dict. XX. p. 233 (1816).—LAFR. Rev. Zool. 1844, p. 167.—BP. Consp. I, p. 271 (1850).
Cichlherminia densirostris BP. Compt. Rend. XXXVIII. p. 2 (1854).—SHARPE, Cat. Bds. Brit. Mus. VI, p. 330 (1881).
Margarops densirostris SCL. P. Z. S. 1859, p. 336.—GRAY. Handl. Bds. I, p. 259 (1869).—SCL. & SALV. Nom. Avium Neotr. p. 2 (1873).—LAWR. Pr. U. S. Nat. Mus. I, p. 233 (1878).—SCL. P. Z. S. 1879, p. 765.—ALLEN, Bull. Nutt, Orn. Club, V, p. 166 (1880).—CORY, List Bds. W. I. p. 6 (1885).

SP. CHAR. *Male:*—Above dark brown. feathers edged with pale brown; primaries dark brown, margined with reddish brown; inner secondaries tipped with white; throat heavily marked with white on the upper portion, shading into dark brown on the breast, the feathers edged with white; centre of the belly dull white; sides mottled with white and brown; under tail-coverts white, banded with brown; tail dark brown, tipped with white; bill and legs horn color; iris pale yellow.

The sexes are similar.

Length (skin), 10.75; wing, 5.30; tail, 4.30; tarsus, 1.25; bill, 1.10.

HABITAT. Dominica, Martinique, Montserrat, Santa Lucia and Guadeloupe.

Margarops montanus (LAFR.).

Turdus montanus LAFR. Rev. Zool. 1844, p. 167.
Margarops montanus SCL. P. Z. S. 1859, p. 336; ib. 1871, p. 268.—GRAY, Handl. Bds. I, p. 259 (1869).—SCL. & SALV. Nom. Avium Neotr. p. 2 (1873).—LAWR. Pr. U. S. Nat. Mus. I, p. 5: (1878). — LISTER. Ibis,

1880, p. 39.—Allen, Bull. Nutt. Orn. Club, V, p. 166 (1880).—
Cory, List Bds. W. I. p. 6 (1885).

Cichlherminia montana Sharpe, Cat. Bds. Brit. Mus. VI, p. 330 (1881).

Sp. Char. *Male:*—Upper plumage dark olive brown; throat and breast
brown, the feathers edged with white; feathers of the lower breast
dull white, banded with pale brown, the whole giving a mottled
white and brown appearance to the underparts: wings and tail
dark brown; the inner secondaries and some of the coverts tipped
with white; tail-feathers tipped with white; bill and feet dark brown.

The female is somewhat lighter brown than the male on the under
surface.

Length (skin), 9.20; wing, 4.55; tail, 3.75; tarsus, 1; bill, .65.

Habitat.   Martinique, St. Vincent, Dominica, Santa Lucia,
Guadeloupe and Grenada.

Genus **Ramphocinclus** Lafr.

Ramphocinclus Lafr. Rev. Zool. 1843, p. 66.

Ramphocinclus brachyurus (Vieill.).

Turdus brachyurus Vieill. Nouv. Dict. XX, p. 255 (1818).—Gray, Gen.
Bds. I, p. 219 (1844).

Pterodroma mexicanus Less. Ann. Soc. Nat. 2d ser. IX, p. 168 (1838).

Ramphocinclus brachyurus Lafr. Rev. Zool. 1843, p. 66.—Taylor, Ibis,
1864, p. 166.—Baird, Rev. Am. Bds. p. 41 (1864).—Gray, Handl.
Bds. I, p. 254 (1869).—Scl. & Salv. Nom. Avium Neotr. p. 2 (1873).
—Lawr. Pr. U. S. Nat. Mus. I, p. 486 (1878).—Cory, List Bds.
W. I. p. 6 (1885).

Formicarius brachyurus Gray, Gen. Bds. I, p. 211 (1844).

Legriocinclus mexicanus Less. Descr. Mamm. et Ois. p. 278 (1847).

Campylorhynchus brachyurus Gray, Gen. Bds. III, App. p. 7 (1849).

Zoothera cinclops Br. Consp. I, p. 253 (1850).

Cinclocerthia brachyurus Scl. P. Z. S. 1855, p. 214.

Rhamphocinclus brachyurus Scl. P. Z. S. 1859, p. 338;   —Sharpe, Cat.
Bds. Brit. Mus. VI, p. 325 (1881).

Sp. Char. *Male*—Top of the head dark brown, rest of upper surface dark brown, showing a tinge of chocolate brown on the back; lores and below the eye black, shading into brown on the ear-coverts; throat and breast pure white; belly white; sides of the body chocolate brown; wings and tail dark brown; bill dark brown, almost black; legs dark olive brown; iris reddish brown.

The sexes are similar.

Length (skin), 8.50; wing, 4.25; tail, 3.80; tarsus, 1.25; bill, 1.

Habitat. Martinique.

Genus Cinclocerthia Gray.

Cinclocerthia Gray, List Gen. Bds. p. 17 (1840).

Cinclocerthia ruficauda (Gould).

Stenorhynchus ruficaudus Gould, P. Z. S. 1835, p. 186.
Cinclocerthia ruficauda Gray, List Gen. Bds. p. 17 (1840).—Scl. Cat. Am. Bds. p. 7 (1862).—Lawr. Pr. U. S. Nat. Mus. I. p. 486 (1878). —Sharpe, Cat. Bds. Brit. Mus. VI, p. 320 (1881).—Cory, List Bds. W. I. p. 6 (1885).
Ramphocinclus tremulus Lafr. Rev. Zool. 1843, p. 67.—Scl. P. Z. S. 1855, p. 213.
Herminierus guadeloupensis Less. Rev. Zool. 1843. p. 325.
Herminierus infaustus Less. *t. c.* p. 325.
Thriothorus l'herminieri Less. *t. c.* p. 326.
Formicarius tremulus Gray, Gen. Bds. I, p. 211 (1844).

Sp. Char. *Male:*—Above ashy brown, shading into rufous brown on the back and rump; lores and ear-coverts dark brown; a patch in front of the eye brownish black; chin and throat very pale brown, becoming reddish brown on the belly; tail rufous brown; quills dark brown, edged with rufous brown.

The sexes are apparently similar.

Length (skin), 9.30; wing, 4; tail, 3.70; tarsus, 1; bill, 1.30.

Habitat. Guadeloupe and Dominica.

Cinclocerthia macrorhyncha Scl.

Cinclocerthia macrorhyncha Scl. P. Z. S. 1866, p. 320; *ib.* 1871, p. 268.— Gray, Handl. Bds. I, p. 263 (1869).—Scl. & Salv. Nom. Avium Neotr. p. 2 (1873).—Allen, Bull. Nutt. Orn. Club, V, p. 166 (1880). —Sharpe, Cat. Bds. Brit. Mus. VI, p. 325 (1881).—Cory, List Bds. W. I. p. 6 (1885).

Sp. Char. *Male:*—General plumage above ashy; forehead dark brown; feathers in front of the eye, including lores and ear-coverts dark brown; throat dull white, shading into ashy on the breast, and showing a tinge of rufous on the sides and under tail-coverts; the rufous slightly perceptible on the abdomen, varying in different specimens; wings dull brown, the coverts ashy; tail brown, an olive tinge on the upper surface; legs greenish; iris dull yellow.

The sexes are similar.

Length (skin), 9.30; wing, 4.20; tail, 3.20; tarsus, 1.20; bill, 1.35.

Habitat. Santa Lucia.

Cinclocerthia gutturalis (Lafr.).

Ramphocinclus gutturalis Lafr. Rev. Zool. 1843, p. 67.—Bp. Consp. I, p. 223 (1850).

Formicarius gutturalis Gray. Gen. Bds. I, p. 211 (1844).

Campylorhynchus gutturalis Gray, Gen. Bds. III, App. p. 7 (1849).

Cinclocerthia gutturalis Scl. P. Z. S. 1855, p. 214.—Gray, Handl. Bds. I, p. 263 (1869).—Scl. & Salv. Nom. Avium Neotr. p. 2 (1873).— Lawr. Pr. U. S. Nat. Mus. II, p. 351 (1879).—Sharpe, Cat. Bds. Brit. Mus. VI, p. 324 (1881).—Cory, List Bds. W. I. p. 6 (1885).

Sp. Char. *Male:*—Upper parts brown, darkest on the head; underparts dull brownish white, the white showing clearest on the throat and belly, but never entirely free from a grayish tinge; wings and tail brown; bill and feet dark brown; iris gray.

The sexes are similar.

Length (skin), 9.25; wing, 4.50; tail, 4; tarsus, 1.25.

It is possible that at some seasons the under surface may be differently colored, but in all the specimens before me, the underparts are marked with a dull mixture of brown and white.

Habitat. Martinique.

Genus Galeoscoptes Caban.

Galeoscoptes Cabanis, Mus. Hein. I, p. 82 (1850).

Galeoscoptes carolinensis (Linn.).

Muscicapa carolinensis Linn. Syst. Nat. I, p. 328 (1766).

Turdus carolinensis Licht.—D'Orn. in La Sagra's Hist. Nat. Cuba, Ois. p. 51 (1840).—Gundl. J. f. O. 1861, p. 324 (Cuba).

Galeoscoptes carolinensis Cab. Mus. Hein. I, p. 82 (1850).—Gundl. Repert. Fisico-Nat. Cuba, I, p. 230 (1865); *ib.* J. f. O. 1872, p. 407 (Cuba).—Cory, List Bds. W. I. p. 6 (1885).

Mimus carolinensis Brewer, Pr. Bost. Soc. Nat. Hist. VII, p. 307 (1860) (Cuba).—Cory, Bds. Bahama I. p. 51 (1880).

Mimus (Galeoscoptes) carolinensis Bryant, Pr. Bost. Soc. Nat. Hist. XI, p. 69 (1867).

Occasional in the Bahama Islands and Cuba.

GENUS **Mimus** BOIE.

Mimus BOIE, Isis, 1826, p. 972.

Mimus polyglottus (LINN.).

Turdus polyglottus LINN. Syst. Nat. I, p. 293 (1766).
Orpheus polyglottus D'ORB. in La Sagra's Hist. Nat. Cuba, Ois. p. 53
    (1840).
Mimus polyglottus BREWER, Pr. Bost. Soc. Nat. Hist. VII, p. 307 (1860).—
    GUNDL. Repert. Fisico-Nat. Cuba, I, p. 230 (1865) (Cuba); *ib.*
    J. f. O. 1872, p. 408 (Cuba).—CORY, List Bds. W. I. p. 6 (1885).

Cuban specimens of this bird are very rare. A specimen in my collection is labelled, in the handwriting of Dr. Gundlach, *Mimus polyglottus cubensis*. It is apparently *M. elegans*, although somewhat larger, and may represent a new race. Perhaps both species are represented there, as I have seen specimens of *polyglottus* labelled "Cuba."

Mimus orpheus (LINN.).

Turdus orpheus LINN. Syst. Nat. I, p. 293 (1766).—VIEILL. Ois. Am.
    Sept. II, p. 12, pl. 68 (1807).—GOSSE, Bds. Jam. p. 144 (1847).
Mimus orpheus GRAY, Gen. Bds. I. p. 221 (1844).—BP. Consp. I, p. 276
    (1850).—MARCH, Pr. Acad. Nat. Sci. Phila. 1863, p. 290.—BAIRD,
    Rev. Am. Bds. p. 50 (1864).—SCL. & SALV. Nom. Avium Neotr.
    p. 3 (1873).—A. & E. NEWTON, Handb. Jamaica, p. 105 (1881).—
    SHARPE, Cat. Bds. Brit. Mus. VI, p. 340 (1881).—CORY, List Bds.
    W. I. p. 6 (1885).
Mimus polyglottus GOSSE, Bds. Jam. p. 144 (1847).—ALBRECHT, J. f. O.
    1862, pp. 194, 201.—HILL, Pr. Acad. Nat Sci. Phila. 1863, p. 304.—
    GUNDL. Anal. Soc. Esp. Hist. Nat. VII, p. 173 (1878).
Mimus polyglottus var. portoricensis BRYANT, Pr. Bost. Soc. Nat. Hist.
    XI, p. 68 (1866).
Mimus polyglottus var. cubanensis BRYANT, *t. c.* p. 68.

SP. CHAR.—Above grayish brown, showing ashy on the back; underparts
    white, showing a tinge of ash on the breast; wings brown, prima-
    ries heavily marked with white, the eighth and ninth almost entirely
    white; tail dark brown, outer feather entirely white, second nearly
    so, showing a brownish line on outer web more or less distinct,
    third feather having outer web brown, inner web white; bill black;
    legs brownish.

Length, 9.50; wing, 4.30; tail, 5; tarsus, 1.20.

HABITAT. Jamaica and Grand Cayman.

Mimus elegans Sharpe.

Mimus polyglottus (var. *bahamensis?*) Bryant, Pr. Bost. Soc. Nat. Hist
     XI, p. 68 (1866).—Gray, Handl. Bds. I, p. 261 (1869).
Mimus orpheus var. *dominicus* Cory, Bds. Bahama I, p. 48 (1880).
Mimus elegans Sharpe, Cat. Bds. Brit. Mus. VI, p. 339 (1881).—Cory,
     List Bds. W. I. p. 6 (1885).

Sp. Char. *Male:*—Above grayish brown, showing ashy on the back;
     underparts white, slightly tinged with ashy on the breast; wings
     brown; all of the primaries heavily marked with, and the eighth
     and ninth almost entirely white; tail brown, having the first two
     and entire inner web of third feathers white; bill black; legs
     brownish.

     The sexes are similar.

     Length, 8.50; wing, 4; tail, 4.20; tarsus, 1.20; bill, .64.

Habitat.   Inagua, Bahama Islands.

Mimus dominicus (Linn.).

Turdus dominicus Linn. Syst. Nat. I, p. 295 (1766).
Turdus merle Müll. Syst. Nat. Anhang, p. 139 (1766).
Mimus dominicus Gray, Gen. Bds. I, p. 221 (1844).—Bp. Consp. I, p.
     276 (1850).—Scl. P. Z. S. 1859, p. 341.—Gray, Handl. Bds. I, p.
     262 (1869).—Sharpe, Cat. Bds. Brit. Mus. VI, p. 341 (1881).—
     Cory, Bds. Haiti & San Domingo, p. 21 (1885); *ib.* List Bds. W.
     I. p. 6 (1885).
Mimus polyglottus var. *dominicus* Bryant, Pr. Bost. Soc. Nat. Hist. XI,
     p. 63 (1866).
Mimus orpheus dominicus Cory, Bull. Nutt. Orn. Club, VI, p. 151 (1881).

Sp. Char. *Male:*—Above grayish brown, showing ashy on the back;
     underparts white, slightly tinged with ashy on the breast; wings
     brown; all of the primaries heavily marked with, and the eighth
     and ninth almost entirely white; tail dark brown, having the first
     two and inner web of third feathers white; bill black; legs brown-
     ish.

     Sexes are similar.

     Length, 8.50; wing, 4; tail, 4.18; tarsus, 1.20; bill, .64.

Habitat.   Haiti and San Domingo.

This species is very closely allied to *M. orpheus*, and perhaps
should not be separated from it.

Mimus gilvus (Vieill.).

Turdus gilvus Vieill. Ois. Am. Sept. II, p. 15 (1807).
Mimus gilvus Jard. Ann. Nat. Hist. 2nd ser. XX, p. 329 (1847).— Bp.

Consp. I, p. 276 (1850).— SCL. P. Z. S. 1859, p. 342.—SCL. &
SALV. Nom. Avium Neotr. p. 3 (1873).—LAWR. Pr. U. S. Nat.
Mus. I, p. 187 (1878).—ALLEN, Bull. Nutt. Orn Club, V, p. 166
(1880).—SHARPE, Cat. Bds. Brit. Mus. VI, p. 350 (1881).—RIDGW.
Pr. U. S. Nat. Mus. VII, p. 172 (1884).—CORY, List Bds. W. I. p. 6
(1885).
Mimus melanopterus LAWR. Ann. Lyc. N. Y. V, p. 35, pl. 2 (1849).—
SCL. Cat. Am. Bds. p. 9 (1862).—FINSCH, P. Z. S. 1870, p. 553.
Mimus columbianus CAB. Mus. Hein. I, p. 82 (1850).
Mimus gracilis CAB. Mus. Hein. I, p. 83 (1850).—BAIRD, Rev. Am. Bds.
p. 54 (1864).—LAWR. Ann. Lyc. N. Y. IX, p. 91 (1868).

SP. CHAR.—Above grayish brown, ashy on the rump and forehead;
underparts dull ashy white; flanks streaked slightly with brownish,
wings brown, edged with dull white; under wing-coverts marked
with brown; tail dark brown, all the feathers tipped with white,
central feathers very slightly, sometimes apparently not at all, the
white increasing to the outer feathers, which show a patch of white
on tip of inner web, about three quarters of an inch in length,
extending to a less extent to the outer web; bill and feet black.

Length (skin), 8.75; wing, 4.45; tail, 4; tarsus, 1.25.

Common in St. Vincent, Grenada, Santa Lucia, and St.
Thomas.

Mimus gundlachi CABAN.

Mimus gundlachi CAB. J. f. O. 1855, p. 470.—SCL. P. Z. S. 1859, p. 342.—
BAIRD, Rev. Am. Bds. p. 59 (1864).—BREWER, Pr. Bost. Soc. Nat.
Hist. VII, p. 307 (1860).—GUNDL. Repert. Fisico-Nat. Cuba, I, p.
230 (1865).—SHARPE, Cat. Bds. Brit. Mus. VI, p. 344 (1881).—
CORY, List Bds. W. I. p. 6 (1885).
Mimus bahamensis BRYANT, Pr. Bost. Soc. Nat. Hist. VII, p. 114 (1859).—
BAIRD, Rev. Am. Bds. p. 52 (1864).—CORY, Bds. Bahama I. p. 48
(1880); ib. List Bds W. I. p. 6 (1885).—SHARPE, Cat. Bds. Brit.
Mus. VI, p. 334 (1881).
Scotiomimus bahamensis BRYANT, Pr. Bost. Soc. Nat. Hist. XI, p. 68
(1866).—GRAY, Handl. Bds. I. p. 262 (1869).
Mimus gundlachii GRAY, Handl. Bds. I, p. 262 (1869).—GUNDL. J. f. O.
1872, p. 409.

SP. CHAR. *Male:*—Much larger than *M. polyglottus*, and the white tail-
feathers wanting. Above pale rufous brown, the rufous tint most
marked on the rump and upper tail-coverts; below pale ash, streaked
with fine lines of brown, becoming broader upon the sides;
wings rufous brown, feathers slightly edged with pale rufous;
wing-coverts tipped with white, forming two narrow bars; tail

dark brown, slightly tipped with dull white, wanting on the two
middle feathers; legs bluish black; bill black; iris yellow.
The female resembles the male.

Length, about 11; wing, 5; tail, 5; tarsus, 1.60; bill, .90.

HABITAT.   Bahama Islands and Cuba.

After a careful examination and comparison of a series of
twenty-two of the so-called *M. bahamensis* and three specimens
of *M. gundlachi*, I can not find any difference sufficient to
characterize them as distinct species.   One specimen from
Cuba has more white on the tail-feathers than any from the
Bahama Islands, but some of the latter show the white fully as
much as the other Cuban examples.   A large series from Cuba
would determine the matter more satisfactorily.

Mimus hillii MARCH.

Mimus hillii MARCH, Pr. Acad. Nat. Sci. Phila. 1863, p. 291.—BAIRD,
   Rev. Am. Bds. p. 52 (1864).—GRAY, HANDL. Bds. I, p. 262 (1869).—
   SCL. & SALV. Nom. Avium Neotr. p. 3 (1873).—SHARPE, Cat. Bds.
   Brit. Mus. VI, p. 343 (1881).—CORY, List Bds. W. I. p. 6 (1885).
Mimus orpheus HILL, Pr. Acad. Nat. Sci. Phila. 1863, p. 304.—GRAY,
   Handl. Bds. I, p. 262 (1869).
Mimus hilli A. & E. NEWTON, Handb. Jamaica, p. 105 (1881).

SP. CHAR.—General appearance the same as that of *M. gundlachi*, differing
   from it by being slightly browner on the head, and somewhat paler
   on the underparts, with more white on the end of the tail-feathers.
   Length (skin), 11.20; wing, 5; tail, 5.75; tarsus, 1.55.

HABITAT.   Jamaica.

Very closely allied to *M. gundlachi*, and perhaps ought not
to be separated from it.

FAMILY SYLVIIDÆ.

GENUS Polioptila SCL.

Polioptila SCLATER, P. Z. S. 1855, p. 11.

Polioptila lembeyi (GUNDL.).

Culicivora lembeyi GUNDL. Ann. N. Y. Lyc. 1858, p. 273.—BREWER, Pr.
   Bost. Soc. Nat. Hist. VII, p. 306 (1860).—ALBRECHT, J. f. O. 1861,
   p. 211.

Polioptila lembeyii BAIRD, Rev. Am. Bds. p. 68 (1864).
Polioptila lembeyei GUNDL. Repert. Fisico-Nat. Cuba, I, p. 231 (1865);
  ib. J. f. O. 1872, p. 410.—GRAY, Handb. Bds. I, p. 237 (1869).
Polioptila lembeyi ED. BWR. & RIDGW. Hist. N. Am. Bds. I, p. 78 (1874).
  —CORY, List Bds. W. I. p. 6 (1885).—SHARPE, Cat. Bds. Brit. Mus.
  X, p. 444 (1885).

SP. CHAR.—A narrow black line commences at the top of the eye, extending backwards, bordering the ear-coverts; above bluish gray;
  underparts ashy white, the white clearest on the abdomen; tail-
  feathers narrow and long, having the shafts dark brown, outer
  feather white, except the basal half of inner web, which is dark
  brown, second having the terminal third white and outer web narrowly tipped with white, third feather tipped with white, rest of
  tail-feathers brownish black; wings brownish black, the feathers
  edged with white, no white on the edges of the first two primaries.
  Length (skin), 4.58; wing, 1.50; tail, 2; tarsus, .70; bill, .35.
HABITAT.   Cuba.

Polioptila cærulea (LINN.).

Motacilla cærulea LINN. Syst. Nat. I, p. 337 (1766).
Culicivora cærula D'ORB. in La Sagra's Hist. Nat. Cuba, Ois. p. 90
  (1840).—BREWER, Pr. Bost. Soc. Nat. Hist. VII, p. 306 (1860) (Cuba).
  —GUNDL. J. f. O. 1861, p. 407 (Cuba).
Polioptila cærulea GUNDL. J. f. O. 1861, p. 324; *ib.* 1872, p. 409.—CORY,
  Bds. Bahama I. p. 52 (1880); *ib.* List Bds. W. I. p. 6 (1883); *ib.* Auk, p.
  501, 1886 (Grand Cayman).

Common in the Bahama Islands; breeds.   Numerous records
from Cuba.   Mr. Ridgway separates the resident Bahama bird
under the name of *P. cærulea cæsiogaster.*   They are somewhat
paler in coloration.*

FAMILY TROGLODYTIDÆ.

GENUS Thryothorus VIEILL.

Thryothorus VIEILLOT, Analyse, p. 45 (1816).

Thryothorus martinicensis SCL.

Thryothorus martinicensis SCL. P. Z. S. 1866, p. 321.—SCL. & SALV. Nom.
  Avium Neotr. p. 7 (1873).—LAWR. Pr. U. S. Nat. Mus. I, p. 352
  (1878).—SHARPE, Cat. Bds. Brit. Mus. VI, p. 228 (1881).—CORY, List
  Bds. W. I. p. 7 (1885).
Hylemathrous martinicensis GRAY, Handl. Bds. I, p. 191 (1869).

---

* *Polioptila cærulea cæsiogaster* RIDGWAY, Manual N. A. Birds, p. 569 (1887).

Sp. Char. *Male:*—Upper parts dark brown, very narrowly lined on the
back; feathers of the wings and tail banded with narrow lines;
under surface pale rufous brown. Resembles *T. grenadensis*, but
is darker.

The sexes are similar.

Length (skin), 5; wing, 2.15; tail, 2.10; tarsus, .80; bill, .10.

HABITAT.   Martinique.

Thryothorus rufescens Lawr.

Thryothorus rufescens Lawr. Ann. N. Y. Acad. Sci. I, p. 47 (1878); *ib.*
Pr. U. S. Nat. Mus. I, p. 486 (1878).— Sharpe, Cat. Bds. Brit. Mus.
VI, p. 228 (1881).— Cory, List Bds. W. I. p. 7 (1885).

"*Male.* Entire plumage rufous, much deeper in color above, of a lighter
and brighter shade underneath; tail dark rufous, regularly and
closely crossed with narrow bars of black; the coloring of the underpart of the tail is duller, but is barred in a similar manner; inner webs of quills blackish brown, outer webs and both webs of the
innermost secondaries dark rufous, with distinct narrow bars of
black; upper mandible dark brown, the under yellowish-white; feet
pale brown.

Length, 4⅝ in.; wing, 2⅛; tail, 1⅞; tarsus, 11-16; bill from front
9-16; from rictus ⅜." (Lawr. orig. deser.)

HABITAT.   Dominica and Guadeloupe.

Thryothorus musicus Lawr.

Thryothorus musicus Lawr. Ann. N. Y. Acad. Sci. I. p. 148 (1878); *ib.*
Pr. U. S. Nat. Mus. I. p. 486 (1878).— Sharpe, Cat. Bds. Brit. Mus.
VI, p. 223 (1881).— Cory, List Bds. W. I. p. 7 (1885).

"*Male.* Above of a dark ferruginous, somewhat darker on the crown and
brighter on the rump; lores, and a line running back from the eye,
white tinged with rufous; the exposed portions of the wings are
dark rufous, conspicuously barred with black; the inner webs of
the primaries are blackish-brown; under wing-coverts white; the
tail-feathers are dark rufous, barred with black; the entire back and
upper tail-coverts are marked inconspicuously with narrow transverse dusky lines; the feathers of the rump have concealed white
shaft-stripes, which become wider towards the ends of the feathers;
the feathers of the back also have the basal portion of their shafts
marked with white; the throat, breast, and middle of the abdomen
are white, the latter tinged with rufous, the sides are light ferruginous; the under tail-coverts are rufous, each feather marked with
a subterminal round black spot; upper mandible, black; the under

whitish, with the end dusky; tarsi and toes light brownish flesh
color.

"Length (fresh), 5¼ in.; wing 2½; tail 1 13-16; tarsus ⅞." (LAWR.,
orig. descr.)

HABITAT. St. Vincent.

Thryothorus grenadensis LAWR.

Thryothorus grenadensis LAWR. Ann. N. Y. Acad. Sci. I, p. 161 (1878);
ib. Pr. U. S. Nat. Mus. I, p. 486 (1878).—SHARPE, Cat. Bds. Brit.
Mus. VI, p. 228 (1881).—CORY, List Bds. W. I. p. 7 (1885).

"*Female.* Upper plumage of a rather bright ferruginous, a little inclining to brownish on the head and hind neck, and brighter on the
rump; lores whitish tinged with rufous; a light rufous stripe extends over the eye to the hind neck; tail dull rufous, barred with
black; the primary quills have their outer webs of a dull light
rufous, with broad black bars; the inner webs are brownish-black;
the wing-coverts and tertials are rufous with narrower black bars;
under wing-coverts pale rufous; the throat is very pale rufous, inclining to whitish; the breast light rufous; the middle of the abdomen is of a rather paler shade; the sides and under tail-coverts are
of a bright darker ferruginous; the upper mandible brownish-
black; the under pale yellow, dusky at the tip; tarsi and toes hazel
brown.

"Length (fresh), 4¾ inches; wing 2¼; tail 1½; tarsus ⅞; bill from
front, 11-16." (LAWR. orig. descr.)

HABITAT. Grenada.

Thryothorus mesoleucus SCL.

Thryothorus mesoleucus SCL. P. Z. S. 1876, p. 14.—ALLEN, Bull. Nutt.
Orn. Club, V, p. 166 (1880).—SHARPE, Cat. Bds. Brit. Mus. VI, p.
223 (1881).—CORY, List Bds. W. I. p. 7 (1885).

SP. CHAR.— Top of head brown, the feathers delicately edged with lighter
brown, giving a faint mottled appearance to the crown; back rufous
brown, the rufous showing brightest on the rump; wings and tail
brown delicately banded with brownish black; sides of the head
and neck buff, shading into buffy white on the throat and breast;
abdomen and crissum pale rufous; bill pale.

Length (skin), 4.05; wing, 1.95; tail, 1.50; tarsus, .70; bill, .80.

HABITAT. Santa Lucia.

Family MNIOTILTIDÆ.

Genus **Mniotilta** Vieill.

Mniotilta Vieillot. Analyse, p. 45 (1816).

Mniotilta varia (Linn.).

Motacilla varia Linn. Syst. Nat. I. p. 333 (1766).
Mniotilta varia Gosse, Bds. Jam. p. 134 (1847).—Lemb. Aves Cuba, p.
    68 (1850).—Gundl. J. f. O. 1855, p. 475; *ib.* Repert. Fisico-Nat.
    Cuba, I, p. 232 (1865) (Cuba).—Brewer, Pr. Bost. Soc. Nat. Hist.
    VII, p. 306 (1860) (Cuba).—Albrecht, J. f. O. 1862, p. 193 (Ja-
    maica).— A. & E. Newton, Ibis, 1859, p. 143 (St. Croix).—Bry-
    ant, Pr. Bost. Soc. Nat. Hist. VII, p. 110 (1859) (Bahamas).— Scl.
    P. Z. S. 1861, p. 70 (Jamaica).—March, Pr. Acad. Nat. Sci. Phila.
    1863, p. 293 (Jamaica).—Gundl. Anal. Soc. Esp. Hist. Nat. VII,
    p. 177 (1878) (Porto Rico).—Cory, Bds. Bahama I. p. 54 (1880);
    ib. Bds. Haiti & San Domingo, p. 23 (1885).—A. & E. Newton,
    Handb. Jamaica, p. 105 (1881).
Sylvicola (Mniotilta) varia Bryant, Pr. Bost. Soc. Nat. Hist. XI, p. 91
    (1867) (San Domingo).

Bahama Islands and Greater Antilles. Recorded also from the
Lesser Antilles.

Genus **Compsothlypis** Cab.

Compsothlypis Cabanis. Mus. Hein. I, p. 20 (1851).

Compsothlypis americana (Linn.).

Parus americanus Linn. Syst. Nat. I. p. 190 (1766).
Sylvia americana D'Orb. in La Sagra's Hist. Nat. Cuba. Ois. p. 69
    (1840).—A. & E. Newton, Ibis, 1859, p. 143 (St. Croix).
Parula americana Gosse, Bds. Jam. p. 154 (1847).—Cass. Proc. Acad.
    Nat. Sci. Phila. 1860, p. 376 (St. Thomas).—Gundl. J. f. O. 1861,
    p. 326 (Cuba); *ib.* 1872, p. 411 (Cuba).—Albrecht, J. f. O. 1862,
    p. 192 (Jamaica).— March, Pr. Acad. Nat. Sci. Phila. 1863, p.
    293 (Jamaica).—Gundl. Anal. Soc. Esp. Hist. Nat. VII, p. 176
    (1878) (Porto Rico).—Cory, Bds. Bahama I. p. 55 (1880); *ib.* Bull.
    Nutt. Orn. Club, VI, p. 151 (1881) (San Domingo); *ib.* Bds. Haiti
    & San Domingo, p. 24 (1885).

Bahamas and Greater Antilles; recorded from some of the
Lesser Antilles.

GENUS **Protonotaria** BAIRD.

Protonotaria BAIRD, Bds. N. Am. p. 239 (1858).

Protonotaria citrea (BODD.).

Motacilla citrea BODD. Tab. pl. 704 (1783).
Protonotaria citrea GUNDL. J. f. O. 1861, p. 324; *ib.* 1862, p. 178; *ib.* 1872,
    p. 411; *ib.* Repert. Fisico-Nat. Cuba, I, p. 231 (1865) (Cuba).
    —BAIRD, Rev. Am. Bds. p. 173 (1864).—CORY, List Bds. W. I.
    p. 7 (1885).

Accidental in Cuba.

GENUS **Helmitherus** RAF.

Helmitherus RAFINESQUE, Journ de Phys. LXXXVIII, p. 417 (1819).

Helmitherus vermivorus (GMEL.).

Motacilla vermivora GMEL. Syst. Nat. I, p. 95 (1788).
Vermivora pennsylvanica GOSSE, Bds. Jam. p. 150 (1847).—ALBRECHT,
    J. f. O. 1862, pp. 194, 201 (Jamaica).— MARCH, Pr. Acad. Nat. Sci.
    Phila. 1863, p. 293 (Jamaica).
Helinaia vermivorus LEMB. Aves Cuba, p. 35 (1850).
Helmitheros vermivorus GUNDL. J. f. O. 1855, p. 476; *ib.* 1861, pp. 326,
    409 (Cuba).
Helinaia vermivora BREWER, Pr. Bost. Soc. Nat. Hist. VII, p. 307 (1860).
    (Cuba).
Helmitherus vermivorus GUNDL. Repert. Fisico-Nat. Cuba, I, p. 232
    (1865); *ib.* J. f. O. 1872, p. 412 (Cuba).—CORY, List Bds. W. I. p. 7
    (1885).
Helminthotherus vermivorus A. & E. NEWTON, Handb. Jamaica, p. 105
    (1881).

Recorded from Cuba and Jamaica.

Helmitherus swainsoni AUD.

Sylvia swainsoni AUD. Orn. Biog. II, p. 563 (1834).
Helmitherus swainsoni BAIRD, Rev. Am. Bds. p. 180 (1864).—GUNDL.
    Repert. Fisico-Nat. Cuba, I, p. 232 (1865); *ib.* J. f. O. 1872, p.
    412 (Cuba).—CORY, List Bds. W. I. p. 7 (1885).
Helonaea swainsoni NEWTON, P. Z. S. 1879. p. 552 (Jamaica).
Helminthotherus swainsoni A. & E. NEWTON, Handb. Jamaica, p. 105
    (1881).
Helinaia swainsoni MERRIAM, Auk, II, p. 377 (1885) (Jamaica).

Recorded from Cuba and Jamaica.

Genus Helminthophila Ridgw.

Helminthophila Ridgway, Bull. Nutt. Orn. Club, VII, p. 53 (1882).

Helminthophila chrysoptera (Linn.).

Motacilla chrysoptera Linn. Syst. Nat. I. p. 333 (1766).
Helinaia chrysoptera Brewer, Pr. Bost. Soc. Nat. Hist. VII, p. 307 (1860)
    (Cuba).
Helminthophaga chrysoptera Gundl. J. f. O. 1861. p. 326; *ib.* 1862, p. 177;
    ib. 1872, p. 411; *ib.* Repert. Fisico-Nat. Cuba, I, p. 232 (1865)
    (Cuba).—Cory, List Bds. W. I. p. 7 (1885).

Accidental in Cuba.

Helminthophila bachmani (Aud.).

Sylvia bachmani Aud. Orn. Biog. II, p. 483 (1834).
Helinaia bachmanii Lemb. Aves Cuba, p. 36 (1850).
Helminthophaga bachmani "Cab." Gundl. J. f. O. 1885, p. 475; *ib.* :861,
    pp. 326, 409; *ib.* Repert. Fisico-Nat. Cuba, I, p. 232 (1865) (Cuba).
    —Cory, List Bds. W. I. p. 7 (1885).
Helinaia bachmani Brewer, Pr. Bost. Soc. Nat. Hist. VII, p. 307 (1860)
    (Cuba).

Accidental in Cuba.

Helminthophila peregrina (Wils.).

Sylvia peregrina Wils. Am. Orn. IV, p. 83 (1811).
Helinaia peregrina Brewer, Pr. Bost. Soc. Nat. Hist. VII, p. 307 (1860)
    (Cuba).
Helminthophaga peregrina Gundl. J. f. O. 1861, p. 326; *ib.* 1862, p. 177;
    ib. 1872, p. 412; *ib.* Repert. Fisico-Nat. Cuba, I, p. 232 (1865)
    (Cuba).—Cory, List Bds. W. I. p. 7 (1885).

Accidental in Cuba. Bahama Islands? A specimen in my
cabinet is labelled "Bahama I."; the collector is unknown.

Genus Dendroica Gray.

Dendroica Gray, Gen. Bds. App. 8 (1842).

Dendroica tigrina (Gmel.).

Motacilla tigrina Gmel. Syst. Nat. I, p. 985 (1788).
Sylvia maritima D'Orb. in La Sagra's Hist. Nat. Cuba, Ois. p. 70 (1840).
Certhiola maritima Gosse, Bds. Jam. p. 87 (1847).

Rhimamphus maritimus GUNDL. J. f. O. 1853. p. 474; *ib.* 1861, p. 409
(Cuba).
Sylvicola maritima BRYANT, Pr. Bost. Soc. Nat. Hist. VII, p. 110 (1859)
(Bahamas).—BREWER, Pr. Bost. Soc. Nat. Hist. VII, p. 307 (1860)
(Cuba).
Dendroeca tigrina A. & E. NEWTON, Ibis, 1859, p. 144 (St. Croix).—SCL.
P. Z. S. 1861, p. 71 (Jamaica).—ALBRECHT, J. f. O. 1862, p. 193
(Jamaica).—CORY, Bds. Bahama I. p. 63 (1880); *ib.* Bull. Nutt.
Orn. Club, VI, p. 151 (1881); *ib.* Bds. Haiti & San Domingo, p. 25
(1885).
Dendroica trigrina GUNDL. J. f. O. 1861, p. 326 (Cuba).—MARCH, Pr.
Acad. Nat. Sci. Phila. 1863, p. 293 (Jamaica).
Perissoglossa tigrina GUNDL. Repert. Fisico-Nat. Cuba, I, p. 233 (1865);
ib. J. f. O. 1872. p. 412 (Cuba); *ib.* Anal. Soc. Esp. Hist. Nat. VII,
p. 178 (1878) (Porto Rico).—CORY, List Bds. W. I. p. 7 (1885).

Recorded from Bahama Islands, Greater Antilles, and St.
Croix.

Dendroica æstiva (GMEL.).

Motacilla æstiva GMEL. Syst. Nat. I, p. 996 (1788).
Rhimamphus æstivus Bp.? GUNDL. J. f. O. 1885, p. 472 (Cuba)?—CAB.
J. f. O. 1860, p. 326 (Cuba).
Sylvicola æstiva? BREWER, Pr. Bost. Soc. Nat. Hist. VII. p. 307 (1860)
(Cuba)?—FINSCH. P. Z. S. 1870, p. 564 (Trinidad).
Dendroeca æstiva? TAYLOR, Ibis, 1864, p. 81 (Trinidad).—CORY, Bds.
Bahama, I. p. 56 (1880).

Cuba? and the Bahama Islands?
It is doubtful if *D. æstiva* occurs in the West Indies, as in
some plumages it is difficult to distinguish from the closely allied
forms which occur there.

Dendroica petechia (LINN.).

Motacilla petechia LINN. Syst. Nat. I, p. 334 (1766).
Sylvia petechia LATH. Gen. Syn. II, p. 535 (1790).—VIEILL. Ois. Am.
Sept. II, p. 32, (1807).
Sylvicola æstiva GOSSE. Bds. Jam. p. 157 (1847).
Dendroeca æstiva A. & E. NEWTON, Ibis, 1859, p. 143.
Dendroeca petechia SCL. Cat. Am. Bds. p. 32 (1862).—ALBRECHT. J. f. O.
1862, p. 193.—GRAY, Handl. Bds. I, p. 240 (1869).—SCL. & SALV.
Nom. Avium Neotr. p. 9 (1873).—GUNDL. Anal. Soc. Esp. Hist.
Nat. VII. p. 182 (1878).—CORY, Bds. Bahama I. p. 57 (1880); *ib.*
List Bds. W. I. p. 8 (1885).—A. & E. NEWTON, Handb. Jamaica, p.

106 (1881).—Ridgw. Pr. U. S. Nat. Mus. VII, p. 172 (1884).—
Coues, Key N. Am. Bds. p. 297 (1884).—Sharpe, Cat. Bds. Brit.
Mus. X, p. 277 (1885).
Sylvicola petechia Bryant, Pr. Bost. Soc. Nat. Hist. XI, p. 67 (1867).
Dendrœca petechia e. *jamaicensis* Sund. Oefv. K. Vet. Akad. Förh.
1869, p. 607.
Dendroica petechia Cassin, Pr. Acad. Nat. Sci. Phila. 1860, pp. 192,
376.—March, Pr. Acad. Nat. Sci. Phila. 1863. p. 292.—Baird, Rev.
Am. Bds. p. 199 (1864).—Bd. Bwr. & Ridgw. Hist. N. Am. Bds.
I, p. 216 (1874).

Sp. Char. *Male:*—Underparts bright yellow, streaked with dull rufous on
the breast and sides; forehead yellowish, shading into olive green
on the top of the head; a tinge of rufous on the concealed portions
of the feathers on the forehead; back olive green; wings and tail
brown, edged with yellowish; under surface of tail having the appearance of bright yellow, the feathers tipped with olive green; the
upper surface of tail-feathers having the inner webs yellow.

Female:—Somewhat greener than the male; more yellow on
the rump and tail-coverts; no rufous on the head.

Length (skin), 4.50; wing, 2.50; tail, 1.60; tarsus, .74.

Habitat.   Jamaica.   Accidental in the Bahama Islands.

Dendroica petechia gundlachi.

(?) *Motacilla albicollis* Gmel. Syst. Nat. I, p. 983 (1788).
Rhimamphus æstivus Cab. J. f. O. 1855, p. 472.
Sylvicola petechia Brewer, Pr. Bost. Soc. Nat. Hist. VII, p. 307 (1860).
Dendroica albicollis Cassin, Pr. Acad. Nat. Sci. 1860, p. 192.—Lawr
Ann. N. Y. Lyc. 1860, p. 18.—Gundl. J. f. O. 1861, p. 326.
Dendroica gundlachi Baird, Rev. Am. Bds. p. 197 (1864).—Gundl.
J. f. O. 1872, p. 414.
Dendroica gundlachi Gundl. Repert. Fisico-Nat. Cuba, I, p. 234 (1865)
—Gray, Handl. Bds. I, p. 241 (1869).—Sharpe, Cat. Bds. Brit.
Mus. X, p. 278 (1885).
Dendrœca petechia d. *cubana* Sund. Oefv. K. Vet. Akad. Förh. 1869
p. 608.
Dendroica petechia var. *gundlachi* Bd. Bwr. & Ridgw. Hist. N. Am. Bds.
I, p. 216 (1874).
Dendrœca petechia var. *gundlachi* Cory. Bds. Bahama I. p. 58 (1880).
Dendrœca petechia gundlachi Coues, Bds. Colo. Vall. p. 255 (1878).—
Cory, List Bds. W. I. p. 8 (1885).

Sp. Char. *Male:*—Lower part of throat streaked; above yellowish green;
crown showing no signs of rufous, or only a faint tinge; feathers
yellowish, brighter towards the bill.

Female:—Similar to the male, but somewhat paler, and showing less yellow on the tail.

Length (skin), 4.8; wing, 2.45; tail, 2.15; tarsus, .83.

HABITAT. Cuba. Accidental in the Bahama Islands.

Dendroica petechia ruficapilla.

Motacilla ruficapilla GMEL. Syst. Nat. I, p. 971 (1788).
Sylvicola ruficapilla BP. Consp. I, p. 307 (1850).
Dendroica ruficapilla BAIRD, Rev. Am. Bds. p. 291 (1864).
Dendræca ruficapilla GRAY, Handl. Bds. I, p. 240 (1869).—SHARPE, Cat. Bds. Brit. Mus. X, p. 275 (1885).
Dendroica petechia var. *ruficapilla* BD. BWR. & RIDGW. Hist. N. Am. Bds. I, p. 217 (1874).
Dendræca petechia var. *ruficapilla* LAWR. Pr. U. S. Nat. Mus. I, p. 486 (1878).
Dendræca petechia ruficapilla CORY, List Bds. W. I. p. 8 (1885).

Length, 4.75; wing, 2.6; tail, 2.10; tarsus, .82.

This form approaches very closely to *D. petechia*, but lacks the distinct rufous crown. Throat streaked heavily; the under tail-coverts are also streaked; otherwise like *D. petechia*.

HABITAT. Barbuda, Antigua, Porto Rico, and St. Thomas.

Dendroica petechia melanoptera.

Dendræca petechia var. *melanoptera* LAWR. Pr. U. S. Nat. Mus. I, p. 453 (1878).
Dendræca petechia melanoptera CORY, List Bds. W. I. p. 8 (1885).
Dendræca melanoptera SHARPE, Cat. Bds. Brit. Mus. X, p. 279 (1885).

Length, 4.50; wing, 2.30; tail, 1.85; tarsus, .69.

This form resembles *petechia ruficapilla*, as would be expected, but varies in having the wing-coverts black, and it is somewhat smaller in size; the rufous streaks are narrower and darker. The female lacks the rufous crown and stripes on the under surface.

HABITAT. Guadeloupe and Dominica.

Dendroica capitalis LAWR.

Dendræca petechia c. *barbadensis* SUND. Oefv. K. Vet. Akad. Förh. 1869, p. 608.
Dendræca capitalis LAWR. Pr. Acad. Nat. Sci. Phila. 1868, p. 359.—GRAY. Handl. Bds. III, Index, p. 202 (1871).—COUES, Key N. Am. Bds. p.

46   CORY *on the Birds of the West Indies.*

297 (1884).—CORY, List. Bds. W. I. p. 8 (1885).—SHARPE, Cat. Bds.
Brit. Mus. X. p. 280 (1885).
Dendroica capitalis BD. BWR. & RIDGW. Hist. N. Am. Bds. I, p. 271
(1874).

SP. CHAR. *Male:*—Top of the head dark rufous brown, extending to the
nape, but not reaching the eye; upperparts greenish yellow; wings
and tail brown, edged with yellow; inner webs of the tail-feathers
broadly edged with bright yellow; underparts yellow, streaked with
rufous brown.
   Female:—Entire upper surface olive green; entire under surface
pale yellow; tail as in the male.
   Length (skin), 4; wing, 2.45; tail, 1.75; tarsus, .75.
HABITAT. Barbadoes.

Dendroica rufigula BAIRD.

Dendroica rufigula BAIRD, Rev. Am. Bds. p. 204 (1864).
Dendroeca rufigula GRAY, Handl. Bds I, p. 241 (1869).—LAWR. Pr. U. S.
Nat. Mus. I, p. 486 (1878).—CORY, List Bds. W. I. p. 8 (1885).—
SHARPE, Cat. Bds. Brit. Mus. X. p. 285 (1885).
Dendroica vieilloti var. *rufigula* BD. BWR. & RIDGW. Hist. N. Am. Bds.
I, p. 217 (1874).
Dendroeca vieilloti rufigula COUES, Bds. Colo. Vall. p. 256 (1878).
Dendroeca vieilloti (pt.) SALV. & GODM. Biol. Centr. Amer. Aves, I,
p. 125 (1880).

SP. CHAR. *Male:*—Head and throat rufous brown; upper parts greenish
yellow; wings and tail brown, broadly edged with yellow; underparts bright yellow, streaked with rufous on the breast and flanks;
axillaries and under wing-coverts bright yellow.
   Length (skin), 5; wing, 2.25; tail, 2; tarsus, .75.
HABITAT. Martinique.

Dendroica eoa (GOSSE).

Sylvicola eoa GOSSE, Bds. Jam. p. 158 (1847).—BP. Consp. I, p. 309
(1850).—ALBRECHT, J. f. O. 1862, p. 201.
Dendroeca eoa SCL. P. Z. S. 1861, p. 71 (?)—GRAY, Handl. Bds. I, p. 240
(1869).—SUND. Oefv. K. Vet. Akad. Förh. 1869, p. 609—A. &
E. NEWTON, Handb. Jamaica, p. 106 (1881).—COUES, Key N. Am.
Bds. p. 297 (1884).—CORY, List Bds. W. I. p. 8 (1885).—SHARPE,
Cat. Bds. Brit. Mus. X, p. 266 (1885).
Dendroica eoa BAIRD, Rev. Am. Bds. p. 195 (1864).—BD. BWR. & RIDGW.
Hist. N. Am. Bds. I, p. 218 (1874).

"*Male:*—Upper parts olive, approaching to yellow on the rump; sides of
head marked with a band of orange, extending from the ear to the
beak, and meeting both on the forehead and on the chin. Wings
(quills and coverts) blackish, with yellowish edges. Tail blackish-
olive, with yellow edges; the outermost two feathers on each side
have the greatest portion of the inner webs pale yellow. Underparts pale yellow. The crown, rump, tertials, belly, and
under tail-coverts sparsely marked with undefined spots of pale
orange.

"*Female:*—Nearly as in the male, but the deep orange is spread
over the whole cheeks, chin, throat, and breast. The head and back
are dusky gray, tinged with olive, and patched with the fulvous much
more largely, but irregularly, as if *laid* upon the darker hue.
Length, 5 inches; expanse, 7.60; wing, 2.70; tail, 1.90; rictus
nearly .60; tarsus, .90; middle toe, .50. Iris dark hazel; feet horn-
color; beak pale horn; culmen and tip darker." (*Gosse,* l. c.).

HABITAT. Jamaica.

Mr. Sharpe considers *D. eoa* to be a hybrid between *D. blackburniæ* and *D. petechia* or *D. æstiva*. The type specimens
are in the British Museum.

Dendroica cærulescens (GMEL.).

Motacilla cærulescens GMEL. Syst. Nat. I, p. 960 (1788).

Sylvia cærulescens D'ORB. in La Sagra's Hist. Nat. Cuba, Ois. p. 63
(1840).

Sylvicola pannosa GOSSE, Bds. Jam. p. 162 (1847).

Sylvicola canadensis GOSSE, Bds. Jam. p. 162 (1847).—SALLÉ, P. Z. S.
1857, p. 231 (San Domingo).—BRYANT, Pr. Bost. Soc. Nat. Hist.
VII, p. 110 (1859) (Bahamas).—BREWER, Pr. Bost. Soc. Nat. Hist.
VII, p. 307 (1860) (Cuba).

Rhimamphus canadensis GUNDL. J. f. O. 1855, p. 473; 1861, p. 408 (Cuba).

Dendrœca pannosa ALBRECHT, J. f. O. 1862, p. 193 (Jamaica).

Dendroica canadensis GUNDL. J. f. O. 1861, p. 396 (Cuba).—MARCH, Pr.
Acad. Nat. Sci. Phila. 1863, p. 293 (1863) (Jamaica).

Dendrœca canadensis SCL. P. Z. S. 1861, p. 70 (Jamaica).—ALBRECHT, J.
f. O. 1862, p. 193.

Dendroica cærulescens BAIRD, Rev. Am. Bds. p. 186 (1864) (?).—GUNDL.
Report. Fisico-Nat. Cuba, I, p. 233 (1865): *ib.* J. f. O. 1872, p. 413
(Cuba); *ib.* Anal. Soc. Esp. Hist. Nat. VII, p. 179 (1878) (Porto
Rico).

Sylvicola (Dendrœca) canadensis BRYANT, Pr. Bost. Soc. Nat. Hist. XI,
p. 91 (1867).

Dendrœca cærulescens CORY, Bds. Bahama I. p. 58 (1880); *ib.* Bull. Nutt.
Orn. Club, VI, p. 151 (1881); *ib.* Bds. Haiti & San Domingo, p. 26

(1885); *ib.* List Bds. W. I. p. 8 (1885).—A. & E. Newton, Handb. Jamaica, p. 106 (1881).—Tristram, Ibis, 1884, p. 168.

Common in the Bahamas and Greater Antilles in winter.

Dendroica coronata (Linn.).

Motacilla coronata Linn. Syst. Nat. I, p. 333 (1766).
Sylvia coronata D'Orb in La Sagra's Hist. Nat. Cuba, Ois. p. 60 (1840).
Rhimamphus coronatus Gundl. J. f. O. 1855, p. 473; *ib.* 1861, p. 408 (Cuba).
Sylvicola coronata Gosse, Bds. Jam. p. 155 (1847).—Sallé, P. Z. S. 1857, p. 231 (San Domingo).—Bryant, Pr. Bost. Soc. Nat. Hist. VII, p. 110, (1859) (Bahamas).—Brewer, Pr. Bost. Soc. Nat. Hist. VII, p. 307 (1860) (Cuba).—Albrecht, J. f. O. 1862, p. 201 (Jamaica).
Dendroica coronatus Gundl. J. f. O. 1861, p. 326 (Cuba).
Dendroica coronata March, Pr. Acad Nat. Sci. Phila. 1863, p. 292 (Jamaica).—Gundl. Repert. Fisico-Nat. Cuba, I, p. 233 (1865) (Cuba); *ib.* Anal. Soc. Esp. Hist. Nat. VII, p. 180 (1878) (Porto Rico).
Sylvicola (Dendrœca) coronata Bryant, Pr. Bost. Soc. Nat. Hist. XI, p. 91 (1867).
Dendrœca coronata Cory, Bds. Bahama I, p. 59 (1880); *ib.* Bull. Nutt. Orn. Club, VI, p. 151 (1881); *ib.* Bds. Haiti & San Domingo, p. 30 (1885); *ib.* List. Bds. W. I. p. 8 (1885).—A. & E. Newton, Handb. Jamaica, p. 109 (1881).

Common in winter in the Bahamas and Greater Antilles.

Dendroica maculosa (Gmel.).

Motacilla maculosa Gmel. Syst. Nat. I. p. 984 (1788).
Sylvia maculosa D'Orb in La Sagra's Hist. Nat. Cuba, Ois. p. 72 (1840).
Rhimamphus maculosus Gundl. J. f. O. 1855, p. 474 (Cuba).
Sylvicola maculosa Bryant, Pr. Bost. Soc. Nat. Hist. VII, p. 110 (1859) (Bahamas).—Brewer, Pr. Bost. Soc. Nat. Hist. VII, p. 307 (1860) (Cuba).
Dendroica maculosa Gundl. J. f. O. 1861, p. 326; *ib.* 1872, p. 415; *ib.* Repert. Fisico-Nat. Cuba, I, p. 234 (1865) (Cuba); *ib.* Anal. Soc. Esp. Hist. Nat. VII, p. 183 (1878) (Porto Rico).
Dendrœca maculosa Cory, Bds. Bahama I. p. 62 (1880); *ib.* Bds. Haiti & San Domingo, p. 29 (1885); *ib.* List Bds. W. I. p. 8 (1885).

Occasional winter visitant in the Greater Antilles and the Bahama Islands.

Dendroica cærulea (WILS.).

Sylvia cærulea WILS. Am. Orn. II. p. 141 (1810).
Dendroica cærulea GUNDL. J. f. O. 1861, p. 326; *ib.* 1872, p. 411; *ib.*
    Repert. Fisico-Nat. Cuba, I, p. 234 (1865) (Cuba).—BAIRD, Rev.
    Am. Bds. p. 191 (1864).
Rhimamphus cæruleus GUNDL. J. f. O. 1862, p. 177 (Cuba).
Dendrœca cærulea CORY, List Bds. W. I. p. 8 (1885).

Cuba. No other West India Record.

Dendroica pennsylvanica (LINN.).

Motacilla pennsylvanica LINN. Syst. Nat. I. p. 333 (1766).
Sylvicola icterocephala BRYANT, Pr. Bost. Soc. Nat. Hist. VII, p. 110
    (1859) (Bahamas).
Dendroica pennsylvanica BAIRD, Rev. Am. Bds. p. 191 (1864).
Dendrœca pennsylvanica CORY, Bds. Bahama I. p. 62 (1880); *ib.* List
    Bds. W. I. p. 8 (1885).

Bahama Islands in winter.

Dendroica striata (FORST.).

Muscicapa striata "FORSTER, Phil. Trans. LXII, 383."
Sylvia striata LEMB. Aves Cuba, p, 33 (1850).
Rhimamphus striatus GUNDL. J. f. O. 1855. p. 475; *ib.* 1861, p. 409 (Cuba).
Sylvicola striata BRYANT, Pr. Bost. Soc. Nat. Hist. VII, p. 110 (1859)
    (Bahamas).—BREWER *ib.* p. 307 (Cuba).
Dendroica striatus GUNDL., J. f. O. 1861, p. 326 (Cuba).
Dendroica striata GUNDL. Repert. Fisico-Nat. Cuba, I, p. 234 (1865);
    ib. J. f. O. 1872, p. 414 (Cuba); *ib.* Anal. Soc. Esp. Hist. Nat. VII,
    p. 181 (1878) (Porto Rico).
Dendrœca striata CORY, Bds. Bahama I. p. 61 (1880); *ib.* List Bds. W. I.
    p. 8 (1885).—A. & E. NEWTON, Handb. Jamaica, p. 106 (1881).

Common in winter in the Bahama Islands. Recorded from
Cuba, Porto Rico, and Jamaica.

Dendroica pharetra (GOSSE).

Sylvicola pharetra GOSSE, Bds. Jam. p. 163 (1857).—BP. Consp. I. p.
    309 (1850).—OSBURN, Zool. 1859, p. 6660.
Dendrœca pharetra SCL. P. Z. S. 1861, p. 71.—ALBRECHT, J. f. O. 1862,
    p. 193.—GRAY, Handl. Bds. I, p. 241 (1869).—SCL. & SALV. Nom.
    Avium Neotr. p. 9 (1873).—A. & E. NEWTON, Handb. Jamaica, p.
    106 (1881).—COUES, Key N. Am. Bds. p. 297 (1884).—CORY, List
    Bds. W. I. p. 8 (1885).—SHARPE, Cat. Bds. Brit. Mus. X, p. 332
    (1885).

Dendroica pharetra BAIRD, Rev. Am. Bds. p. 192 (1864).—ED. BWR. &
     RIDGW. Hist. N. Am. Bds. I, p. 220 (1874).

SP. CHAR. Male :—Entire plumage dull white and black, in general appear-
     ance resembling *Mniotilta varia* at the first glance. Throat white,
     the feathers narrowly tipped with black, giving a dotted appearance ;
     the black marking becomes heavier on the breast and belly ; top
     of head heavily streaked with black and white ; rump and upper tail
     coverts olive brown ; wings and tail brown, showing a faint olive
     tinge ; under wing-coverts white.
        Female:—Similar to the male, but duller in coloration ; less black
     on the under surface ; more brown on the lower back, rump and
     tail.
        Length (skin), 4.40; wing. 2.30; tail, 2.05; tarsus, .72.
HABITAT.   Jamaica.

Dendroica blackburniæ (GMEL.).

Motacilla blackburniæ GMEL. Syst. Nat. I, p. 977 (1788).
Sylvicola blackburniæ BRYANT, Pr. Bost. Soc. Nat. Hist. VII, p. 110
     (1859) (Bahamas).
Dendroica blackburniæ BAIRD, Rev. Am. Bds. p. 189 (1864).
Dendroeca blackburniæ CORY, Bds. Bahama I. p. 60 (1880) ; *ib.* List Bds.
     W. I. p. 8 (1885).

   Accidental in the Bahama Islands in winter.

Dendroica dominica (LINN.).

Motacilla dominica LINN. Syst. Nat. I, p. 334 (1766).
Sylvia pensilis D'ORB. in La Sagra's Hist. Nat. Cuba, Ois. p. 65 (1840).
Sylvicola pensilis GOSSE, Bds. Jam. p. 156 (1847).—SALLE, P. Z. S. 1857,
     p. 231 (San Domingo).—BREWER, Pr. Bost. Soc. Nat. Hist. VII, p.
     307 (1860) (Cuba).—ALBRECHT, J. f. O. 1862, p. 201 (Jamaica).
Rhimamphus pensilis GUNDL. J. f. O. 1855, p. 474 ; *ib.* 1861, p. 408 (Cuba).
Dendroica superciliosa GUNDL. J. f. O. 1861, p. 326 (Cuba).—MARCH, Pr.
     Acad. Nat. Sci. Phila. 1863, p. 293 (Jamaica).
Dendroica dominica GUNDL. Repert. Fisico-Nat. Cuba. I, p. 235 (1865) ;
     ib. J. f. O. 1872, p. 415 (Cuba).—BRACE, Pr. Bost. Soc. Nat. Hist.
     XIX, p. 240 (1877) (Bahamas).—GUNDL. Anal. Soc. Esp. Hist.
     Nat. VII, p. 184 (1878) (Porto Rico).
Dendroeca dominica CORY, Bds. Bahama I. p. 65 (1880) ; *ib.* Bds. Haiti &
     San Domingo, p. 27 (1885) ; *ib.* List Bds. W. I. p. 8 (1885).—A. & E.
     NEWTON, Handb. Jamaica. p. 106 (1881).

   Common in winter in the Bahamas and Greater Antilles ; possibly resident in Jamaica.

Dendroica adelaidæ BAIRD.

Dendroica adelaidæ BAIRD, Rev. Am. Bds. p. 212 (1864).
Sylvicola (Dendrœca) adelaidæ BRYANT, Pr. Bost. Soc. Nat. Hist. X,
 p. 251 (1866).
Dendrœca adelaidæ SUND. Oefv. K. Vet. Akad. Förh. Stockh. 1869, p.
 615.—GRAY, Handl. Bds. II. p. 241 (1870).—SCL. & SALV. Nom.
 Avium Neotr. p. 9 (1873).—GUNDL. Anal. Soc. Esp. Hist. Nat. VII,
 p. 185 (1878).—RIDGW. Pr. U. S. Nat. Mus. V, pp. 525, 526 (1883).
 —COUES, Key N. Am. Bds. p. 297 (1884).—CORY, List Bds. W. I.
 p. 8 (1885).—SHARPE, Cat. Bds. Brit. Mus. X, p. 306 (1885).
Dendroica gracie var. *adelaidæ* BD. BREW. & RIDGW. Hist. N. Am. Bds.
 I, p. 220 (1874).

Sp. CHAR. *Male:*—"Entire upper parts, and sides of neck as far forward
 as the eyes, uniform ash gray. Beneath, including edge of bend of
 wing, bright yellow; lining of wings, axillaries, and crissum,
 white. A broad yellow line from bill to eye, with the eyelids yellow;
 forehead and sides of vertex black. A black loral line. Wings
 with two conspicuous white bands; the quills and tail-feathers
 blackish, edged externally with whitish, internally with purer
 white. There lateral tail-feathers with a quadrate terminal white
 patch on inner web. Bill black. Legs pale yellowish." (BAIRD, l. c.).
   Length, 4.7; wing, 2.1; tail, 2.05; tarsus, .65.

HABITAT. Porto Rico.

Dendroica adelaidæ delicata.

Dendrœca adelaidæ SCL. P. Z. S. 1871, p. 269.—ALLEN, Bull. Nutt. Orn.
 Club, V, p. 166 (1880).
Dendrœca adelaidæ delicata RIDGW. Pr. U. S. Nat. Mus. V, p. 525 (1882).
 —CORY, List Bds. W. I. p. 8 (1885).
Dendrœca delicata SHARPE, Cat. Bds. Brit. Mus. X, p. 306 (1885).

   General appearance of *D. adelaidæ*, but differs in having brighter
yellow on the superciliaries and underparts; the yellow superciliary line
is broader, occupying the whole forehead except a narrow central line,
and the back more plumbeous; it is also slightly larger.
   Length, 4.45; wing, 2.10; tail, 2.10; tarsus, .70.

HABITAT. Santa Lucia.

Dendroica virens (GMEL.).

Motacilla virens GMEL. Syst. Nat. I, p. 985 (1788).
Rhimamphus virens GUNDL. J. f. O. 1855, p. 474 (Cuba).
Sylvicola virens BREWER Pr. Bost. Soc. Nat. Hist. VII, p. 307 (1860)
 (Cuba).

Dendroica virens BAIRD, Rev. Am. Bds. p. 182 (1864).—GUNDL. J. f. O.
    1861, p. 426; *ib.* 1872, p. 413; *ib.* Repert. Fisico-Nat. Cuba, 1, p.
    233 (1865) (Cuba).
Dendroeca virens LAWR. Pr. U. S. Nat. Mus. I, p. 54 (1878) (Dominica).—
    A. & E. NEWTON, Handb. Jamaica, p. 106 (1881).—CORY, List Bds.
    W. I. p. 8 (1885).

Recorded from Cuba, Jamaica, and Dominica.

Dendroica kirtlandi BAIRD.

Sylvicola kirtlandi BAIRD, Ann. N. Y. Lyc. V, p. 217 (1852).
Dendroica kirtlandii BAIRD, Rev. Am. Bds. p. 206 (1864).
Dendroeca kirtlandi CORY, Bds. Bahama I. p. 66 (1880); *ib.* List Bds. W.
    I. p. 8 (1885).

Common in winter at New Providence and Andros, Bahama
Islands; probably ranges as far south as Long Island; no other
record. It is possible that it is resident and breeds in the
Bahama Islands.

Dendroica pityophila (GUNDL.).

Sylvicola pityophila GUNDL. Ann. N. Y. Lyc. 1855, p. 160.—BREWER, Pr.
    Bost. Soc. Nat. Hist. VII, p. 307 (1860).
Rhimamphus pityophilus GUNDL. J. f. O. 1857, p. 240.
Dendroica pityophila BAIRD, Rev. Am. Bds. p. 208 (1864).—BD. BWR. &
    RIDGW. Hist. N. Am. Bds. I, p. 221 (1874).
Dendroeca pityophila GUNDL. Repert. Fisico-Nat. Cuba, I, p. 234 (1865).
    —GRAY, Handb. Bds. I, p. 241 (1869).—COUES, Key N. Am. Bds. p.
    297 (1884).—CORY, List Bds. W. I. p. 8 (1885).—SHARPE, Cat. Bds.
    Brit. Mus. X, p. 322 (1885).

SP. CHAR. *Male:*—"Above, including sides of head and neck, uniform
    plumbeous gray; the forehead, vertex and loral region olive green;
    chin and fore-neck bright yellow, extending on the middle of jugu-
    lum, and bordered by black streaks towards lower part of neck, most
    conspicuous on sides of breast. Beneath dull white, the insides of
    wings more ashy, the flanks something like the back. Two dull
    ashy white bands across the wing-coverts; the quill- and tail-feathers
    edged with paler ash than the ground color. Lateral tail-feather
    with a whitish patch on the inner web, running forward to a point
    along the shaft, including the whole web at the end; second feather
    with a more restricted patch of the same." (BAIRD, l. c.)
    Length, 4.50; wing, 2.30; tail, 2.20; tarsus, .56; bill, .45.

HABITAT. Cuba.

Dendroica vigorsii (AUD.).

Sylvia pinus WILS. Am. Orn. III, p. 25 (1811) (Nec LATHAM, 1790).
Sylvicola (Dendrœca) pinus BRYANT, Pr. Bost. Soc. Nat. Hist. XI, p. 67
(1867).
Dendrœca pinus CORY, Bds. Bahama I. p. 69 (1880); *ib.* Bds. Haiti & St.
Domingo, p. 33 (1885); *ib.* List. Bds. W. I. p. 8 (1885).
Sylvia vigorsii AUD. Orn. Biog. I, 153 (1835).
Dendroica vigorsii STEJN. Auk, II, 343 (1885).

Common in winter in the Bahama Islands and San Domingo ;
breeds in San Domingo.

Dendroica discolor (VIEILL.).

Sylvia discolor VIEILL. Ois. Am. Sept. II, p. 37 (1807).—LEMB. Aves
Cuba, p. 32 (1850) (Cuba).
Sylvicola discolor GOSSE, Bds. Jam. p. 159 (1847).—BRYANT, Pr. Bost.
Soc. Nat. Hist. VII, p. 110 (1859) (Bahamas); *ib.* X, p. 251 (1866)
—BREWER, *ib.* VII, p. 307 (1860) (Cuba).
Rhimamphus discolor GUNDL. J. f. O. 1855, p. 474 (Cuba).
Dendrœca discolor A. & E. NEWTON, Ibis, 1859, p. 144 (St. Croix); *ib.*
Handb. Jamaica, p. 106 (1881).—SCL. P. Z. S. 1861, p. 71 (Ja-
maica).— CORY, Bds. Bahama I. p. 64 (1880); *ib.* Bull. Nutt. Orn.
Club, VI, p. 151 (1881); *ib.* Bds. Haiti & San Domingo, p. 31
(1885); *ib.* List Bds. W. I. p. 8 (1885).
Dendroica discolor GUNDL. J. f. O. 1861, p. 326; *ib.* 1872, p. 416; *ib.* Re-
pert. Fisico-Nat. Cuba, I, p. 235 (1865) (Cuba).—MARCH, Pr. Acad.
Nat. Sci. Phila. 1863, p. 293 (Jamaica).—GUNDL. Anal. Soc. Esp.
Hist. Nat. VII, p. 186 (1878) (Porto Rico).
Sylvicola (Dendrœca) discolor BRYANT, Pr. Bost. Soc. Nat. Hist. XI, p.
91 (1867).

Winters in the Bahamas, the Greater and some of the Lesser
Antilles.

Dendroica palmarum (GMEL.).

Motacilla palmarum GMEL. Syst. Nat. I, p. 951 (1788).
Sylvia palmarum D'ORB. in La Sagra's Hist. Nat. Cuba, Ois. p. 61 (1840).
Rhimamphus ruficapillus GUNDL. J. f. O. 1855, p. 473; *ib.* 1861, p. 408
(Cuba).
Sylvicola palmarum SALLÉ, P. Z. S. 1857, p. 231 (San. Domingo).—BRY-
ANT, Pr. Bost. Soc. Nat. Hist. VII, p. 110 (1859) (Bahamas).
Dendroica palmarum GUNDL. J. f. O. 1861, p. 326; *ib.* 1872, p. 415; *ib.*
Repert. Fisico-Nat. Cuba, I. p. 234 (1865) (Cuba); *ib.* Anal. Soc.
Esp. Hist. Nat. VII, p. 183 (1878) (Porto Rico)

Dendrœca palmarum SCL. P. Z. S. 1861, p. 71 (Jamaica).—ALBRECHT,
    J. f. O. 1862, p. 93 (Jamaica).—CORY, Bds. Bahama I. p. 68 (1880).
    —A. & E. NEWTON, Handb. Jamaica. p. 106 (1881).
Sylvicola (Dendrœca) palmarum BRYANT, Pr. Bost. Soc. Nat. Hist. XI,
    p. 91 (1867).

Common in winter in the Bahama Islands and Greater Antilles.

Dendroica plumbea LAWR.

Dendrœca plumbea LAWR. Ann. N. Y. Acad. Sci. I, p. 47 (1878); *ib.* Pr.
    U. S. Nat. Mus. I. p. 486 (1878).—CORY. List Bds. W. I. p. 8 (1885).
    —SHARPE, Cat. Bds. Brit. Mus. X, p. 333 (1885).

SP. CHAR. *Male:*—General plumage above dark plumbeous; a superciliary
    stripe of white from the bill; a spot of white on the lower eyelid;
    lores very dark brown, almost black; underparts mixed with ashy
    and dull white; outer tail-feather tipped with white on the inner
    web; next feather showing a smaller spot; next two narrowly tip-
    ped with white; middle and greater wing-coverts tipped with white,
    forming two wing-bands.

    Female:—Above dark olive: underparts grayish, tinged with olive;
    showing a pale yellowish wash on the throat, breast and middle of
    the abdomen.

    Length (skin), 5.20; wing, 2.45; tail, 2.25; tarsus, .72.

HABITAT. Guadeloupe and Dominica.

GENUS Leucopeza SCL.

Leucopeza SCLATER, P. Z. S. 1876, p. 14.

Leucopeza semperi SCL.

Leucopeza semperi SCL. P. Z. S. 1876, p. 14.—ALLEN, Bull. Nutt. Orn.
    Club, V, p. 166 (1880).—CORY, List Bds. W. I. p. 8 (1885).—SHARPE,
    Cat. Bds. Brit. Mus. X, p. 228 (1885).

SP. CHAR. *Male:*—General plumage above dark bluish gray; slightly
    brownish on lower back and rump; sides of head and ear-coverts
    slightly paler; throat and breast grayish white, shading into brown-
    ish on the belly; crissum. axillaries and under wing-coverts ashy
    gray, edged with dull white.

    Length, 5.70; wing, 2.60; tail, 2.20; tarsus, 88; bill, .68.

HABITAT. Santa Lucia.

Genus Catharopeza Scl.

Catharopeza Sclater, Ibis, 1880, p. 73.

Catharopeza bishopi (Lawr.).

Leucopeza bishopi Lawr. Ann. N. Y. Acad. Sci. I, p. 151 (1878); *ib.* Pr. U. S. Nat. Mus. I. p. 486 (1878).—Sharpe, Cat. Bds. Brit. Mus. X, p. 228 (1885).
Catharopeza bishopi Scl. Ibis, 1880, p. 73.—Lister, Ibis, 1880, p. 40.— Cory, List Bds. W. I, p. 8 (1885).

Habitat. St. Vincent.

Sp. Char. *Male:*—"The general plumage is smoky black; rather darker on the head; the sides are blackish cinereous; a circle of pure white surrounds the eye; a large roundish spot on the middle of the throat; the upper part of the breast, and the middle of the abdomen, are dull white, somewhat mixed with blackish on the throat and with cinereous on the abdomen; a very small spot on the chin, and the tips of the feathers on the upper part of the throat are dull white; the black on the upper part of the breast has the appearance of a broad band, separating the white of the throat from that of the lower part of the breast; the under tail-coverts are cinereous-black at base, ending largely with dull white; wings and tail black, the outer two tail-feathers have a small white spot, triangular in shape, on the inner webs at the end; bill black; tarsi and toes very pale yellowish-brown, perhaps much lighter colored in the living bird, nails also pale.

Length (fresh), 5¾ in.; wing, 2¾; tail, 2½; tarsus, ⅞. Two specimens marked as females do not differ in plumage from the males." (Lawr. l. c.)

Habitat. St. Vincent.

Genus Seiurus Swains.

Seiurus Swainson, Zool. Journ. III, p. 171 (1827).

Seiurus aurocapillus (LINN.).

Motacilla aurocapilla LINN. Syst. Nat. I. p. 334 (1766).
Seiurus aurocapillus D'ORB. in La Sagra's Hist. Nat. Cuba, Ois. p. 55
    (1840).
Seiurus aurocapillus GOSSE, Bds. Jam. p. 152 (1847).—SALLÉ, P. Z. S.
    1857, p. 321 (San Domingo).—MARCH, Pr. Acad. Nat. Sci. Phila.
    1863, p. 294 (Jamaica).—GUNDL. Repert. Fisico-Nat. Cuba, I, p. 325
    (1865) (Cuba); ib. Anal. Soc. Esp. Hist. Nat. VII. p. 175 (1878)
    (Porto Rico).—CORY, Bds. Bahama I. p. 70 (1880); ib. Bull. Nutt.
    Orn. Club, VII, p. 151 (1881); ib. Bds. Haiti & San Domingo,
    p. 34 (1885); ib. List Bds. W. I. p. 8 (1885).
Henicocichla aurocapilla GUNDL. J. f. O. 1855, p. 471; ib. 1861, pp. 326,
    407 (Cuba).—SCL. P. Z. S. 1861, p. 70 (Jamaica).—ALBRECHT, J. f.
    O. 1862, p. 192 (Jamaica).
Siurus aurocapillus A. & E. NEWTON, Ibis, 1859, p. 142 (St. Croix); ib.
    Handb. Jamaica, p. 105 (1881).
Enicocichla aurocapillus BREWER, Pr. Bost. Soc. Nat. Hist. VII, p. 306
    (1860) (Cuba).

Ranges in winter throughout the West Indies.

Seiurus noveboracensis (GMEL.).

Motacilla noveboracensis GMEL. Syst. Nat. I, p. 958 (1788).
Siurus nævius LAWR. Pr. U. S. Nat. Mus. I, p. 233 (Antigua), p. 453
    (Guadeloupe), p. 54 (Dominica) (1878).
Seiurus sulfurascens D'ORB. in La Sagra's Hist. Nat. Cuba, Ois. p. 57
    (1840) (Cuba).
Seiurus noveboracensis GOSSE, Bds. Jam. p. 151 (1847).—MARCH, Pr.
    Acad. Nat. Sci. Phila. 1863, p. 294 (Jamaica).—GUNDL. Repert. Fisico-
    Nat. Cuba, I, p. 235 (1865); ib. J. f. O. 1872, p. 416 (Cuba); ib.
    Anal. Soc. Esp. Hist. Nat. VII, p. 175 (1878) (Porto Rico).—CORY,
    Bds. Bahama I. p. 71 (1880); ib. Bull. Nutt. Orn. Club, VI, p. 151
    (1881); ib. List Bds. W. I. p. 8 (1885).
Seiurus gossii BP. Consp. I, p. 306 (1850) (Jamaica).
Henicocichla sulphurascens GUNDL. J. f. O. 1855, p. 471; ib. 1861, p. 407
    (Cuba).
Henicocichla noveboracensis GUNDL. J. f. O. 1855, p. 471; ib. 1861, pp. 326,
    407 (Cuba).—SCL. P. Z. S. 1861, p. 70 (Jamaica).—ALBRECHT, J. f.
    O. 1862, p. 192 (Jamaica).
Siurus noveboracensis A. & E. NEWTON, Ibis, 1859, p. 145 (St. Croix); ib.
    Handb. Jamaica, p. 105 (1881).
Enicocichla noveboracensis BREWER, Pr. Bost. Soc. Nat. Hist. VII, p. 306
    (1860) (Cuba).

The present species ranges in winter throughout the West
Indies.

Seiurus motacilla (VIEILL.).

Turdus motacilla VIEILL. Ois. Am. Sept. II. p. 9 (1807).
Henicocichla motacilla CAB. J. f. O. 1857, p. 240 (Cuba).—GUNDL. J. f. O.
1861, p. 326 (Cuba).
Henicocichla major CAB. J. f. O. 1857, p. 240 (Cuba).
Enicocichla major BREWER, Pr. Bost. Soc. Nat. Hist. VII, p. 306 (1860)
(Cuba).
Henicocichla ludoviciana SCL. P. Z. S. 1861, p. 70 (Jamaica).—ALBRECHT,
J. f. O. 1862. p. 192 (Jamaica).
Seiurus ludovicianus GUNDL. Repert. Fisico-Nat. Cuba, I. p. 236 (1865);
ib. J. f. O. 1872, p. 417 (Cuba).
Siurus motacilla LAWR. Pr. U. S. Nat. Mus. I, pp. 233, 486 (1878) (An-
tigua).
Seiurus motacilla CORY, Bds. Haiti & San Domingo, p. 35 (1885); *ib.*
List Bds. W. I. p. 8 (1885).

Winters in the Greater Antilles; probably occurs throughout
the West Indies.

GENUS Geothlypis CABAN.

Geothlypis CABANIS, Arch. für Naturg. I, pp. 316, 449 (1847).

Geothlypis formosa (WILS.).

Sylvia formosa WILS. Am. Orn. III, p. 85 (1811).
Myiodioctes formosus LEMB. Aves Cuba, p. 37 (1850) (Cuba).—GUNDL.
J. f. O. 1861, p. 326 (Cuba).
Myiotomus formosus GUNDL. J. f. O. 1855, p. 472 (Cuba).
Setophaga formosa BREWER, Pr. Bost. Soc. Nat. Hist. VII, p. 307 (1860)
(Cuba).
Oporornis formosus GUNDL. Repert. Fisico-Nat. Cuba, I. p. 236 (1865);
ib. J. f. O. 1872. p. 417 (Cuba).—CORY, List Bds. W. I. p. 8 (1885).
Geothlypis formosa RIDGW. Pr. U. S. Nat. Mus. VIII, p. 354 (1885).
Accidental in Cuba.

Geothlypis rostrata BRYANT.

Geothlypis rostratus BRYANT, Pr. Bost. Soc. Nat. Hist. XI, p. 67 (1866).—
CORY, Bds. Bahama I. p. 73 (1880); *ib.* List Bds. W. I. p. 9 (1885).
Trichas rostrata GRAY, Handl. Bds. I, p. 242 (1869).
Geothlypis trichas var. *rostrata* BD. BWR. & RIDGW. Hist. N. Am. Bds. I,
p. 296 (1874).
Geothlypis rostrata SHARPE, Cat. Bds. Brit. Mus. X. p. 355 (1885).

SP. CHAR. *Male:*—Above bright olive green; a broad band of black passing from the sides of the neck, over the forehead, including the eye,

and extending to the nostril, just touching the lower mandible, the black bordered posteriorly with pearl gray, becoming deeper gray upon the crown; underparts bright yellow, the flanks shaded with olive; quills brown, with the outer webs olive green; third primary longest.

Female:—The black band wanting; plumage slightly paler; a pale ash-colored line from over the eye to sides of the neck; crown showing a trace of brown; otherwise resembling the male.

Length, 5.50; wing, 2.70; tail, 2.36; tarsus, .92; bill, .72.

HABITAT.   New Providence, Bahama Islands.

Geothlypis trichas· (LINN.).

Turdus trichas LINN. Syst. Nat. I, p. 293 (1766).

Sylvia trichas D'ORB. in La Sagra's Hist. Nat. Cuba, Ois. p. 67 (1840) (Cuba).

Trichas marylandica GOSSE, Bds. Jam. p. 148 (1847).—BRYANT, Pr. Bost. Soc. Nat. Hist. VII, p. 110 (1859) (Bahamas).

Trichas marilandica BREWER, Pr. Bost. Soc. Nat. Hist. VII. p. 307 (1860) (Cuba).

Geothlypis trichas GUNDL. J. f. O. 1855, p. 472; *ib.* 1861, p. 326; *ib.* 1872, p. 417; *ib.* Repert. Fisico-Nat. Cuba, I, p. 236 (1865) (Cuba).—SCL. P. Z. S. 1861, p. 70 (Jamaica).—ALBRECHT, J. f. O. 1862, p. 192 (Jamaica).—MARCH, Pr. Acad. Nat. Sci. Phila. 1863, p. 293 (Jamaica).—GUNDL. Anal. Soc. Esp. Hist. Nat. VII, p. 187 (1878) (Porto Rico).—CORY, Bds. Bahama I. p. 71 (1880); *ib.* Bull. Nutt. Orn. Club, VI, p. 151 (1881); *ib.* Bds. Haiti & San Domingo, p. 36 (1885); *ib.* List Bds. W. I. p. 9 (1885).—A. & E. NEWTON, Handb. Jam. p. 106 (1881).

Geothlypis restricta MAYNARD, Am. Ex. and Mart. Dec. 15, 1886.

Common in winter in the Bahama Islands and Greater Antilles.

GENUS Microligea CORY.

Microligea CORY, Auk, I, p. 290 (1884).

Microligea palustris CORY.

Ligea palustris CORY, Auk, I, p. 1 (1884); *ib.* Bds. Haiti & San Domingo, p. 38 (1885).—SHARPE, Cat. Bds. Brit. Mus. X, p. 349 (1885).

Microligea palustris CORY, Auk, I, p. 290 (1884); *ib.* List Bds. W. I. p. 9 (1885).

SP. CHAR. *Male:*—Crown, nape and upper portion of back siaty plumbeous; rest of back and upper surface of wings and tail yellowish
green; throat, breast and sides grayish plumbeous, showing a dull
olive tinge on the sides, darkest on the flanks; the middle of the
throat showing a slight grayish tinge, and the middle of the belly
showing distinctly white; outer webs of primaries and most of the
secondaries yellowish green, giving the wing a general greenish
appearance; inner webs of primaries dark brown, apparently slate
color in some lights; under surface of tail dull green; eyelids
white.

Female:—In general appearance like the male, but differs from it
by underparts being tinged with olive, mixing with the gray, and
top of the head green, showing the slate color faintly.

Length, 5.50; wing, 2.50; tail, 2.50; tarsus, .75; bill, .50; middle toe, .40.

HABITAT.   San Domingo.

GENUS Teretistris CABAN.

Teretistris CABANIS, "J. f. O. 1855, p. 475."

Teretistris fernandinæ (LEMB.).

Anabates fernandinæ LEMB. Aves Cuba, p. 66 (1850).—GUNDL. Journ.
Bost. Soc. Nat. Hist. VI, p. 317 (1852).
Helmitherus blandus Br. Consp. I, p. 314 (1850).
Teretistris fernandinæ CAB. J. f. O. 1855, p. 475.—BREWER, Pr. Bost. Soc.
Nat. Hist. VII, p. 307 (1860).—GUNDL. Repert. Fisico-Nat. Cuba, I,
p. 236 (1865); *ib.* J. f. O. 1872, p. 418.—GRAY, Handl. Bds. I, p. 384
(1869).—CORY, List Bds. W. I. p. 9 (1885).—SHARPE, Cat. Bds.
Brit. Mus. X, p. 368 (1885).
Teretristis fernandinæ BAIRD, Rev. Am. Bds. p. 234 (1864).—SCL. &
SALV. Nom. Avium Neotr. p. 11 (1873).

SP. CHAR. *Male:*—Top of the head bright olive green, this color extending to the upper back; rest of upper parts ash-gray; throat and
sides of the head bright yellow, tinged with olive on the cheeks
and ear-coverts; eyelids bright yellow; rest of underparts ash-gray,
whitish on the middle of the belly, and tinged with olive on the
flanks and sides; a slight tinge of olive on the carpus; under wing-
coverts white, slightly tinged with yellow.

Length, 4.85; wing, 2.20; tail, 1.95; tarsus, .75.

HABITAT.   Western part of Cuba.

Teretistris fornsi GUNDL.

Teretistris fornsi GUNDL. Ann. N. Y. Lyc. Nat. Hist. VI, p. 274 (1858).—
BREWER, Pr. Bost. Soc. Nat. Hist. VII. p. 307 (1860).—ALBRECHT,

J. f. O. 1861, p. 211; *ib.* J. f. O. 1862, p. 177; *ib.* Repert. Fisico-Nat. Cuba. I, p. 236 (1865); *ib.* J. f. O. 1872, p. 418.—CORY, List Bds. W. I. p. 9 (1885).—SHARPE, Cat. Bds. Brit. Mus. X, p. 368 (1885).
Teretristis fornsii BAIRD, Rev. Am. Bds. p. 235 (1864).
Teretistris fornsii GRAY, Handl. Bds. I, p. 384 (1869).
Teretristis fornsi SCL. & SALV. Nom. Avium Neotr. p. 11 (1873).

SP. CHAR. *Male:*—Top of head and upper parts pale ash-gray; a faint indication of yellow on the extreme forehead; sides of the head (including the eye), throat, and underparts yellow, becoming pale on the belly and ashy white on the flanks and crissum; wings and tail pale brown, the feathers pale edged; a tinge of yellow on the carpus and under wing-coverts.

　　Female:—Similar to the male, but less yellow on the underparts; ashy white on the belly.

　　Length, 4.60; wing, 2.15; tail, 1.95; tarsus, .72.
HABITAT. Eastern portion of Cuba.

GENUS Sylvania NUTTALL.

Sylvania "NUTT. Man. Orn. 1832."

Sylvania mitrata (GMEL.).

Motacilla mitrata GMEL. Syst. Nat. I, p. 977 (1788).
Setophaga mitrata D'ORB. in La Sagra's Hist. Nat. Cuba, Ois. p. 89 (1840) (Cuba).—BREWER, Pr. Bost. Soc. Nat. Hist. VII, p. 307 (1860) (Cuba).
Myioctomus mitratus GUNDL. J. f. O. 1855, p. 472; *ib.* 1861, p. 407 (Cuba); *ib.* 1872, p. 419 (Cuba).
Myiodioctes mitratus GUNDL. J. f. O. 1861, p. 326; *ib.* 1872, p. 419; *ib.* Repert. Fisico-Nat. Cuba, I, p. 237 (1865) (Cuba).—A. & E. NEWTON, Handb. Jam. p. 106 (1881).
Sylvania mitratus CORY, List Bds. W. I. p. 9 (1885).
　　Accidental in Cuba and Jamaica.

GENUS Setophaga SWAINS.

Setophaga SWAINSON, Zool. Journ. III, p. 360 (1827).

Setophaga ruticilla (LINN.).

Muscicapa ruticilla LINN. Syst. Nat. I, p. 326 (1766).—D'ORB. in La Sagra's Hist. Nat. Cuba, Ois. p. 87 (1840) (Cuba).
Setophaga ruticilla GOSSE, Bds. Jam. p. 164 (1874) (Jamaica).—GUNDL. J. f. O. 1855, p. 472; *ib.* 1861, p. 326; *ib.* 1872, p. 419; *ib.* Repert. Fisico-Nat. Cuba, I, p. 237 (1865) (Cuba).—SALLE, P. Z. S. 1857,

p. 231 (San Domingo).—A. & E. NEWTON, Ibis. 1859, p. 144 (St.
Croix).—BRYANT, Pr. Bost. Soc. Nat. Hist. VII, p. 111 (1859)
(Bahamas).—BREWER, Pr. Bost. Soc. Nat. Hist. VII, p. 307 (1860)
(Cuba).—SCL. P. Z. S. 1861, p. 72 (Jamaica).—ALBRECHT, J. f. O.
1862, p. 194 (Jamaica).—MARCH, Pr. Acad. Nat. Sci. Phila. 1863, p.
293 (Jamaica).—LAWR. Ann. Lyc. N. Y. VIII, p. 97 (1864) (Sombre-
ro Is.); ib. Pr. U. S. Nat. Mus. I, p. 486 (1878) (Lesser Antilles).—
SCL. P. Z. S. 1876, p. 14 (Sta. Lucia).—GUNDL. Anal Soc. Esp.
Hist. Nat. VII, p. 187 (1878) (Porto Rico).—CORY, Bds. Bahama I.
p. 75 (1880); ib. Bull. Nutt. Orn Club, VI, p. 151 (1881) (Haiti);
ib. Bds. Haiti & San Domingo, p. 40 (1885).—A. & E. NEWTON,
Handb. Jam. p. 106 (1881).—TRISTRAM, Ibis, 1884, p. 168 (San
Domingo).

The present species probably occurs in most of the West India
Islands. It is recorded from the Bahamas, all of the Greater,
and some of the Lesser Antilles.

FAMILY CŒREBIDÆ.

GENUS **Certhiola** SUNDEV.

Certhiola SUNDEV. Vet. Akad. Handl. Stockholm, p. 99 (1835).

Certhiola bahamensis REICH.

Parus bahamensis SELIGM, Samml. ausl. Vögel. III, p. t. xviii (1753).
Certhia bahamensis BRISS. Orn. III, p. 620 (1760).
Certhia flaveola var. β. LINN. Syst. Nat. I, p. 187 (1766).
Certhia flaveola var. γ. GMEL. Syst. Nat. I, p. 479 (1788).
Certhia flaveola var. γ. LATH. Ind. Orn. I, p. 297 (1790).—BECHST. Lath.
  Uebers. IV, p. 188.
Certhiola flaveola GRAY, Gen. Bds. I, p. 102 (1844).—BP. Consp. I, p.
  402 (1850).—BAIRD, Bds. N. Am. p. 924 (1858).
Certhiola bahamensis REICH. Handb. I, p. 253 (1853).—CASSIN, Pr. Acad.
  Nat. Sci. Phila. 1864, p. 271.—CAB. J. f. O. 1865. p. 412.—BRYANT,
  Pr. Bost. Soc. Nat. Hist. XI, p. 66 (1866).—GRAY, Handl. Bds. I,
  p. 120 (1869).—FINSCH, Verhandl. Zool. Botan. Gesells. Wien.
  XXI, p. 752 (1871).—SCL. & SALV. Nom. Avium Neotr. p. 16
  (1873).—BD. BWR. & RIDGW. Hist. N. Am. Bds. I, p. 428
  (1874).—CORY, Bds. Bahama I. p. 76 (1880); ib. List Bds. W. I.
  p. 9 (1885).—COUES, Key N. Am. Bds. p. 317 (1884).—RIDGW. Pr.
  U. S. Nat. Mus. VIII, pp. 27, 29 (1885).
Certhiola bairdii CAB. J. f. O. 1865, p. 412.

SP. CHAR. *Male:*—Above black, with a slight grayish tinge; a superciliary
line of white, from bill to nape; throat ashy white; breast bright
yellow, extending upon the sides of the abdomen, and shading into
gray upon the flanks; crissum white, wing-feathers slightly edged
with dull white; a white patch at the base of the primaries, forming
a bar on the wings; edge of the carpus bright yellow; tail, color
of the back, tipped with white, wanting upon the middle, and largest
upon the two outer feathers.

Female:—Slightly paler than the male, but otherwise resembling it.

Length, 4.50; wing, 2.60; tail, 1.90; tarsus; 70; bill, .54.

HABITAT.   Bahamas.

Certhiola portoricensis (BRYANT).

Cœreba flaveola VIEILL. Ency. Méth. 1820, p. 611.
Nectarinia flaveola MORITZ. Wiegm. Arch. für. Naturg. II, p. 387 (1836).
Certhiola flaveola SCL. Cat. Am. Bds. p. 54 (1862) (St. Thomas).—
  TAYLOR, Ibis, 1864, p. 166.—CASSIN, Pr. Acad. Nat. Sci. Phila.
  1864. p. 271.
Certhiola flaveola var. *portoricensis* BRYANT, Pr. Bost. Soc. Nat. Hist.
  X, p. 252 (1866).
Certhiola sti. thomæ SUND. Consp. 1869. p. 621 (?).
Certhiola portoricensis SUND. Consp. 1869, p. 622.—FINSCH, Verhandl.
  Zool. Botan. Gesells. Wien, XXI, p. 760 (1871).—SCL. & SALV.
  Nom. Avium Neotr. p. 16 (1873).—BD. BWR. & RIDGW. Hist. N.
  Am. Bds. I, p. 427 (1874).—GUNDL. Anal. Soc. Esp. Hist. Nat. VII,
  p. 216 (1878).—RIDGW. Pr. U. S. Nat. Mus. VII, p. 172 (1884); *ib.*
  VIII, pp. 28, 29 (1885).—CORY, List Bds. W. I. p. 9 (1885).

SP. CHAR.—Back dark slate color, showing an olive tint in some specimens;
in others the back almost black; rump olive yellow; breast color of
rump, showing more olive on the abdomen; throat gray; second,
third, fourth, and fifth primaries banded at base with white, sixth
primary nearly so, rest of primaries showing white on the webs at
the base.

Length (skin), 4.25; wing, 2.30; tail, 1.45; tarsus. .72; bill, .50.

HABITAT.   Porto Rico and St. Thomas.

Certhiola sancti-thomæ RIDGW.

Certhiola portoricensis FINSCH, Verhandl. Zool. Botan. Gesells. Wien,
  XXI, p. 760 (1871).—BAIRD, Am. Nat. VII. p. 671 (1873) and
  authors from St. Thomas and St. John.
Certhiola sancti-thomæ RIDGW. Pr. U. S. Nat. Mus. VIII. pp. 28, 29
  (1885).—CORY, List Bds. W. I. p. 9 (1885).

Sp. Char.—Very close to *C. portoricensis*, but separated from it by having
the back lighter slate color, and throat lighter gray.

Measurements practically the same as those of *C. portoricensis*.

Habitat. St. Thomas, and St. John, W. I.

This is a somewhat doubtful species, and requires further
investigation. Specimens in my collection from St. John and
St. Thomas show the dark back of *C. portoricensis*, while
others show the gray tinge, representing *sancti-thomæ*. I have
also a specimen of *C. portoricensis* which has the back nearly
as gray as any from St. Thomas. A specimen from St. Thomas
also agrees with one from Port Rico, in the color of the throat,
although other specimens have the throat lighter. It is possible
that some of the specimens in question may be incorrectly labelled,
as several of them were obtained by purchase.

Certhiola bananivora (Gmel.).

Motacilla bananivora Gmel. Syst. Nat. I. p. 951 (1788).

Certhiola—(?) Sallé, P. Z. S. 1857. p. 233.

Certhiola bananivora Bryant, Pr. Bost. Soc. Nat. Hist. XI, p. 95 (1865).
—Ed. Bwr. & Ridgw. Hist. N. Am. Bds. I, p. 427 (1874).—
Cory, Bds. Haiti, and San Domingo, p. 41 (1885); *ib.* List Bds.
W. I. p. 9 (1885).

Certhiola clusiæ "Herz Von Wurtlemb. Hartl. Naumannia, II, Heft.
2, p. 56 (1852) (sine descr.)."—Finsch, Verhandl. Zool. Botan.
Gesells. Wien, XXI, p. 771 (1871).—Scl. & Salv. Nom. Avium
Neotr. p. 17 (1873).—Cory, Bull. Nutt. Orn. Club, VI, p. 151
(1881).

Sp. Char. *Male:*—Upper surface, including head, cheeks, wings, and tail,
dull black; a superciliary white stripe, extending from the base
of the upper mandible to the nape; throat dark slaty color; underparts bright yellow, becoming grayish olive upon the sides and
thighs; rump and carpus bright yellow; an edging of white upon the
basal portion of primaries on the outer webs, very narrow upon
the first, the whole nearly concealed by the coverts, forming a
narrow white wing-band; bill and feet black; tail slightly tipped
with dull white on the outer feathers.

The sexes are similar.

Length, 4.40; wing, 2.40; tail, 1.60; tarsus, .60; bill, .50.

Habitat. San Domingo.

Young birds of this species have the superciliary stripe yellow,
and the back more gray. Specimens in my collection show all

intermediate stages, from the yellow one, some having it half white, half yellow, while others show but a faint spot of yellow in front of the eye. The color of the throat also varies slightly at different seasons and ages.

Certhiola bartholemica (Sparrm.).

Certhia bartholemica Sparrm. Mus. Carls. fasc. III, No. 57 (1788).— Bechst. Lath. Uebers. I, p. 611. (1793).
Cæreba flaveola Vieill. Ency. Méth. p. 611 (1820).
Certhiola bartholemica Reich. Handb. Scans. p. 253 (1853).—Sundev. Kütisk. Framställ. in K. Vet. Akad. Handl. II, No. 3, p. 10 (1857); *ib.* Vet. Akad. Förh. 1869, p. 622.—Finsch, Verhandl. Zool. Botan. Gesells. Wien, XXI, p. 763 (1872).—Scl. & Salv. Nom. Avium Neotr. p. 16 (1873).—Bd. Bwr. & Ridgw. Hist. N. Am. Bds. I, p. 428 (1874). —Cory, List. Bds. W. I. p. 9 (1885).—Ridgw. Pr. U. S. Nat. Mus. VIII, pp. 28, 30 (1885).

Sp. Char.—Forehead dull gray; throat dark plumbeous; superciliary stripe extending backward, commencing above the eye; white marking near base of primaries very small; lower part of rump dull yellowish green.
    Length (skin), 395; wing, 2.35; tail, 1.70.
    Habitat.   St. Bartholemew.

Certhiola saccharina Lawr.

Certhiola saccharina Lawr. Am. N. Y. Acad. Sci. I, p. 151 (1878); *ib.* Pr. U. S. Nat. Mus. I, p. 487 (1878).—Cory, List. Bds. W. I. p. 9 (1885).—Ridgw. Pr. U. S. Nat. Mus. VIII, pp. 28, 30. (1885).

Sp. Char.—Throat very dark slate color, much darker than in *C. porto-ricensis*, and extending lower; underparts brighter yellow; the white marking on the primaries somewhat heavier; rump yellowish green; back very dark slate color, not quite as dark as in *C. portoricensis*.
    Length (skin), 4; wing, 2.30; tail, 1.50; tarsus, .58.
    Habitat.   St. Vincent, and Grenada.

Certhiola flaveola (Linn.).

Certhia flaveola Linn. Syst. Nat. I, p. 187 (1766).—Gmel. Syst. Nat. I, p. 497 (1788).—Vieill. Ency. Méth. p. 611 (1820).—Denny, P. Z. S. 1847, p. 39.
Certhiola flaveola Gosse, Bds. Jam. p. 84 (1847).—Scl. Cat. Am. Bds. p. 54 (1862).—Albrecht, J. f. O. 1862, p. 196.—March, Pr. Acad. Nat. Sci. Phila. 1863, p. 296.—Gray, Handl. Bds. I, p. 120 (1869).

—FINSCH, Verhandl. Zool. Botan. Gesells. Wien, XXI, p. 756 (1871).
—SCL. & SALV. Nom. Avium Neotr. p. 16 (1873).—BD. BWR. &
RIDGW. Hist. N. Am. Bds. I, p. 427 (1874).—A. & E. NEWTON,
Handb. Jamaica. p. 103 (1881).—CORY, List. Bds. W. I. p. 9 (1885).
—RIDGW. Pr. U. S. Nat. Mus. VIII, pp. 28-30 (1885).

SP. CHAR.—General appearance of *C. portoricensis*, but having the throat
much darker gray. Upper parts of breast showing an olive tinge;
the yellow of the breast is duller than in *C. portoricensis*, and somewhat ochraceous; outer webs of primaries heavily marked with white,
extending fully half their length, inner webs showing much white
at the base, and narrowly edged with the same; secondaries broadly
marked with white on the inner webs; rump yellow, as bright as
the belly.

Length (skin), 4; wing, 2.32; tail, 1.60; tarsus, .58.

HABITAT. Jamaica.

Certhiola newtoni BAIRD.

Certhiola flaveola A. & E. NEWTON, Ibis, 1859, p. 67.—SCL. Cat. Am.
Bds. p. 54 (1862) (St. Croix).—SUNDEV. Vet. Akad. Förh. 1869,
p. 623 (St. Croix).
Certhiola bartholemica FINSCH, Verhandl. Zool. Botan. Gesells. Wien.
XXI, p. 763 (1871) (St. Croix).—SCL. & SALV. Nom. Avium Neotr.
p. 16 (1873) (St. Croix).—CORY, List Bds. W. I. p. 9 (1884).
Certhiola newtoni BAIRD, Am. Nat. VII, p. 611 (1873)—BD. BWR. &
RIDGW. Hist. N. Am. Bds. I, p. 427 (1884).—RIDGW. Pr. U. S. Nat.
Mus. VIII, pp. 28-30 (1885).

SP. CHAR.—Similar to *C. flaveola.* "White patch of wing more quadrate
on each quill; transverse; not tapering off gradually and uniformly
behind; not reaching the shaft on outer primary. Breast without
ochraceous; rump olivaceous yellow; the color different from that
of the belly." (BD. BWR. & RIDGW. Hist. N. Am. Bds.)

HABITAT. St. Croix.

Certhiola dominicana TAYLOR.

Certhiola dominicana TAYLOR, Ibis, 1864, p. 167.—GRAY, Handl. Bds. I.
p. 120 (1869).—FINSCH, Verhandl. Zool. Botan. Gesells. Wien, XXI.
p. 787 (1871).—SCL. & SALV. Nom. Avium Neotr. p. 17 (1873).—
BD. BWR. & RIDGW. Hist. N. Am. Bds. I, p. 428 (1874).—LAWR.
Pr. U. S. Nat. Mus. I, p. 487 (1878).—CORY, List. Bds. W. I. p. 9
(1885).—RIDGW. Pr. U. S. Nat. Mus. VIII, p. 30 (1885).
Certhiola frontalis BAIRD, MSS. Bd. Bwr. & Ridgw. Hist. N. Am. Bds. I,
p. 428 (1874).

SP. CHAR.—Superciliary stripe lacking, or extremely indistinct in front of the eye; frontal region dull grayish black; back smoky black, sometimes showing a slight olive tinge when held in the light; throat dark slate color; lower part of rump showing olive green; a delicate penciling of white on the outer webs of primaries.

Length (skin), 4.85; wing, 2.50; tail, 1.60; tarsus, .65.

HABITAT. Dominica, Antigua, Barbuda, Nevis, St. Eustatius, Guadeloupe, and Saba.

C. sundevalli Ridgw. is probably a phase of plumage of this species, the yellow superciliary stripe changing with age, as in C. bananivora.

Certhiola barbadensis BAIRD.

Certhiola martinicana SCL. P. Z. S. 1874, p. 174.

Certhiola barbadensis BAIRD, Am. Nat. VII, p. 612 (1873).—BR. BWR. & RIDGW. Hist. N. Am. Bds. I, p. 428 (1874).—CORY, List Bds. W. I. p. 9 (1885).—RIDGW. Pr. U. S. Nat. Mus. VIII, pp. 28, 30 (1886).

SP. CHAR.—"Upper part of throat slate black, bordered laterally by a gray rictal patch, and below by a yellowish white patch; separating the black from the yellow of the jugulum. Upper parts as in C. dominicana, but superciliary stripe broadest and most sharply defined anteriorly." (RIDGW. Pr. U. S. Nat. Mus. VIII, p. 28 (1885.)

Length, 3.75; wing, 2.40; tail, 1.75.

HABITAT. Barbadoes.

Certhiola martinicana REICH.

Certhia martinicana s. saccharivora BRISS. Orn. III, p. 611 (1860).

Certhia flaveola var. β. LINN. Syst. Nat. I, p. 187 (1766).—GMEL. Syst. Nat. I, p. 479 (1788).

Certhiola martinicana REICH. Handb. I, p. 252 (1853).—CASSIN, Pr. Acad. Nat. Sci. Phila. 1864, p. 271.—CAB. J. f. O. 1865, p. 412.—GRAY, Handl. Bds. I, p. 120 (1869).—FINSCH, Verhandl. Zool. Botan. Gesells. Wien, XXI, p. 788 (1871).—SCL. P. Z. S. 1871, p. 269.—SCL. & SALV. Nom. Avium Neotr. p. 17 (1873).—BD. BWR. & RIDGW. Hist. N. Am. Bds. I, p. 428 (1874).—LAWR. Pr. U. S. Nat. Mus. I, p. 487 (1878).—ALLEN, Bull. Nutt. Orn. Club, V, p. 166 (1880).—CORY, List Bds. W. I. p. 9 (1885).—RIDGW. Pr. U. S. Nat. Mus. VIII, pp. 28-30 (1885).

Certhiola albigula BP. Compt. Rend. 1854, p. 259.—TAYOR, Ibis, 1864, p. 167.—NEWTON, Zool. Record, 1864, p. 76.

SP. CHAR.—Sides of the throat grayish black; a patch of white on the middle of the throat to breast; underparts bright yellow, a tinge of olive on the abdomen; lower rump narrowly banded with olive green; upper parts dull slate color; wing-coverts sometimes slightly tipped with white.

Length (skin), 4.15; wing, 2.28; tail, 1.60; tarsus, .64. Another specimen: Length (skin), 4.35; wing, 2.33; tail, 1.70; tarsus, 68.

HABITAT. Santa Lucia and Martinique.

C. finschi Ridgw. is probably a phase of plumage of this species. Some specimens from Martinique in my collection have the superciliary stripe yellow, and also show yellow on the throat. The locality given where the type specimen of *C. finschi* was taken is, as Mr. Ridgway suggests, undoubtedly incorrect. The same variation in coloring on account of age and season is shown in the San Domingo species *C. bananivora.*

Certhiola atrata LAWR.

Certhiola atrata LAWR. Am. N. Y. Acad. Sci. I, p. 150 (1878); *ib.* Pr. U. S. Nat. Mus. I, p. 487 (1878).—LISTER, Ibis, 1880, p. 40.— CORY, List. Bds. W. I. p. 9 (1885).—RIDGW. Pr. U. S. Nat. Mus. VIII*pp. 28, 30 (1885).

SP. CHAR.—Entire plumage dull black; a tinge of olive is perceptible on the underparts, and on the rump.

Length (skin), 4.05; wing, 2.35; tail, 1.50; tarsus, .56.

HABITAT. St. Vincent and Grenada.

Mr. Ridgway expresses the opinion that this is perhaps a melanotic variety of *C. saccharina.*

GENUS Cœreba VIEILL.

Cœreba VIEILLOT, Ois. Am. Sept. 1807.

Cœreba cyanea (LINN.).

Certhia cyanea LINN. Syst. Nat. I, p. 188 (1766).
Certhia cyanogastra LATH. Ind. Orn. I, p. 295 (1790).
Cœreba cyanea VIEILL. Ois. Dos. pls. 41, 42, 43, et Gal. Ois. pl. 176 (1820-26).—MAX. Beitr. III, p. 761 (1831).—Br. Consp. I, p. 399 (1850).— THIENEM. J. f. O. 1857, p. 152.—BURM. Syst. Ueb. III, p. 150.—SCL. Cat. Am. Bds. p. 52 (1862).—GRAY, Handl. Bds. I, p. 116 (1869).— SCL. & SALV. Nom. Avium Neotr. p. 16 (1873).—BD. BWR. & RIDGW. Hist. N. Am. Bds. I, p. 425 (1874).—CORY, List Bds. W. I. p. 9 (1885).
Cœreba cyanea D'ORB. in La Sagra's Hist. Nat. Cuba. Ois. p. 124 (1840). —LEMB. Aves Cuba. p. 131 (1850).

Diceum aterrimum, Lesson Traité d'Orn. I, p. 303 (1831).—PUCHERAN, Rev. Zool. 1846, p. 134.—SALV. and GODM. C. A. Ave. I, p. 251?
Certhiola atrata SCLATER, Cat. Bds. Brit. 'Ius. Vol. XI, p. 47.

Arbelorhina cyanea CAB. in Schomb. Guian. III, p. 675 (1848); *ib.* J. f. O.
1856, p. 98; *ib.* 1874, p. 139.—GUNDL. Repert. Fisico-Nat. Cuba,
I, p. 291 (1865).—BREWER, Pr. Bost. Soc. Nat. Hist. VIII, p. 306
(1860).

SP. CHAR. *Male:*—Top of head bright pale blue; a stripe of black passing
from the upper mandible and encircling the eye; sides of the head,
lower back, and entire underparts dark purplish blue, wings and
upper back black; inner webs of primaries and secondaries bright
yellow; sides and flanks greenish.

Female:—Entire upper parts bright green; underparts green, the shafts
of the feathers showing dull white, giving a finely pencilled appearance to the throat and breast; central portion of belly showing a
pale yellowish tinge.

Length (skin), 4; wing, 2.60; tail, 1.30; tarsus, .50; bill, 50.

Dr. Gundlach writes me that this species is abundant in many
portions of the Island of Cuba.

GENUS **Glossiptila** SCL.

Glossiptila SCLATER, P. Z. S. 1856, p. 269.

Glossiptila ruficollis (GMEL.).

Motacilla campestris LINN. Syst. Nat. I, p. 329 (1766).
Tanagra ruficollis "GMEL. Syst. Nat. II."
Tachyphonus rufigularis LAFR. Rev. Zool. 1846, p. 320.
Tanagrella ruficollis GOSSE, Bds. Jam. p. 236 (1847).—GRAY, Gen. Bds.
III, App. p. 17 (1849).—BP. Consp. I, p. 236 (1850).
Pyrrhulagra ruficollis BP. Consp. I, p. 493 (1850) (excl. syn.).
Neornis cærulea HARTL. Nachtr. z. Verz. Mus. Brem. p. 8 (descr. nulla).
Glossiptila ruficollis SCL. P. Z. S. 1856, p. 269.—ALBRECHT, J. f. O. 1862,
p. 196.—MARCH, Pr. Acad. Nat. Sci. Phila. 1863, p. 296.—BAIRD,
Rev. Am. Bds. I, p. 163 (1864).—GRAY, Handl. Bds. I, p. 120 (1869).
—SCL. & SALV. Nom. Avium Neotr. p. 17 (1873).—A. & E. NEW-
TON, Handb. Jamaica, p. 104 (1881).—CORY, List Bds. W. I. p. 9
(1885).

Sp. Char. *Male:*—General plumage dull blue; a stripe of dull black from the bill to the eye, showing slightly on the forehead; a large patch of rufous on the throat; quills and tail dark brown, feathers edged with blue; bill black; feet horn color.

Female:—Top of head bluish gray, shading into grayish olive on the back; wings edged with pale brown; underparts gray, faintly streaked; tail brown.

Length (skin), 5; wing, 3; tail, 1.75; tarsus, .58.

Habitat. Jamaica.

Genus Chlorophanes Reich.

Chlorophanes " Reich. Handb. p. 234 (1853). '

Chlorophanes spiza (Linn.).

Certhia spiza Linn. Syst. Nat. I, p. 186 (1766).
Certhia spiza var.? Gmel. Syst. Nat. I, p. 476 (1788).
Cœreba atricapilla Vieill. Nouv. Dict. XIV, p. 50. (1817).
Cœreba spiza Max. Beitr. III, p. 771 (1831).
Cœreba atricapilla Bp. Consp. I, p. 400 (1850).
Dacnis atricapilla Scl. Contr. Orn. p. 108 (1851).
Chlorophanes atricapilla "Reich. Handb. p. 234 (1853)."—Scl. Cat. Am. Bds. p. 52 (1862).—Baird, Rev. Am. Bds. p. 163 (1864).—Gray, Handl. Bds. I, p. 118 (1869).—Scl. & Salv. Nom. Avium Neotr. p. 16 (1873) (Cuba).—Cory, List Bds. W. I. p. 9 (1885).
Dacnis spiza Cab. Mus. Hein. I, p. 95 (1850).—Burm. Syst. Ueb. III, p. 152.
Nectarina mitrata Licht. Doubl. p. 15.

Sp. Char. *Male:*—Head and cheeks black, rest of plumage, including throat, bright bluish green; quills and tail dark brown, edged with greenish; under surface of wing steel gray.

Female:—Entire plumage light green, brightest on the back, and palest on the underparts; under surface of wing dull white.

Length (skin), 5; wing, 2.75; tail, 2; tarsus, .75.

A male bird of this species in my cabinet is labelled Cuba, and Messrs. Sclater and Salvin (l. c.) record it from there. It is probable that if the localities given are correct, the specimens in question were escaped cage birds.

Family HIRUNDINIDÆ.

Genus Progne Boie.

Progne Boie, Isis, 1826, p. 971.

Progne dominicensis (GMEL.).

Hirundo dominicensis GMEL. Syst. Nat. I. p. 1025 (1788).—VIEILL. Ois.
    Am. Sept. p. 59 (1807).
Hirundo albiventris VIEILL. Nouv. Dict. Hist. Nat. XIV, p. 533 (1817).
Progne dominicensis BOIE, Isis, 1826, p. 971.—GOSSE, Bds. Jam. p. 69
    (1847).—BP. Consp. I, p. 337 (1850).—ALBRECHT, J. f. O. 1862,
    p. 194.—MARCH, Pr. Acad. Nat Sci. Phila. 1863. p. 295.—TAYLOR,
    Ibis, 1864, p. 166.—BAIRD, Rev. Am. Bds. p. 279 (1864).—GUNDL.
    J. f. O. 1872, p. 419; *ib.* Anal. Soc. Esp. Hist. Nat. VII, p. 196 (1878).
    —SCL. & SALV. Nom. Avium Neotr. p. 14 (1873).—LAWR. Pr. U. S.
    Nat. Mus. I. p. 487 (1878).— LISTER, Ibis, 1880, p. 40.—A. & E.
    NEWTON, Handb. Jamaica, p. 107 (1881).— CORY, Bds. Haiti & San
    Domingo, p. 44 (1885); *ib.* List Bds. W. I. p. 10 (1885).—SHARPE,
    Cat. Bds. Brit. Mus. X, p. 176 (1885).
Hirundo (Progne) dominicensis BRYANT, Pr. Bost. Soc. Nat. Hist. XI, p.
    94 (1866).
Progne subis. var. *dominicensis* BD. BWR. & RIDGW. Hist. N. Am. Bds. I,
    p. 328 (1874).

SP. CHAR. *Male:*—Entire upper surface, throat, and sides steel blue, show-
    ing purplish reflections in some lights; rest of underparts white;
    quills and tail dark brown, the feathers having a faint bluish tinge
    on the outer webs; crissum dull white; bill and feet black.
     Female:—Upper surface as in the male; throat and sides ashy
    brown; otherwise resembling the male.
     Length, 7; wing, 5.60; tail. 3.10; tarsus, .50; bill, .50.
HABITAT.   San Domingo and Antilles.

Progne subis (LINN.).

Hirundo subis LINN. Syst. Nat. p. 192 (1758).
Hirundo purpurea LINN. Syst. Nat. I, p. 344 (1766).
Progne purpurea D'ORB. in La Sagra's Hist. Nat Cuba, Ois. p. 94 (1840).—
    GUNDL. J. f. O. 1856, p. 3; *ib.* 1861, p. 328 (Cuba).—BREWER, Pr.
    Bost. Soc. Nat. Hist. VII, p. 306 (1860) (Cuba).
Progne cryptoleuca BAIRD, Rev. Am. Bds. p. 277 (1864).—GUNDL. J. f. O.
    1872, p. 431.—CORY, List Bds. W. I. p. 10 (1885).
Progne subis var. *cryptoleuca* BD. BWR. & RIDGW. Hist. N. Am. Bds. I, p.
    322 (1874).
       Cuba.

GENUS **Petrochelidon** CABAN.

Petrochelidon CABANIS, Mus Hein. I, 1850-51, p. 47.

Petrochelidon fulva (VIEILL.).

Hirundo fulva VIEILL. Ois. Am. Sept. I. p. 62 (1807).—MARCH, Pr.
Acad. Nat. Sci. Phila. 1863, p. 295.
Cecropsis fulva BOIE, Isis, 1828, p. 315.
Hirundo melanogaster DENNY, P. Z. S. 1847, p. 38.
Hirundo pœciloma GOSSE, Bds. Jam. p. 64 (1847).—OSBORN, P. Z. S.
1865, p. 63.
Hirundo coronata LEMB. Aves Cuba, p. 45 (1850).—GUNDL. Journ. Bost.
Soc. Nat. Hist. VII, p. 318 (1852).
Herse fulva BR. Consp. I. p. 341 (1850).
Petrochelidon fulva CAB. Mus. Hein. I. p. 47 (1850).—BREWER, Pr. Bost.
Soc. Nat. Hist. VII, p. 306 (1860).—SCL. Cat. Am. Bds. p. 40
(1862).—ALBRECHT, J. f. O. 1862. p. 194.—BAIRD, Rev. Am. Bds.
p. 291 (1864).—GRAY, Handl. Bds. I, p. 71 (1869).—SCL. & SALV.
Nom. Avium Neotr. p. 14 (1873).—GUNDL. J. f. O. 1874, p. 133;
ib. Anal. Soc. Esp. Hist. Nat. VII, p. 198 (1878).—A. & E. NEWTON,
Handb. Jamaica, p. 107 (1881).—CORY, Bull. Nutt. Orn. Club. VI. p.
152 (1881); *ib.* Bds. Haïti & San Domingo. p. 47 (1885): *ib.* List
Bds. W. I. p. 10 (1885).—SHARPE, Cat. Bds. Brit. Mus. X, p. 195
(1885).
Petrochelidon pœciloma BAIRD, Rev. Am. Bds. p. 292 (1864).—GRAY,
Handl. Bds. I, p. 71 (1869).—GUNDL. J. f. O. 1874. p. 311.
Hirundo (Petrochelidon) fulva BRYANT, Pr. Bost. Soc. Nat. Hist. X,
p. 252 (1866).

SP. CHAR. *Male:*—Throat and sides of the breast pale rufous brown, the
color passing around the neck in a narrow line at the nape; belly
and crissum dull white, the latter showing a rufous tinge; top of the
head bluish black, the color nearly encircling the eye; forehead
and rump dark rufous brown; back bluish black, streaked with
white; wings and tail dark brown; bill and feet black.

The sexes are apparently similar.

Length, 4.70: wing, 4; tail, 1.85; tarsus, .40; bill, .27.

HABITAT. Antilles.

GENUS Tachycineta CABAN.

Tachycineta CABANIS, Mus. Hein. I, p. 48 (1850).

Tachycineta bicolor (VIEILL.).

Hirundo bicolor VIEILL. Ois. Am. Sept. I. p. 61 (1807).—CORY, List Bds.
W. I, p. 10 (1885).
Tachycineta bicolor GUNDL. J. f. O. 1856, p. 4: *ib.* 1861. p. 330 (Cuba);
ib. 1874, p. 113 (Cuba).—CORY, Bds. Bahama I. p. 80 (1880).
SHARPE, Cat. Bds. Brit. Mus. X, p. 117 (1885).

Petrochelidon bicolor BREW. Pr. Bost. Soc. Nat. Hist. VII. p. 306 (1860)
(Cuba).

Accidental in Cuba and Bahama Islands.

Tachycineta euchrysea (GOSSE).

Hirundo euchrysea GOSSE, Bds. Jam. p. 68 (1847).—MARCH, Pr. Acad. Nat.
Sci. Phila. 1863. p. 295.—SCL. & SALV. Nom. Avium Neotr. p. 14
(1873).—CORY, List Bds. W. I. p. 10 (1885).—SHARPE, Cat. Bds.
Brit. Mus. X, p. 170 (1885).
Herse euchrysea Br. Bonsp. I, p. 34 (1850).
Petrochelidon euchrysea SCL. P. Z. S. 1861, p. 72; *ib.* Cat. Am. Bds. p. 39
(1862).—ALBRECHT. J. f. O. 1862, p. 194.
Callichelidon euchrysea BAIRD, Rev. Am. Bds. p. 304 (1864).—GRAY.
Handl. Bds. I, p. 72 (1869).—A. & E. NEWTON, Handb. Jamaica, p.
107 (1881).

SP. CHAR. *Male:*—Entire upper surface including head bright golden
green; a slight bluish tinge perceptible on the forehead, when held
in the light; underparts white; wings and tail brown, showing a
tinge of bronzy green on the upper surface.
Female similar to male.
Length (skin), 4.50, wing, 4.25; tail, 2.25.
HABITAT.   Jamaica.

Tachycineta sclateri (CORY).

Hirundo euchrysea (var. *dominicensis?*) BRYANT, Pr. Bost. Soc. Nat. Hist.
XI, p. 95 (1866).
Callichelidon euchrysea var. *dominicensis* GRAY, Handl. Bds. I, p. 72
(1869).
Hirundo sclateri CORY, Auk, I, p. 2 (1884): *ib.* Bds. Haiti & San Do-
mingo, p. 45 (1855); *ib.* List. Bds. W. I. p. 10 (1855).—SHARPE,
Cat. Bds. Brit. Mus. X, p. 171 (1885).

SP. CHAR.—Above bright bluish green, showing a golden color in some
lights, becoming decidedly blue on the forehead: upper surface of
wings and tail showing a tinge of dull blue, brightest on the tail;
underparts pure white; primaries brown; bill and legs very dark
brown.
The sexes are similar.
Length, 5; wing, 4.60; tail, 2.
HABITAT.   San Domingo.

GENUS Chelidon FORST.

[*Chelidon* FORSTER. Syn. Cat. Brit. Bds. p. 55 (1817).

Chelidon erythrogastra (BODD.).

Hirundo erythrogastra "BODD. Tabl. P. E. 45 (1873)."—CORY, List. Bds.
    W. I. p. 10 (1885).—SHARPE, Cat. Bds. Brit. Mus. X, p. 137 (1885).
Hirundo americana LEMB. Aves Cuba, p. 44 (1850).
Hirundo rufa GUNDL. J. f. O. 1855, p. 3; *ib.* 1861, p. 328 (Cuba).—BREW.
    Pr. Bost. Soc. Nat. Hist. VII. p. 306 (1860).
Hirundo horreorum A. & E. NEWTON, Ibis, 1859, p. 66 (St. Croix); *ib.*
    Handb. Jam. p. 107 (1881).—SUNDV. Oefv. K. Vet. Akad. p. 584
    (1869) (St. Bartholemew).—GUNDL. J. f. O. 1872, p. 431 (Cuba).—
    LAWR. Pr. U. S. Nat. Mus. I, p. 455 (1878) (Guadeloupe).—CORY,
    Bds. Bahama I. p. 78 (1880).

Recorded from Bahama Islands, Greater Antilles, St. Croix
and Guadeloupe, and St. Bartholemew.

GENUS Callichelidon BRYANT.

Callichelidon (BRYANT, MSS.) BAIRD, Rev. Am. Bds. p. 303 (1864).

Callichelidon cyaneoviridis (BRYANT).

Hirundo cyaneoviridis BRYANT, Pr.Bost. Soc. Nat. Hist. VII, p. 111 (1859).
    —BAIRD, Rev. Am. Bds. p. 303 (1864).—SALV. Ibis, 1874. p, 307.—
    CORY, Bds. Bahama I. p. 79 (1880); *ib.* List Bds. W. I. p. 10 (1885).
Callichelidon cyaneoviridis GRAY, Handl. Bds. I, p. 72 (1869).—BD. BWR. &
    RIDGW. Hist. N. Am. Bds. I, p. 327 (1874).
Tachycineta cyaneoviridis SHARPE, Cat. Bds. Brit. Mus. X, p. 121 (1885).

SP. CHAR. *Male:*—Above velvet green, shading into steel blue, with purple
    reflections upon the rump and wings: a black stripe from the nostrils
    to the eye; underparts pure white: tail forked, the inner webs of the
    outer feathers edged with dull white.
        Female:—Resembles the male, but the plumage much duller, and
    showing traces of dusky; bill and feet black.
        Length, 6.40; wing, 4.40; tail, 3.10; tarsus, .42; bill, .15.

HABITAT. Bahamas.

Genus Clivicola FORST.

Clivicola FORST. Syn. Cat. Brit. Bds. 55 (1817).

Clivicola riparia (FORST.).

Hirundo riparia LINN. Syst. Nat. I, p. 192 (1758).—LEMB. Aves Cuba,
    p. 47 (1850).
Cotyle riparia Gundl. J. f. O. 1856, p. 5; *ib.* 1861. p. 330; *ib.* 1874. p. 114
    (Cuba).—BREW. Pr. Bost. Soc. Nat. Hist. VII, p. 306 (1860) (Cuba)
    A. & E. NEWTON, Handb. Jamaica, p. 107 (1881).—CORY. List Bds.
    W. I. p. 10 (1885).

Cuba; Jamaica; probably wanders throughout the Antilles.

FAMILY VIREONIDÆ.

GENUS Vireo VIEILL.

Vireo VIEILLOT, Ois. Am. Sept. I, p. 83 (1807).

Vireo modestus SCL.

Vireo noveboracensis GOSSE, Bds. Jam. p. 192 (1847).
Vireo modestus SCL. P. Z. S. 1860, p. 462.—ALBRECHT, J. f. O. 1862, p.
194.—MARCH, Pr. Acad. Nat. Sci Phila. 1863. p. 294.—BAIRD, Rev.
Am. Bds. p. 362 (1864).—SCL. & SALV. Nom. Avium Neotr. p. 12
(1873). —A. & E. NEWTON, Handb. Jamaica, p. 106 (1881).—
GADOW Cat. Bds. Brit. Mus. VIII, p. 303 (1883).—CORY, List Bds.
W. I. p. 10 (1885).
Vireonella modestus GRAY, Handl. Bds. I, p. 382 (1869).

SP. CHAR. *Male:*—Upper plumage dull olive green; throat pale; belly
dull yellowish brown; primaries and secondaries brown, edged with
dull greenish on the outer webs; coverts edged with dull yellowish
white, forming an imperfect wing-band; tail-feathers narrowly edged
with olive green.
The sexes are similar.
Length (skin), 4.50; wing, 2.40; tail, 2; tarsus, .65.
HABITAT. Jamaica.

Vireo latimeri BAIRD.

Vireo latimeri BAIRD, Rev. Am. Bds. p. 364 (1864).—BRYANT, Pr. Bost.
Soc. Nat. Hist. X, p. 252 (1866).—SCL. & SALV. Nom Avium Neotr.
p. 12 (1873).—GUNDL. Anal. Soc. Esp. Hist. Nat. VII. p. 135 (1878);
ib. J. f. O. 1878, p. 158.—GADOW, Cat. Bds. Brit. Mus. VIII, p. 304
(1883).—CORY, List Bds. W. I. p. 10 (1885).
Vireonella latimeri GRAY, Handl. Bds. I, p. 382 (1869).

SP. CHAR. *Male:*—Top of head grayish brown; back olive green; underparts showing throat dull white, shading into bright yellow on the
breast and belly; wings and tail brown, feathers edged with pale
greenish.
Length (skin), 4.25; wing, 2.25; tail, 1.75; tarsus, .75.
HABITAT. Porto Rico.

Vireo alleni CORY.

Vireo alleni CORY, Auk, iii. pp. 500, 501 (1886).

Sp. Char.—Above dull olive, showing a dull yellow tinge on the forehead; a stripe of yellow from the upper mandible to the eye, the yellow showing on the upper and lower eyelids; entire under surface dull yellow, tinged with olive on the flanks and sides; two distinct yellowish-white wing bands; quills dark brown, most of the feathers edged with yellowish-green on the outer webs; tail brown, showing faint olive edgings on the outer webs; bill horn color; legs dark brown or slaty brown.

Length, 4.10; wing, 2.30; tail, 1.85; tarsus, .75; bill, .45.

Habitat.   Grand Cayman.

Vireo crassirostris (Bryant).

Lanivireo crassirostris Bryant, Pr. Bost. Soc. Nat. Hist. VII, p. 112 (1859).—Cory, Bds. Bahama I. p. 83 (1880).

Vireo crassirostris Baird, Rev. Am. Bds. p. 368 (1864).—Scl. & Salv. Nom. Avium Neotr. p. 12 (1873).—Gadow, Cat. Bds. Brit. Mus. VIII, p. 300 (1883).—Cory, List Bds. W. I. p. 10 (1885).

Vireonella crassirostris Gray, Handl. Bds. I, p. 382 (1869).

Sp. Char. *Male:*—Above yellowish olive; a streak of olive from the nostril, encircling the eye; underparts yellowish; wings and tail brown, the feathers edged with greenish, the former showing two white bands.

Length, 5; wing, 2.40; tail, 1.90; tarsus, .84; bill, .40.

Habitat.   Bahamas.

This species is nearly allied to *V. noveboracensis*, but differs from it in being slightly larger, and in having the entire underparts an almost uniform color—pale yellow, or yellowish white.[*]

Vireo noveboracensis (Gmel.).

Muscicapa noveboracensis Gmel. Syst. Nat. I, p. 947 (1788).

Vireo noveboracensis Gundl. J. f. O. 1855, p. 469 (Cuba); *ib.* 1861, p. 404 (Cuba); *ib.* 1872, p. 484 (Cuba).—Brewer, Pr. Bost. Soc. Nat. Hist. VII, p. 307 (1860) (Cuba).—Gadow, Cat. Bds. Brit. Mus. VII, p. 300 (1883).—Cory, List Bds. W. I. p. 10 (1885).

Vireo (Lanivireo) noveboracensis Gundl. J. f. O. 1861, p. 324 (Cuba).

Accidental in Cuba.

Vireo gundlachi Lemb.

Vireo gundlachi Lemb. Aves Cuba, p. 29 (1850).—Cab. J. f. O. 1855, p. 468.—Brewer, Pr. Bost. Soc. Nat. Hist. VII, p. 307 (1860).—Gundl. Report. Fisico-Nat. Cuba, I, p. 228 (1865).—Scl. & Salv. Nom. Avium Neotr. p. 12 (1873).—Gadow, Cat. Bds. Brit. Mus. VIII, p. 304 (1883).—Cory, List Bds. W. I. p. 10 (1885).

Vireonella gundlachi Baird, Rev. Am. Bds. p. 369 (1864).—Gray, Handl. Bds. I, p. 382 (1869).—Bd. Bwr. & Ridgw. Hist. N. Am. Bds. I, p. 382 (1874).

---

[*] *Vireo crassirostris flavescens*, a new sub-species described by Ridgway, Man. N. A. Bds. p. 476, 1887, "Yellow beneath, and olive green above." It occurs on Rum Cay and Conception I., Bahamas. It is very closely allied to *V. alleni*.

Sp. Char.—Wings short and rounded; upper plumage grayish olive; lores and circle around the eye yellowish; underparts dull yellow; two narrow wing-bands; wings and tail brown edged with grayish olive, pale on the secondaries; quills narrowly edged on inner webs with dull white; bill dull horn color.

Length (skin), 5; wing, 2.15; tail, 2.10; tarsus, .78; bill. .43.

Habitat. Cuba.

Vireo flavifrons Vieill.

Vireo flavifrons Vieill. Ois. Am. Sept. I, p. 85 (1807).—Gundl. J. f. O. 1855, p. 468 (Cuba); *ib.* 1861, p. 404 (Cuba); *ib.* 1872, p. 403 (Cuba) —Brewer, Pr. Bost. Soc. Nat, Hist. VII, p. 307 (1860) (Cuba).— Cory, List Bds. W. I. p. 10 (1885).

Accidental in Cuba.

Vireo calidris (Linn.).

Motacilla calidris Linn. Syst. Nat. I, p. 329 (1766).
Vireosylvia olivacea Gosse, Bds. Jam. p. 194 (1847).
Vireo altiloquus Gamb. Pr. Acad. Nat. Sci. Phila. 1848, p. 127.—Baird, Bds. N. Am. p. 354 (1858).
Vireosylvia altiloqua Cassin, Pr. Acad Nat. Sci. Phila. 1851, p. 152; *ib.* Pr. Acad. Nat. Sci. Phila. 1860, p. 375.—Newton, Ibis, 1859, p. 145. —Albrecht, J. f. O. 1862, p. 195.—Scl. & Salv. P. Z. S. 1864, p. 348.
Vireo altiloqua Sallé, P. Z. S. 1857, p. 231.
Vireosylvia calidris Baird, Rev. Am. Bds. p. 329 (1864).—Scl. & Salv. P. Z. S. 1875. p. 234.—Lawr. Pr. U. S. Nat. Mus. I, p. 233 (1878).— A. & E. Newton, Handb. Jamaica, p. 106 (1881).
Vireo calidris Bryant, Pr. Bost. Soc. Nat. Hist. XI, p. 93 (1866).—Salv. & Godm. Biol. Centr. Amer. Aves, p. 186 (1881).—Gadow, Cat. Bds. Brit. Mus. VIII, p. 293 (1883).—Cory, Bds. Haiti & San Domingo, p. 49 (1885); *ib.* List Bds. W. I. p. 10 (1885).
Phyllomanes calidris Gundl. Anal. Soc. Esp. Hist. Nat. VII, p. 168 (1878); *ib.* J. f. O. 1878, p. 158.
Vireosylvia calidris var. *dominicana* Lawr. Pr. U. S. Nat. Mus. I, p. 55 (1878).

Sp. Char. *Male:*—Crown grayish, but showing a slight olive tinge; upper parts dull olive green; a buff superciliary line and a dusky stripe through the eye; a narrow dusky maxillary line half way down the sides of the throat; sides pale yellowish-olive; lining of wings and under tail-coverts pale yellow; tail olive.

The sexes are similar.

Length, 5.80; wing, 3.20; tail, 2.50; tarsus, .68; bill, .60.

Habitat. Jamaica, San Domingo, and Antilles.

Vireo calidris barbatula.

Vireo gilvus D'ORB. In La Sagra's Hist. Nat. Cuba, Ois. p. 43 (1840) (?)
Phyllomanes barbatulus CAB. J. f. O. 1855, p. 467 (Cuba); *ib.* GUNDL.
    1861, p. 324 (Cuba); *ib.* 1872, p. 401 (Cuba).—BREWER, Pr. Bost.
    Soc. Nat. Hist. VII, p. 307 (1860) (Cuba).
Vireo olivaceus THIENEM. J. f. O. 1857, p. 147 (Cuba)?
Vireosylvia altiloqua BRYANT, Pr. Bost. Soc. Nat. Hist. VII, p. 113 (1859)
    (Bahamas).—ALBRECHT, J. f. O. 1861, p. 206 (Cuba).
Vireo calidris var. *barbatulus* BD. BWR. & RIDGW. Hist. N. Am. Bds. I,
    p. 36 (1874).
Vireo altiloquus var. *barbatulus* CORY, Bds. Bahama I. p. 82 (1880).
Vireo calidris barbatula CORY, List Bds. W. I. p. 10 (1885).

    The North American variety of *V. calidris* occurs in the
Bahama Islands, and is recorded from Cuba.

Vireo olivaceus (LINN.).

Muscicapa olivacea LINN. Syst. Nat. I. p. 327 (1766).
Vireo olivaceus BREWER, Pr. Bost. Soc. Nat. Hist. VII, p. 387 (1860)
    (Cuba).
Phyllomanes olivaceus GUNDL. J. f. O. 1872, p. 403 (Cuba); *ib.* 1878, p.
    158 (Porto Rico); *ib.* Anal. Soc. Esp. Hist. Nat. VII, p. 169 (1878)
    (Porto Rico).
Vireo olivacea CORY, List Bds. W. I. p. 10 (1885).

    V. olivaceus is claimed to have occurred in Cuba and Porto
Rico.

Vireo solitarius (WILS.).

Muscicapa solitaria WILS. Am. Orn. II. p. 43 (1810).
Vireo solitarius GUNDL. J. f. O. 1854, p. 468 (Cuba); *ib.* 1872, p. 403
    (Cuba).—BREWER, Pr. Bost. Soc. Nat. Hist. VII, p. 307 (1860)
    (Cuba).—CORY, List Bds. W. I. p. 10 (1885).
Vireo (Lanivireo) solitarius GUNDL. J. f. O. 1861, p. 324 (Cuba).

    Accidental in Cuba.

GENUS Laletes SCL.

Laletes SCLATER, P. Z. S. 1861, p. 72.

Laletes osburni SCL.

Laletes osburni SCL. P. Z. S. 1861, p. 72, pl. 14.—ALBRECHT, J. f. O. 1862,
    p. 195.—SCL. & SALV. Nom. Avium Neotr. p. 12 (1873).—A. & E.

NEWTON, Handb. Jamaica, p. 106 (1881).—GADOW, Cat. Bds. Brit.
Mus. VIII, p. 313 (1883).—CORY, List Bds. W. I. p. 10 (1885).
Laletes osburnii BAIRD, Rev. Am. Bds. p. 383 (1864).—GRAY, Handl.
Bds. I, p. 384 (1869.)

SP. CHAR. *Male:*—Top of head grayish olive, becoming olive green on the
back; underparts dull greenish yellow; wings and tail pale brown,
narrowly edged with olive; under wing-coverts yellowish white.
   Length (skin), 5.25; wing, 3; tail, 2.50; tarsus, .75.

HABITAT. Jamaica.

FAMILY AMPELIDÆ.

GENUS Dulus VIEILL.

Dulus VIEILLOT, Analyse, p. 42, No. 131, 1816.

Dulus dominicus (LINN.).

Tanagra dominica LINN. Syst. Nat. I, p. 316 (1766).—GMEL. Syst. Nat.
   I, p. 894 (1788).
Dulus palmarum "VIEILL. Nouv. Dict. X, p. 438 (1817)."—BP. Consp.
   I, p. 331 (1850).
Dulus dominicus STRICKL. Contr. Orn. p. 103 (1851).—SALLÉ, P. Z. S.
   1857, p. 231.—BAIRD, Rev. Am. Bds. p. 403 (1864).—BRYANT, Pf.
   Bost Soc. Nat. Hist. XI, p. 92 (1866).—GRAY, Handl. Bds. I, p.
   365 (1869).—SCL. & SALV. Nom. Avium Neotr. p. 13 (1873).—
   CORY, Bull. Nutt. Orn. Club, VI, p. 152 (1881); *ib.* Bds. Haiti &
   San Domingo, p. 51 (1885); *ib.* List Bds. W. I. p. 11 (1885).—
   SHARPE, Cat. Bds. Brit. Mus. X, p. 218 (1885).

SP. CHAR. *Male:*—Above dull olive brown; throat dull white; feathers
   of the throat and underparts dark brown in the centre, broadly
   edged with dull rufous white, giving the bird a heavily streaked appearance; rump green; primaries and secondaries dark brown, the
   outer webs edged with green, the inner webs becoming very pale
   on the edges; tail dark brown, the feathers very narrrowly edged
   with green; iris orange.
      The sexes are similar.
      Length, 6.20; wing, 3.50; tail, 3.10; tarsus, .80; bill, 55.

HABITAT. Haiti and San Domingo.

Dulus nuchalis SWAINS.

Dulus nuchalis SWAINS. Anim. in Menag. p. 345 (1837); *ib.* Classif. Bds.
   II, p. 238 (1837).—STRICKL. Contr. Orn. p. 104 (1851).—BAIRD,

Rev. Am. Bds. I, p. 403 (1864).--SCL. & SALV. Nom. Avium
Neotr. p. 13 (1873).—BOUC. Cat. Avium, p. 224 (1876).—CORY,
List Bds. W. I. p. 11 (1885).—SHARPE, Cat. Bds. Brit. Mus. X, p.
219 (1885).

This species? is described as somewhat smaller than *Dulus
dominicus*, and showing a white patch on the nape; it is otherwise similar. The exact habitat is unknown.

HABITAT. "Antilles."

GENUS Ampelis LINN.

Ampelis LINN. Syst. Nat. I, p. 297 (1766).

Ampelis cedrorum (VIEILL.).

Ampelis garrulus var. β. LINN. Syst. Nat. I, p. 297 (1766).
Bombycilla cedrorum VIEILL. Ois. Am. Sept. I. p. 88 (1807).—CAB. J. f.
    O. 1856, p. 3 (Cuba); *ib.* 1859, p. 350; GUNDL. J. f. O. 1861, p. 328
    (Cuba).
Ampelis carolinensis GOSSE, Bds. Jam. p. 197 (1847).—ALBRECHT, J. f. O.
    1862, p. 202 (Jamaica).
Bombycilla carolinensis BREWER, Pr. Bost. Soc. Nat. Hist. VII, p. 307
    (1860) (Cuba).
Ampelis cedrorum MARCH, Pr. Acad. Nat. Sci. Phila. 1863, p. 294 (Jamai-
    ca).—GUNDL. Repert. Fisico-Nat. Cuba, I, p. 240 (1865); *ib.* J. f.
    O. 1872, p. 430 (Cuba).—A. & E. NEWTON, Handb. Jamaica, p. 107
    (1881).—CORY, List Bds. W. I. p. 11 (1885).

Recorded from Cuba and Jamaica.

FAMILY TANAGRIDÆ.

GENUS Euphonia DESM.

Euphonia DESMAREST, Hist. Nat. des Tanagras, etc. p. 19 (1805).

Euphonia musica (GMEL.).

L'Organiste de S. Domingue. "BUFF. Pl. Enl. 809, fig. 1."
Pipra musica GMEL. Syst. Nat. I, p. 1004 (1788).
Tanagra musica "VIEILL. Enc. Meth. p. 787."
Euphonia musica GRAY, Gen. Bds. I, p. 367 (1846).—Br. Consp. I. p. 232
    (1850).—SALLÉ. P. Z. S. 1857, p. 231.—BRYANT, Pr. Bost. Soc. Nat.
    Hist. XI, p. 92 (1866).—SCL. & SALV. Nom. Avium Neotr. p. 17
    (1873).—CORY, Bull. Nutt. Orn. Club, VI, p. 152 (1881); *ib.* Bds.

Haiti & San Domingo, p. 61 (1885); *ib.* List Bds. W. I. p. 11 (1885).
—TRISTRAM, Ibis, 1884. p. 168.
Euphone musica LEMB. Aves Cuba. p. 42 (1850)?
Euphona musica GUNDL. J. f. O. 1855, p. 476.

SP. CHAR. *Male:*—Crown light blue, the color extending upon the nape,
and slightly upon the sides of the neck; forehead, underparts, and
rump brownish-orange; throat, cheeks, back, and tail bluish black,
showing purple reflections, the purple very prominent on the back:
a line of purplish black separating the blue and orange of the head
and forehead: primaries dark brown, becoming pale on the edges of
the inner webs; bill and feet black.

Female:—Underparts yellowish green, becoming yellowish on the
throat; cheeks and line above the forehead dull black; head and
nape, extending upon the sides of the neck, light blue; forehead
orange brown; back, rump, and wing-coverts olive green; tail dull
black, showing a tinge of green upon the feathers; primaries as in
the male, except showing an almost indistinct greenish edging upon
the outer webs.

Immature Male:—Forehead pale orange; top of the head grayish
blue; back olive green, blotched with dark blue; rump brownish
orange; wings and tail black, some of the tertiaries and coverts edged
with olive green; underparts olive green, marked with brownish on
the throat; dark orange, shaded with greenish, on the belly and crissum; bill and feet black.

Length, 4.40; wing, 2.60; tail, 1.80; tarsus .50; bill, .25.

HABITAT.   Haiti and San Domingo.   No species of *Euphonia*
has as yet been taken in Cuba.

Euphonia flavifrons (SPARRM.).

Emberiza flavifrons SPARRM. Mus. Carls. IV, No. 92 (♀).
Tanagra flavifrons LATH.   Ind. Orn. Suppl. p. 47 (♀). — VIEILL. Enc.
    Méth. p. 775.
Euphone organiste DESM. Hist. Nat. Tan. pls. 19. 20 (1805).—VIEILL.
    Gal. Ois. Suppl. pl. s. n. (♂ & ♀).
Cyanophonia musica BP. Rev. Zool. 1851, p. 138.
Euphonia flavifrons SCL. P. Z. S. 1856. p. 271; SUNDEV. Oefv. K. Vet.
    Akad. Förh. 1869. p. 583.—SCL. & SALV. Nom. Avium Neotr. p. 17
    (1873).—LAWR. Pr. U. S. Nat. Mus. I, pp. 56, 190, 269, 354, 455
    (1878).—ALLEN. Bull. Nutt. Orn. Club, V, p. 166 (1880.)—CORY.
    List Bds. W. I. p. 11 (1885).

SP. CHAR. *Male:*—Top of head bright blue, extending to the nape;
forehead bright yellow, separated from the blue by a bluish black
line; back green, shading into yellow on the rump; underparts

yellowish green; cheeks bluish black; outer webs of primaries narrowly edged with green, wanting on the first; bill dark.

Female:—Similar to the male but paler in coloration, cheek-marking dark olive and much less distinct.

Length, 4.20; wing, 2.30; tail, 1.40.

HABITAT. St. Bartholemew, Martinique, Guadeloupe, Dominica, St. Vincent, Grenada, and St. Lucia.

Euphonia jamaica (LINN.).

Fringilla jamaica LINN. Syst. Nat. I, p. 323 (1766).

Euphonia jamaica GOSSE, Bds. Jam. p. 238 (1847).—BP. Consp. I, p. 233 (1850).—ALBRECHT, J. f. O. 1862, p. 196.—SCL. Cat. Am. Bds. p. 60 (1862).—MARCH, Pr. Acad. Nat. Sci. Phila. 1863. p. 296.— SCL. & SALV. Nom. Avium Neotr. p. 18 (1873).—A. & E. NEWTON, Handb. Jamaica, p. 104 (1881).—CORY, List Bds. W. I. p. 11 (1885).

Pyrrhuphonia jamaica BP. Rev. Zool. 1851, p. 157.—GRAY, Handl. Bds. II, p. 79 (1870).

Euphonia jamaicensis SCL. P. Z. S. 1856, p. 280; *ib.* 1861, p. 73.

SP. CHAR. *Male:*—Above slaty blue; throat and breast gray; belly, abdomen, and flanks showing bright yellow; crissum dull white; lining of wing whitish, showing yellow on the axillaries.

Female:—Head and neck bluish gray; back yellowish green; wings showing yellowish green on the outer webs; under surface dull gray; a faint greenish tinge on the flanks.

Length, 4.30; wing, 2.50; tail, 1.45.

HABITAT. Jamaica.

Euphonia sclateri BP.

Euphonia sclaterii "BP. Mus. Par."

Euphonia sclateri SUNDEV. Oefv. K. Vet. Akad. Förh. 1869, p. 596.— GRAY, Handl. Bds. II, p. 77 (1870).—SCL. & SALV. Nom. Avium Neotr. p. 17 (1873).—BOUC. Cat. Avium, p. 240 (1876).—GUNDL. J. f. O. 1874. p. 311; *ib.* 1878, p. 159; *ib.* Anal. Soc. Esp. Hist. Nat. VII, p. 191 (1878).—CORY, List Bds. W. I. p. 11 (1885).

SP. CHAR. *Male:*—Forehead dull orange yellow, bordered by a narrow band of dark blue, succeeded by light blue, which color covers the entire top of the head to the nape; cheeks and ear-coverts very dark blue, almost black; wings and tail black with bluish reflections; back bluish black, distinctly blue when held in the light; rump yellow, showing a faint brownish tinge; throat yellow; breast and rest of underparts dull orange yellow, showing a slight brownish tinge on the crissum.

Length (skin), 4; wing, 2.35; tail, 1.65; tarsus, 58; bill, .25.

HABITAT. Porto Rico.

GENUS **Calliste** BOIE.

Calliste BOIE, Isis, 1826, p. 978.

Calliste versicolor LAWR.

Calliste versicolor LAWR. Ann. N. Y. Acad. Sci. I, p. 153 (1878); *ib.* Proc. U. S.
    Nat. Mus. I, pp. 190, 487 (1878).—CORY, List Bds. W. I. p. 11 (1885);
    ib. Ibis, p. 472 (1886).—SCLATER, Cat. Bds. Brit. Mus. XI. p. 113 (1886).
Calliste cucullata SCLATER & SALVIN, Ibis, p. 357 (1879).—CORY, Auk, III, p.
    195 (1886).

SP. CHAR. *Male:*—Top of the head deep chestnut red; upper plumage
golden fawn color; lores black; sides of the head and ear-coverts
dark green; tail black, except the two middle-feathers, which are
bluish green, the rest of tail-feathers and quills black, edged with
bluish green; upper tail-coverts bluish green; underparts pale bluish
lilac when held in the light; feathers of the upper throat tipped
with gray; under tail-coverts cinnamon.

    Female:—Top of the head lighter chestnut than in the male; rest
of upper parts pale green; underparts as in the male, but paler;
under tail-coverts, abdomen, and flanks pale cinnamon.

    Length, 6; wing, 3.30; tail, 2.50

HABITAT. St. Vincent.

GENUS **Spindalis** JARD.

Spindalis "JARD & SELBY. Ill. Orn. U. S. 1836."

Spindalis zena (LINN.).

Fringilla bahamensis BRISS. Orn. III. p. 168.
Fringilla zena LINN. Syst. Nat. I, p. 320 (1766).
Tanagra zena BRYANT, Pr. Bost. Soc. Nat. Hist. VII, p. 111 (1859).
Spindalis zena SCL. P. Z. S. 1856, p. 321.—SCL. & SALV. Nom. Avium
    Neotr. p. 21 (1873).—BOUC. Cat. Avium, p. 244 (1876).—CORY, Bds.
    Bahama I. p. 92 (1880); *ib.* List Bds. W. I. p. 11 (1885).
"*Spindalis pretrei* GRAY, Handl. Bds. II, p. 63 (1870)."

SP. CHAR. *Male:*—Above black; rump, and a broad band over the nape
from side of the neck rufous brown, shading into an orange tinge;
a superciliary stripe, and a stripe on the sides of the throat from
lower mandible and chin white; cheeks black; throat black, shading into brown upon the breast, with a yellow stripe passing from
the chin nearly to the brown of the breast; breast deep yellow,

shading into brown as it nears the throat: belly white, with an olive tint upon the flanks; wings and tail black, edged with white; the tertials, coverts, and base of primaries heavily marked with white; bill black, under mandible bluish; legs black.

Female:—Above olive green; below paler, shading into white on the belly; the sides and flanks pale olive green; the stripe over the eye but faintly indicated, and of an ashy color; wings and tail dark brown, with an olive tinge on the feathers, showing markings of dull white as in the male, but much narrower,

Length, 5.95; wing, 3; tail, 2.50; tarsus, .80; bill, .50.

HABITAT. Bahamas.

Spindalis pretrei (LESS.).

Tanagra pretrei LESS. Rev. Zool. 1839, p. 102.—GRAY, Gen. Bds. I, p. 365 (1846).
Tanagra multicolor et *Tanagra zena* D'ORB. in La Sagra's Hist. Nat. Cuba, Ois. p. 74 (1840).
"*Spindalis zena* et *pretrei* BP. Consp. I, p. 248 (1850.)
Spindalis pretrei CAB. J. f. O. 1855, p. 476.—BREWER, Pr. Bost. Soc. Nat. Hist. VII, p. 307 (1860).—GUNDL. Repert. Fisico-Nat. Cuba, I, p. 237 (1865); *ib.* J. f. O. 1872, p. 419.
Spindalis zena GRAY, Handl. Bds. II, p. 63 (1870).
Spindalis pretrii SCL. & SALV. Nom. Avium Neotr. p. 21 (1873).—BOUC. Cat. Avium, p. 244 (1876).—CORY, List Bds. W. I. p. 11 (1885); *ib.* Revised List Bds. W. I. p. 11 (1886).

SP. CHAR. *Male:*—Head black; a superciliary stripe reaching to the nape, and a stripe reaching from the base of the under mandible down the sides of the throat white; a narrow patch of white on the chin; throat yellow, separated from the white stripe by black; chest and cape chestnut, joining on the sides; back yellowish olive; rump chestnut; underparts grayish white, showing a yellow line down the middle of the belly; wings and tail black, the feathers marked with white; wing-coverts heavily marked with white; a broad patch of chestnut on the carpus; bill and feet dark.

Female:—The black on the head of the male, replaced by dull green; sides of the throat grayish; rump slightly tinged with yellowish; underparts olive gray, palest on the belly.

Length ♂ (skin), 5.60; wing, 3; tail, 2.25.

HABITAT. Cuba.

Spindalis multicolor (VIEILL.).

Tanagra multicolor VIEILL. Enc. Méth. p. 776.
Spindalis multicolor BP. Consp. I, p. 240 (1850).—SALLÉ, P. Z. S. 1857, p. 231.—GRAY, Handl. Bds. II, p. 63 (1870).—SCL. & SALV. Nom.

Avium Neotr. p. 21 (1873).—Bouc. Cat. Avium. p. 244 (1876).—
Cory, Bull. Nutt. Orn. Club, VI, p. 152 (1881): ib. Bds. Haiti &
San Domingo, p. 54 (1885); ib. List Bds. W. I. p. 11 (1885).
Tanagra (Shizampelis) dominicensis Bryant, Pr. Bost. Soc. Nat. Hist.
XI, p. 92 (1866).

Sp. Char. *Male:*—Head black; a superciliary stripe from the forehead to
the nape; a broad stripe of black from the bill, through the eye, to
the neck; chin white, the white extending in a stripe below the
black of the cheek to the neck; rest of throat black, with a yellow
stripe in the centre, reaching the white of the chin; breast chestnut,
shading into yellow upon the underparts and sides; a collar of
bright orange yellow upon the nape, joining the white stripe of
the throat; back olive; rump chestnut; abdomen and crissum white;
tail brownish black, the inner webs of the two outer tail-feathers
broadly marked with white; wings dark brown, with white edgings
to the coverts and secondaries; lesser wing-coverts chestnut; bill
and feet bluish black.

The female is dull colored; olive on the back and yellowish on
the rump; underparts grayish, whitening at the vent.

Length, 6.40; wing, 3.35; tail, 3.30; tarsus, .75; bill, .40.

Habitat.   Haiti and San Domingo.

Spindalis portoricensis (Bryant).

Tanagra (Spindalis) portoricensis Bryant, Pr. Bost. Soc. Nat. Hist. X,
p. 252 (1866).—Sundev. Oefv. K. Vet. Akad. Förh. 1869, p. 596.
Spizampelis portoricensis Gray, Handl. Bds. II, p. 63 (1870).
Spindalis portoricensis Scl. & Salv. Nom. Avium Neotr. p. 21 (1873).—
Bouc. Cat. Avium, p. 244 (1876).—Gundl. Anal. Soc. Esp. Hist.
Nat. VII, p. 188 (1878).—Cory, List Bds. W. I. p. 11 (1885).
Pyrrhulagra portoricensis Gundl. J. f. O. 1874, p. 312.

Sp. Char. *Male:*—Head black; a white superciliary stripe from the nostril to the nape; a white stripe passing down the sides of the throat;
a yellow stripe from the chin to the breast, where it becomes orange
chestnut, separated from the white of the cheek by a black patch,
which nearly reaches the bill; a narrow cape of orange chestnut;
breast yellow, becoming dull white on the abdomen; back green;
wing-coverts showing a patch of chestnut at the carpus; rump and
flanks yellowish green.

Female:—Top of head dull olive green, shading into yellowish
green on the back, brightest on the nape and rump; underparts
ashy, showing dull yellow on the breast; whole under surface indistinctly striped with pale brown.

Length, 6.50; wing, 3.50; tail, 2.50.

Habitat.   Porto Rico.

Spindalis nigricephala (JAMESON).

Tanagra nigricephala JAMESON, Ed. N. Phil. Journ. XIX, p. 213.—GOSSE. Ill. Bds. Jam. pl. 56.
Spindalis bilineatus JARD. & SELB. Ill. Orn. s. n. pl. 9.
Tanagra zena GOSSE, Bds. Jam. p. 231 (1847).
Tanagra zenoides DES MURS, Icon. Orn. pl. 40.
Spindalis nigricephala BP. Consp. I, p. 240 (1850).—SCL. P. Z. S. 1856, p. 230; *ib.* 1861, p. 74: *ib.* Cat. Am. Bds. p. 77 (1862).—ALBRECHT, J. f. O. 1862, p. 196.—MARCH, Pr. Acad. Nat. Sci. Phila. 1863, p. 296.—GRAY, Handl. Bds. II, p. 63 (1870).—SCL. & SALV. Nom. Avium Neotr. p. 21 (1873).—BOUC. Cat. Avium, p. 244 (1876).—A. & E. NEWTON, Handb. Jamaica, p. 104 (1881).—CORY, List Bds. W. I. p. 11 (1885).

SP. CHAR. *Male:*—Head black; a superciliary stripe reaching from the bill to the nape, white; a stripe of white passes down the sides of the throat; chin white, not reaching the orange of the breast; back yellowish green; the central portion of the breast bright orange; rest of underparts greenish yellow, sometimes orange yellow, quills and tail black; most of the primaries, secondaries, and coverts edged with white.

Female:—Top of the head dark olive; light olive green on the back; yellowish green on the rump and upper tail-coverts; throat and cheeks gray; underparts washed with orange yellow, commencing at the upper breast and brightest on the breast and belly; sides and flanks olive green.

Length, 7; wing, 4; tail, 3.

HABITAT. Jamaica.

GENUS Piranga VIEILL.

Piranga VIEILLOT, Ois. Am. Sept. I, p. iv (1807); *ib.* Analyse, p. 32 (1816).

Piranga rubra (LINN.).

Fringilla rubra LINN. Syst. Nat. I, p. 181 (1758).
Muscicapa rubra LINN. Syst. Nat. I, p. 326 (1766).
Tanagra æstiva GMEL. Syst. Nat. I, p. 889 (1788).
Pyranga æstiva D'ORB. La Sagra's Hist. Nat. Cuba, Ois. p. 76 (1840).—BREWER, Pr. Bost. Soc. Nat. Hist. VII, p. 307 (1860) (Cuba).—GUNDL. Repert. Fisico-Nat. Cuba, I, p. 237 (1865); *ib.* J. f. O. 1872, p. 421 (Cuba).—CORY, List Bds. W. I. p. 11 (1885).
Phœnicosoma æstiva GUNDL. J. f. O. 1855, p. 477.

Accidental in Cuba and the Bahama Islands.

Piranga erythromelas VIEILL.

Tanagra rubra LINN. Syst. Nat. I. p. 314 (1766).
Piranga erythromelas VIEILL. Nouv. Dict. XXVIII, p. 293 (1819).
Pyranga rubra D'ORB. in La Sagra's Hist. Nat. Cuba. Ois. p. 78 (1840).
—GOSSE, Bds. Jam. p. 235 (1847).—BREWER. Pr. Bost. Soc. Nat.
Hist. VII. p. 307 (1860) (Cuba).—ALBRECHT. J. f. O. 1862, p. 197
(Jamaica).—MARCH, Pr. Acad. Nat. Sci. Phila. 1863. p. 296
(Jamaica).—GUNDL. Repert. Fisico-Nat. Cuba, I. p. 238 (1865).—
A. & E. NEWTON, Handb. Jamaica, p. 104 (1881).—CORY, List Bds.
W. I. p. 11 (1885).
Phœnicosoma rubra GUNDL. J. f. O. 1855. p. 477 (Cuba).

Cuba and Jamaica: it has also been taken in the Barbadoes,
a specimen so labelled being in the U. S. National Museum.

GENUS Nesospingus SCL.

Nesospingus SCLATER. Ibis, 1885. p. 273.

Nesospingus speculiferus (LAWR.).

Chlorospingus speculiferus LAWR. Ibis, 1875, p. 383, pl. 9.—BOUC. Cat.
Avium, p. 246 (1876).—GUNDL. J. f. O. 1878, p. 159; *ib.* 1882, p.
161; *ib.* Anal. Soc. Esp. Hist. Nat. VII. p. 190 (1878).—CORY, List
Bds. W. I. p. 11 (1885).
Nesospingus speculiferus SCL. Ibis, 1885. p. 273.

SP. CHAR. *Male:*—"Entire upper plumage and sides of the head olive
brown; the feathers of the crown have their centres dark brown
with their margins grayish; the two central tail-feathers are
coloured like the back. the others are light reddish brown and are
closely crossed with nearly obsolete darker bars; quill-feathers dark
brown, first, edged with gray on the outer primaries, the outer webs
of the fourth, fifth, and sixth primaries are marked near their bases
with white. partly concealed by the wing-coverts, the portion
beyond the coverts appearing as a small triangular spot; the under
plumage is grayish white, and has a somewhat mottled appearance,
owing to the darker bases of the feathers showing a little; the sides
are dusky, with a tinge of rufous; under tail-coverts light rufous,
with dusky centres; upper mandible dark brown, the under, pale
brownish white; tarsi and toes brownish black. Length. 6½ inches;
wing, 3½; tail, 2¾; bill, ⅞; tarsus, 1." (LAWR. l. c., orig. descr.)
HABITAT. Porto Rico.

GENUS Phœnicophilus STRICKL.

Phœnicophilus "STRICKLAND, Contr. Orn. p. 104, 1861."

Phœnicophilus palmarum (LINN.).

Turdus palmarum LINN. Syst. Nat. I, p. 295 (1766).
Tachyphonus palmarum "VIEILL. N. D. d'H. N. XXXII. p. 359."
Arremon palmarum GRAY, Gen. Bds. Suppl. p. 16.—BRYANT, Pr. Bost.
    Soc. Nat. Hist. XI. p. 92 (1866).
Dulus palmarum "BP. R. Z. 1851, p. 78."
Dulus poliocephalus "BP. R. Z. 1851. p. 78."
Phœnicophilus palmarum STRICKL. Contr. Orn. p. 104 (1851).—SCL. P.
    Z. S. 1856, p. 84.—GRAY, Handl. Bds. II, p. 72 (1870).—SCL. &
    SALV. Nom. Avium Neotr. p. 25 (1873).—CORY, Bull. Nutt. Orn.
    Club, VI, p. 152 (1881); ib. Bds. Haiti & San Domingo, p. 56
    (1885); ib. List Bds. W. I. p. 12 (1885).—TRISTRAM, Ibis, 1884,
    p. 168.
Phœnicophilus palmarum BOUC. Cat. Avium, p. 247 (1876).

SP. CHAR. Male:—Top of the head and cheeks black; a spot of white on
    each side of the forehead; a white stripe touching the upper eyelid,
    commencing at the centre of the eye, passing backward on the head;
    a patch of white on the lower eyelid; a gray collar on the nape,
    extending upon, and joining the gray of the sides; sides slaty gray;
    throat white, the white extending in a narrow line down the middle
    of belly to the vent; the back, rump, tail, outer webs of secondaries
    and coverts bright yellowish green; quills brown; bill and feet
    bluish black.
        The sexes are similar.
        Length, 6.70; wing. 3.70; tail, 3; tarsus, .85; bill, .70.
HABITAT. Haiti and San Domingo.

Phœnicophilus poliocephalus (BONAP.).

Dulus poliocephalus BP. Rev. Zool. p. 78 (1851).
Phœnicophilus poliocephalus STRICKL. Contr. Orn. p. 194 (1851); SCL. Cat. Bds.
    Brit. Mus. p. 234 (1886).
Phœnicophilus palmarum SCL. P. Z. S. 1856, p. 84.
Phœnicophilus dominicensis CORY, Bull. Nutt. Orn. Club, VI, p. 129 (1881);
    ib. Bds. Haiti and San Domingo, p. 58 (1885); ib. List Bds. W. I. p. 12
    (1885); ib. Auk, III, p. 200 (1886).

SP. CHAR. Male:—Forehead and sides of the head black; a spot of white
    above and below the eye, and on each side of the forehead; chin
    white, extending in two stripes down the sides of the throat to the
    breast, bordering the black of the head; the rest of the head, neck
    and underparts grayish plumbeous; back, wing-coverts, tail and cov-
    erts, and outer edges of wing-feathers bright yellowish green; inner
    webs of primaries and secondaries brown, pale on the edges; legs
    and lower mandible dark slate color; upper mandible black; iris red-
    dish brown.
        The sexes are similar.
        Length, 6.80; wing, 3.50; tail, 2.30; tarsus, .82; bill, .68.
HABITAT. San Domingo.

GENUS **Calyptophilus** CORY.

Calyptophilus CORY. Auk. I. p. 1 (1884).

Calyptophilus frugivorus CORY.

Phœnicophilus frugivorus CORY, Journ. Bost. Zool. Soc. II, No. 4, p. 45
(1883).

Calyptophilus frugivorus CORY, Auk,
I. p. 3 (1884): *ib.* Bds. Haiti and
San Domingo, p. 59 (1885); *ib.*
List Bds. W. I. p. 12 (1885).

SP. CHAR. *Male:*—Top of the head
brown, shading into ashy on the
neck behind the eye; rest of upper
parts, including back and upper surface of wings and tail, brownish
olive; throat white; breast white, becoming ashy upon the sides;
flanks brownish olive, the olive mixing with white upon the crissum;
primaries and secondaries olive brown, the inner webs edged with
very pale brown; a patch of bright yellow under the base of the
wing, extending upon the carpus; eye encircled by a very narrow
line of bright yellow, and a spot of yellow in front of the eye, at the
base of the mandible; upper mandible dark brown; lower mandible
yellowish brown, darkest at the base. Some specimens show a spot
of yellow upon the middle of the breast, but it is not constant. In a
series of fourteen specimens, it is wanting in all but five.

The female is perhaps somewhat duller, and some specimens
appear slightly smaller, but otherwise resembles the male.

Length, 7.50; wing, 3.70; tail, 3.70; tarsus, 1; toe, .82; bill, .75.
HABITAT. San Domingo.

GENUS **Saltator** VIEILL.

Saltator VIEILLOT, Analyse, p. 32 (1816).

Saltator guadeloupensis LAFR.

Saltator guadeloupensis LAFR. Rev. Zool. 1844, p. 167 —Br. Consp. I,
p. 489 (1850).—SCL. Cat. Am. Bds. p. 97 (1862).—TAYLOR, Ibis,
1864, p. 167.—GRAY, Handl. Bds. II. p. 74 (1870).—SCL. & SALV.
Nom. Avium Neotr. p. 26 (1873).—LAWR. Pr. U. S. Nat. Mus. I,
pp. 354, 457 (1878).—CORY, List Bds. W. I. p. 12 (1885).
Saltator martinicensis BR. Consp. I, p. 489 (1850).—GRAY, Handl. Bds.
II, p. 75 (1870).

Sp. Char. *Male:*—Head and back bright olive green, shading into gray on the rump; wings showing the outer webs of the primaries and secondaries green, lacking on the first three primaries, or if showing at all, appearing in a form of a narrow pencilled line; the wing-coverts olive green; throat white, showing a dash of dark brown on either side; a whitish superciliary line; cheeks and ear-coverts olive green; breast and underparts dull buffy white, tinged slightly with olive, and showing faint pencilled lines of pale brown; tail blackish; bill black at the base, pale at the tip.

Female:—Similar to the male; the dark brown streak on the sides of the throat lacking in some specimens, brownish olive in others, but apparently always paler than in the male.

Length (skin), 8; wing, 3.75; tail, 3.50; tarsus, .85; bill, .75.

Habitat. Guadeloupe and Martinique.

Family FRINGILLIDÆ.

Genus Guiraca Swains.

Guiraca Swainson, Zool. Journ. III, p. 350 (1827).

Guiraca cærulea (Linn.).

Loxia cærulea Linn. Syst. Nat. I, p. 306 (1766).
Coccoborus cærulens Lemb. Aves Cuba, p. 61 (1850).—Cab. J. f. O. 1856, p. 9 (Cuba).
Guiraca cærulea Brewer, Pr. Bost. Soc. Nat. Hist. VII, p. 307 (1860) (Cuba).—Gundl. Repert. Fisico-Nat. Cuba, I, p. 285 (1866); *ib.* J. f. O. 1874, p. 126 (Cuba).—Cory, List. Bds. W. I. p. 12 (1885).

Recorded from Cuba.

Genus Habia Reich.

Habia Reich. Av. Syst. Nat. 1850, pl. xxviii.

Habia ludoviciana (Linn.).

Loxia ludoviciana Linn. Syst. Nat. I, p. 306 (1766).
Guiraca ludoviciana Gosse, Bds. Jam. p. 239 (1847).—Brewer, Pr. Bost. Soc. Nat. Hist. VII, p. 307 (1860) (Cuba).—Albrecht, J. f. O. 1862, p. 196 (Jamaica).—A. & E. Newton, Handb. Jamaica, p. 104 (1881).
Coccoborus ludovicianus Lemb. Aves Cuba, p. 59 (1850).

Hedymeles ludoviciana CAB. J. f. O. 1856, p. 9 (Cuba).
Goniaphea ludoviciana GUNDL. Repert. Fisico-Nat. Cuba, I, p. 286 (1866);
     ib J. f. O. 1874, p. 126 (Cuba).
Habia ludoviciana CORY, List Bds. W. I. p. 12 (1885).

Accidental in Cuba and Jamaica.

Habia melanocephala (Swains) is recorded from Cuba (*Hedymeles melanocephala* Cabanis, J. f. O. 1856, p. 9). It has no other West Indian record, and has not been cited by later authors.

GENUS Loxigilla LESS.

Loxigilla LESSON, Traité, p. 443 (1831).

Loxigilla violacea (LINN.).

Loxia violacea LINN. Syst. Nat. I, p. 306 (1766).
Pyrrhula violacea GOSSE, Bds. Jam. p. 254 (1847).
Pyrrhula robinsonii GOSSE, Bds. Jam. p. 259 (1847).—ALBRECHT, J. f. O. 1862, p. 196.
Pyrrhulagra violacea BP. Consp. I, p. 493 (1850).

Loxigilla violacea SALLÉ, P. Z. S. 1857, p. 231. —ALBRECHT, J. f. O. 1862, p. 196.—SCL. Cat. Am. Bds. p. 102 (1862).—MARCH, Pr. Acad. Nat. Sci. Phila. 1863, p. 297.—GRAY, Handl. Bds. II, p. 104 (1370).—SCL. & SALV. Nom. Avium Neotr. p. 27 (1873).— RIDGW. Pr. U. S. Nat. Mus. I, p. 250 (1878). —CORY, Bds. Bahama I. p. 85 (1880); *ib*. Bull. Nutt. Orn. Club. VI. p. 152 (1881).—
A. & E. NEWTON, Handb. Jamaica, p. 104 (1881).—TRISTRAM, Ibis, 1884, p. 168.—CORY, Bds. Haiti & San Domingo, p. 69 (1885); *ib*. List Bds. W. I. p. 12 (1885).
Spermophila violacea BRYANT, Pr. Bost. Soc. Nat. Hist. VII, p. 119 (1859).
Loxia (Pyrrhulagra) violacea BRYANT, Pr. Bost. Soc. Nat. Hist. XI, p. 93 (1866).
Loxigilla violacea β. *bahamensis* RIDGW. Pr. U. S. Nat. Mus. I, p. 250 (1878).

SP. CHAR. *Male:—*Entire plumage black, showing a slight brownish tinge upon the quills; throat, crissum and crescent over the eye reddish brown ; bills and legs black.
     *Female:—*Underparts gray, with a tinge of olive green upon the back ; below ash, lightest upon the belly, showing a tinge of olive upon the breast and sides; quills with fine edgings of dull white;

crissum, a crescent over the eye, and markings upon the chin pale reddish brown, much lighter than in the male; under mandible pale. Immature birds resemble the female.

Length, 5.80; wing, 3; tail, 2.70; tarsus, 90; bill, .50.

HABITAT. Bahamas, Jamaica, Haiti, and San Domingo.

Specimens from different localities often vary in coloration and size, those from Jamaica and San Domingo being somewhat smaller than those from the Bahama Islands. The Jamaica bird differs from the Bahama form, in being somewhat smaller; the red of the throat is lighter, and the under wing-coverts are gray, instead of dull white. It seems to represent a fairly good geographical race. Mr. Ridgway, who first separated them, described the Bahama bird as *L. violacea bahamensis*, but as the type of *L. violacea* came from the Bahamas, the name *bahamensis* becomes a synonym, and the Jamaica form remains as yet unnamed, should it be thought advisable to separate them.

Loxigilla noctis (LINN.).

Fringilla noctis LINN. Syst. Nat. I, p. 320 (1766). —DENNY, P. Z. S. 1847, p. 38.
Pyrrhulagra noctis BP. Consp. I, p. 493 (1850) (excl. syn.).
Loxigilla noctis SCL. Cat. Am. Bds. p. 102 (1862). —TAYLOR, Ibis, 1864, p. 167.—GRAY, Handl.
Bds. II, p. 104 (1870).—SCL. & SALV. Nom. Avium Neotr. p. 27 (1873).—SCL. P. Z. S. 1874, p. 175.—LISTER, Ibis, 1880, p. 40.— GRISDALE, Ibis, 1882, p. 486.—CORY, List Bds. W. I. p. 12 (1885).

SP. CHAR. *Male:*—Entire plumage black; superciliary stripe, and throat chestnut-rufous; under tail-coverts rufous; bill and feet black.

Female:—Upper surface dull reddish brown, brightest on the rump; underparts olive brown; wing-coverts heavily edged with rufous; secondaries tinged with the same color; under mandible brown.

Length (skin), 4.50; wing, 2.70; tail, 1.85; tarsus, .75.

HABITAT. Lesser Antilles.

Loxigilla noctis sclateri.

Loxigilla noctis SCL. P. Z. S. 1871, p. 270.
Loxigilla noctis sclateri ALLEN, Bull. Nutt. Orn. Club, V. p. 166 (1880). —CORY, List Bds. W. I. p. 12 (1885).

Sp. Char. *Male:*—Differs from true *noctis* by having the superciliary line much smaller, almost absent in some specimens, and in lacking the rufous on the under tail-coverts; but the characters are not constant.

Habitat.   Santa Lucia.

Loxigilla anoxantha (Gosse).

Spermophila anoxantha Gosse, Bds. Jam. p. 247 (1847).
Loxigilla anoxantha Scl. P. Z. S. 1861, p. 74; *ib.* Cat. Am. Bds. p. 102 (1866).—Albrecht, J. f. O. 1862, p. 196.—March, Pr. Acad. Nat. Sci. Phila. 1863, p. 297.—Gray, Handl. Bds. II. p. 104 (1870).—Scl. & Salv. Nom. Avium Neotr. p. 27 (1873).—A. & E. Newton, Handb. Jamaica, p. 104 (1881).—Cory, List Bds. W. I. p. 12 (1885).

Sp. Char. *Male:*—Head, throat, and underparts dull black; back, rump and wing-coverts having the feathers edged with bright yellow giving a yellowish appearance to the surface; under tail-coverts chestnut; quills and tail dull brown, slightly edged with yellowish.
   Female:—Entire upper plumage dull green; throat and breast grayish, with a tinge of olive, becoming pale on the belly.
   Length (skin), 4.25; wing, 2.40; tail, 1.40; tarsus, .90.

Habitat.   Jamaica.

Loxigilla portoricensis (Daud.).

Loxia portoricensis Daud. Traité D'Orn. 11, pl. 29.—Sundev. Oefv. K. Vet. Akad. Förh. 1869, p. 597.
Pyrrhula auranticollis Vieill. Enc. Méth. p. 1028.
Pyrrhulagra portoricensis Br. Consp. I, p. 492 (1850).— Gundl. Anal. Soc. Esp. Hist. Nat. VII, p. 308 (1878).
Loxia (Pyrrhulagra) portoricensis Bryant, Pr. Bost. Soc. Nat. Hist. X, p. 254 (1866).

Loxigilla potoricensis Gray, Handl. Bds. II, p. 104 (1870).—Cory, List Bds. W. I. p. 12 (1885).

Sp. Char. *Male:*—A narrow line of black on the forehead; top of the head chestnut rufous, separated at the nape by the black of the back, the black color dividing it like a wedge; throat and under tail-coverts chestnut rufous; rest of plumage black; under wing-coverts dull white; bill and feet black.
   Female.—Similar to the male, possibly somewhat duller in coloration.
   Length (skin), 7.50; wing. 3.50; tail. 3; tarsus, .95.

Habitat.   Porto Rico.

Loxigilla portoricensis grandis.

Loxigilla portoricensis var. *grandis* LAWR. Pr. U. S. Nat. Mus. IV, p. 204 (1881).

Loxigilla portoricensis grandis CORY, List Bds. W. I. p. 12 (1885).

SP. CHAR. *Male:*—Larger than *Loxigilla portoricensis*, and having the rufous chestnut coloring darker.

Length (skin), 8; wing. 4.25; tail, 3.25; tarsus, 1.

HABITAT. St. Christopher.

Dr. Gundlach records "*Cardinalis virginianus*" (Repert. Fisico-Nat. Cuba, I. p. 397, 1886). as occurring in Cuba. It is possible that the specimen in question was an escaped cage bird, although there is no reason why it should not occasionally occur there, being common in Florida.

GENUS **Melopyrrha** BP.

Melopyrrha BONAPARTE, Compt. Rend. XXXVII, p. 924 (1853).

Melopyrrha nigra (LINN.).

Loxia nigra LINN. Syst. Nat. I, p. 306 (1776).

Pyrrhula crenirostris "VIEILL. Ois. Chant. pl. 77 (1805)."

Pyrrhula nigra VIGORS, Zool. Journ. 1827, p. 440.—VIEILL. Gal. Ois. I, p. 65. pl. 57 (1834).—D'ORB. in La Sagra's Hist. Nat. Cuba, Ois. p. 108 (1840).—GUNDL. Journ. Bost. Soc. Nat. Hist. VII, p. 317 (1851).

Melopyrrha nigra CAB. J. f. O. 1856, p. 8.—BREWER, Pr. Bost. Soc. Nat. Hist. VII, p. 307 (1860).—SCL. Cat. Am. Bds. p. 103 (1862).—GUNDL. Repert. Fisico-Nat. Cuba, I. p. 285 (1865); *ib.* J. f. O. 1874, p. 125.—GRAY, Handl. Bds. II, p. 104 (1870).—SCL. & SALV. Nom. Avium Neotr. p. 28 (1873).—CORY, List Bds. W. I. p. 12 (1885).

SP. CHAR. *Male:*—Upper surface dull black, showing a tinge of bluish when held in the light, brightest on the head: under surface blackish, becoming dusky on the belly; primaries broadly edged with white on the basal half of the inner webs, some of the primaries delicately edged with white, showing distinctly on the fourth and fifth; wing-coverts broadly marked with white; carpus and under wing-coverts white; tail brownish black.

The female is similar to the male but slightly duller in plumage.

Length (skin), 5.75; wing, 2; tail, 2.25; tarsus, .68.

HABITAT. Cuba.

GENUS **Loximitris** BRYANT.

Loximitris BRYANT, Pr. Bost. Soc. Nat. Hist. XI, p. 93 (1866).

Loximitris dominicensis (Bryant).

Chrysomitris (Loximitris) dominicensis Bryant, Pr. Bost. Soc. Nat. Hist.
  XI, p. 93 (1866).
Chrysomitris dominicensis Gray, Handl. Bds. II, p. 81 (1870).—Cory,
  Bull. Nutt. Orn. Club, VI, p. 152 (1881).
Loximitris dominicensis Cory, Bds. Haiti & San Domingo, p. 67 (1885);
  ib. List Bds. W. I. p. 12 (1885).

Sp. Char. *Male:*—"Bill light brown color, with the top dusky; whole
            head and throat black; back and scapulars olive; the centre of each
            feather dusky; upper tail-coverts bright olive yellow; wings with
            the quills and coverts blackish brown; the smaller coverts with
            so much of the tips olive as to appear almost wholly of this color;
            the greater coverts and all the quill-feathers, except the first, bor-
            dered externally with the same color, very narrowly on the prima-
            ries, and suddenly wider on the secondaries, but only on the
            posterior half, so that the closed wing presents a distinct blackish
            bar, running nearly across its centre; tail with the centre feather,
            outer web of first, and tips of all, blackish brown, the rest bright
            chrome yellow; beneath yellow, washed with olive on the flanks,
            and brightest on the crissum." (Bryant, l. c., orig. descr.).
            Immature birds are dull olive, mottled with brownish on the
            back, and the underparts yellowish white, streaked with pale
            brown.
            Length, 4.10; wing. 2.60; tail. 1.55; tarsus, .53; bill, .38.

Habitat.   San Domingo.

Genus Pyrrhomitris Bonap.

Pyrrhomitris Bonap. Consp. I, p. 517 (1850).

Pyrrhomitris cucullata (Swains.).

Carduelis cucullata Swains. Zool. Illust. 1820, pl. 7.
Fringilla cubæ Gervais, Mag. Zool. 1835, pl. 44.—Cab. J. f. O. 1856, p.
  10 (Cuba); *ib.* 1857, p. 241 (Cuba); *ib.* Gundl. 1859, p. 295 (Cuba);
  ib. 1861, p. 412 (Cuba); *ib.* 1871, p. 282 (Cuba).
Pyrrhomitris cucullata Bp. Consp. I, p. 517 (1850) (Antilles).—Gundl.
  Repert. Fisico-Nat. Cuba, I, p. 397 (1766) (Cuba); *ib.* Anal. Soc.
  Esp. Hist. Nat. VII, p. 207 (1878) (Porto Rico); *ib.* J. f. O. 1878, p.
  160 (Porto Rico).
Pyrrhomitris cubæ Gray, Handl. Bds. II. p. 82 (1870) (Antilles)?

Sp. Char. *Male:*—Entire head and throat black; back dull reddish orange;
            rump bright orange red; underparts orange red, whitening at the
            vent; under tail-coverts pale orange; primaries having the basal
            half of the outer web pale orange; wings and tail dark brown.

Female:—Entire upper parts grayish olive; rump pale orange; underparts dull gray, whitening near the vent; a patch of orange on the breast; otherwise resembles the male.

Length (skin), 3.50; wing, 2.37; tail, 1.15; tarsus, .50; bill, .40.

Introduced in Cuba and Porto Rico.

Both *Spinus pinus* (Wils.) and *Spinus mexicana* (Swains.) have been recorded from Cuba; the references are as follows:

Chrysomitris pinus GUNDL. J. f. O. 1856. p. 9; *ib* Repert. Fisico-Nat. Cuba, I. p. 397 (1866).

Chrysomitris mexicana GUNDL. Repert. Fisico-Nat. Cuba, I, p. 397 (1866) (Cuba).

GENUS Euetheia REICH.

Euetheia REICHENBACH. Av. Syst. Nat. Knacker. pl. 79, "June 1st, 1850."

Euetheia olivacea (GMEL.).

Emberiza olivacea GMEL. Syst. Nat. I, p. 309 (1788).

Spermophila olivacea GOSSE, Bds. Jam. p. 249 (1847).—ALBRECHT, J. f. O. 1862, p. 196.

Phonipara olivacea SCL. P. Z. S. 1855, p. 159; *ib.* Cat. Am. Bds. p. 107 (1862).—MARCH. Pr. Acad. Nat. Sci. Phila. 1863, p. 297.—SCL. & SALV. Nom. Avium Neotr. p. 29 (1873).—A. & E. NEWTON, Handb. Jamaica, p. 104 (1881).—CORY, Bull. Nutt. Orn. Club, VI, p. 152 (1881); *ib.* Bds. Haiti & San Domingo. p. 65 (1885).

Euetheia lepida CAB. J. f. O. 1856, p. 7.—BREWER, Pr. Bost. Soc. Nat. Hist. VII, p. 307 (1860).—GUNDL. Repert. Fisico-Nat. Cuba, I. p. 284 (1866).—SUNDEV. Oefv. K. Vet. Akad. Förh. 1869, p. 597.—GUNDL. J. f. O. 1874, p. 122; *ib.* Anal. Soc. Esp. Hist. Nat. VII, p. 204 (1878).

Fringilla (Phonipara) olivacea BRYANT, Pr. Bost. Soc. Nat. Hist. XI, p. 93 (1866).

Euetheia olivacea CORY, List Bds. W. I. p. 12 (1885).

SP. CHAR. *Male:*—Above dull olive; a superciliary stripe, and a patch on the chin and upper throat orange yellow, rest of throat black; a narrow line of black bordering the yellow of the throat, reaching to the front of the eye; lower eyelid dull yellow; underparts olivaceous gray; carpus dull yellow; bill and feet dark brown.

Female:—Lacking the black of head and throat in the male; the yellow is much less conspicuous and paler; belly dull gray; the olive of the back duller than in the male.

Length, 4; wing, 2; tail, 1.50; tarsus, .50; bill, .30.

HABITAT. Cuba, Jamaica, Haiti, San Domingo, and Porto Rico.

Euetheia canora (GMEL.).

Loxia canora GMEL. Syst. Nat. 1, p. 858 (1788).
Phonipara canora Br. Consp. 1, p. 494 (1850).—GRAY, Handl. Bds. II. p.
    98 (1870).—SCL. & SALV. Nom. Avium Neotr. p. 29 (1873).—BD.
    BWR. & RIDGW. Hist. N. Am. Bds. II. p. 93 (1874).
Euethia canora BREWER, Pr. Bost. Soc. Nat. Hist. VII, p. 307 (1860).—
    GUNDL. Repert. Fisico-Nat. Cuba, 1, p. 284 (1866); *ib.* J. f. O.
    1874, p. 123.
Euetheia canora CORY, List Bds. W. I. p. 12 (1885).

SP. CHAR. *Male:*—Throat and cheeks black, extending above the eye;
    a broad band of bright yellow extends across the lower throat to
    the sides of the neck, and passes in a narrow line, edging the black,
    to the eye; top of head slaty brown; rest of upper surface bright
    olive green; breast brownish black, shading into pale gray on the
    belly and under tail-coverts.

    Female:—Throat dark chestnut brown, shading into gray on the
    cheeks; yellow collar much paler than in the male; chest and under-
    parts ashy; the rest as in the male.

    Length (skin), 3.75; wing, 2; tail, 2.05; tarsus, .62.

HABITAT. Cuba.

Euetheia bicolor (LINN.).

Fringilla zena LINN. Syst. Nat. I, 10th ed. p. 183 (1758).
Fringilla bicolor LINN. Syst. Nat. I, 12th ed. p. 324 (1766).
Spermophila bicolor GOSSE, Bds. Jam. p. 252 (1847).—BRYANT, Pr. Bost.
    Soc. Nat. Hist. VII, p. 119 (159).—ALBRECHT, J. f. O. 1862, p. 196.
Phonipara bicolor Br. Consp. I, p. 494 (1850).—CASSIN, Pr. Acad. Nat.
    Sci. Phila. 1860, p. 376. — SCL. Cat. Am. Bds. p. 106 (1862). —
    SUNDEV. Oef. K. Vet. Akad. Förh. 1869, p. 596.—SCL. & SALV.
    Nom. Avium Neotr. p. 29 (1873).—CORY, Bds. Bahama I. p. 91
    (1880); *ib.* Bull. Nutt. Orn. Club, VI, p. 152 (1881).—TRISTRAM,
    Ibis, 1884, p. 168.
Phonipara marchii BAIRD, Pr. Acad. Nat. Sci. Phila. 1863, p. 297.—
    A. & E. NEWTON, Handb. Jamaica. p. 104 (1881).
Fringilla (Phonipara) zena var. portoricensis BRYANT, Pr. Bost. Soc. Nat.
    Hist. X, p. 254 (1866).
Fringilla zena var. marchi BRYANT, Pr. Bost. Soc. Nat. Hist. XI, p. 93
    (1867).
Phonipara zena BD. BWR. & RIDGW. Hist. N. Am. Bds. II, p. 93 (1874).
    —RIDGW. Pr. U. S. Nat. Mus. VII, p. 172 (1884).—CORY, Bds.
    Haiti and San Domingo, p. 63 (1885).
Euethia bicolor GUNDL. Anal. Soc. Esp. Hist. Nat. VII, p. 205 (1878).
Euetheia bicolor CORY, List Bds. W. I. p. 12 (1885).

Sp. Char. *Male:*—General plumage dull olive; whole of breast and throat black; a blackish tinge sometimes perceptible on the forehead; belly dull gray, shading into olive on the flanks; surface of wings and tail olive; primaries, secondaries, and tail-feathers brown, showing olive on the outer webs; some males have only a small patch on the chin black, others have nearly the entire undersurface black, the extent of the color varying greatly, perhaps according to the age of the bird.

Female : — Resembles the male, but lacks the black of the throat, which is replaced by dull olivaceous gray.

Length, 4.10; wing, 2.05; tail, 1.60; tarsus, .54; bill, .40.

HABITAT. Bahama Islands and Antilles.

After a careful examination of numerous specimens of the so called *E. marchi,* from Jamaica, and comparing them with a series of some seventy specimens of *E. bicolor,* I fail to see any differences sufficient to separate them. The underparts of *E. marchi* are somewhat browner; the back is also darker; but this stage of plumage occurs in the young and the female of *E. bicolor.*

Euetheia adoxa (Gosse).

Spermophila adoxa Gosse, Bds. Jam. p. 253 (1847).—Albrecht, J. f. O. 1862, p. 196.

Phonipara adoxa " March. P. A. P. 1863, p. 297."—A. & E. Newton, Handb. Jamaica, p. 104 (1881).—Cory, Revised List Bds. W. I. p. 12 (1886).

Sp. Char.—" Irides dark brown; feet purplish; beak horn-colour, nearly black above; whole upper parts olive brown : under parts grayish white; sides and vent fawn color. Length, 4½; expanse, 6½; flexture, 2; tail, 1 6-10 (nearly); rictus, 7-20; tarsus, 7-10; middle toe, 7-10." (Gosse, l. c., orig. descr.)

HABITAT. Jamaica.

I have never seen a specimen of this so-called species ; judging from the description, it would seem possible that it might prove to be the female of *E. bicolor.*

Genus Passerina Vieill.

Passerina Vieillot, Analyse (1816).

Passerina ciris (Linn.).

Emberiza ciris Linn. Kong. Sv. Vet. Akad. Hand. 1750, p. 278, tab. vii, f. 1; *ib.* Syst. Nat. I, p. 313 (1766).

Passerina ciris D'ORB. in La Sagra's Hist. Nat. Cuba, Ois. p. 102 (1840).
—CORY, List Bds. W. I. p. 12 (1885).
?Linaria caniceps D'ORB. in La Sagra's Hist. Nat. Cuba, Ois. p. 107
(1840).—GUNDL. J. f. O. 1871. p. 276 (young male) (Cuba).
Spiza ciris CAB. J. f. O. 1856, p. 8 (Cuba).—BREWER, Pr. Bost. Soc. Nat.
Hist VII. p. 307 (1860) (Cuba).
Cyanospiza ciris GUNDL. Repert. Fisico-Nat. Cuba, I. p. 285 (1866).—
MOORE. Pr. Bost. Soc. Nat. Hist. XIX, p. 247 (1877) (New Provi-
dence).—CORY, Bds. Bahama I. p. 89 (1880).

Accidental in Cuba and Bahamas,

Passerina cyanea (LINN.).

Tanagra cyanea LINN. Syst. Nat. I. p. 315 (1766).
Passerina cyanea D'ORB. in La Sagra's Hist. Nat. Cuba, Ois. p. 100
(1840).—CORY. List Bds. W. I. p. 12 (1885).
Spiza cyanea CAB. J. f. O. 1856. p. 8 (Cuba).—BREWER, Pr. Bost. Soc.
Nat. Hist. VII, p. 307 (1860) (Cuba).
Cyanospiza cyanea GUNDL. Repert. Fisico-Nat. Cuba, I. p. 285 (1866);
ib. J. f. O. 1874. p. 125 (Cuba). — BRACE. Pr. Bost. Soc. Nat. Hist.
XIX. p. 242 (1877) (New Providence). — CORY. Bds. Bahama I.
p. 90 (1880).

Accidental in Cuba and Bahamas.

GENUS **Passer** BRISS.

Passer BRISSON, Orn. 1760.

Passer domesticus (LINN.).

Fringilla domestica LINN. Syst. Nat. I, p. 323 (1766).
Passer domesticus BRACE. Pr. Bost. Soc. Nat. Hist. XIX, p. 242 (1877)
(New Providence).—CORY, Bds. Bahama I. p. 88 (1880); *ib.* List
Bds. W. I. p. 13 (1885).

An introduced species, resident in some of the Bahama
Islands. Cuba?

GENUS **Ammodramus** SWAINS.

Ammodramus SWAINSON. Zool. Journ. III. 1827.

Ammodramus sandwichensis savanna (WILS.).

Fringilla savanna WILS. Am. Orn. III, p. 55 (1811).
Emberiza savanna LEMB. Aves Cuba. p 55 (1850).

Passerculus savanna CAB. J. f. O. 1856, p. 6 (Cuba).—BREWER, Pr. Bost. Soc. Nat. Hist. VII, p. 307 (1860) (Cuba).—GUNDL. Repert. Fisico-Nat. Cuba, I, p. 283 (1866); *ib.* J. f. O. 1874, p. 121 (Cuba).—CORY, Bds. Bahama I. p. 88 (1880); *ib.* List Bds. W. I. p. 13 (1885).

Ammodramus sandwichensis savanna RIDGW. Proc. U. S. N. Mus. VIII, p. 354 (1885).

Recorded from the Bahamas and Cuba.

Ammodramus savannarum (GMEL.).

Fringilla savannarum GMEL. Syst. Nat. I, p. 921 (1788).

Coturniculus tixicrus GOSSE, Bds. Jam. p. 242 (1847).—SCL. P. Z. S. 1861, p. 72 (Jamaica).—ALBRECHT, J. f. O. 1862, p. 196 (Jamaica).—SCL. & SALV. Nom. Avium Neotr. p. 32 (1873) (Jamaica).

Emberiza passerina LEMB. Aves. Cuba, p. 56 (1850)?

Coturniculus passerinus CAB. J. f. O. 1856, p. 6 (Cuba).—BREWER, Pr. Bost. Soc. Nat. Hist. VII, p. 307 (1860) (Cuba).—MARCH, Pr. Acad. Nat. Sci. Phila. 1863, p. 298 (Jamaica).—GUNDL. Repert. Fisico-Nat. Cuba, I, p. 284 (1866); *ib.* J. f. O. 1874, p. 121 (Cuba); *ib.* Anal. Soc. Esp. Hist. Nat. VII, p. 203 (1878) (Porto Rico).—CORY, List Bds. W. I. p. 13 (1885).

Fringilla (Coturniculus) passerina BRYANT, Pr. Bost. Soc. Nat. Hist. X, p. 254 (1866) (Porto Rico).—SUNDEV. Oefv. K. Vet. Akad. Förh. 1869, p. 597 (Porto Rico).

Coturniculus savannarum A. & E. NEWTON, Handb. Jamaica, p. 104 (1881).

Ammodramus australis MAYNARD Am. Ex. & Mart. Jan. 15, 1887.

HABITAT. Jamaica, Cuba, and Porto Rico.

The North American form, *A. savannarum passerinus*, has been taken in the Bahama Islands, specimens of which are in my collection.

Ammodramus maritimus (WILS.).

Fringilla maritima WILS. Am. Orn. IV, p. 68 (1811).

Ammodramus maritimus SWAINS. Zool. Journ. III, p. 328 (1827).—CAB. J. f. O. 1856, p. 7 (Cuba).

Recorded from Cuba.

GENUS Spizella BONAP.

Spizella BONAPARTE, Geog. & Comp. List, p. 33 (1838).

Spizella socialis (WILS.).

Fringilla socialis WILS. Am. Orn. II, p. 127 (1810).

Emberiza pallida LEMB. Aves Cuba, p. 54 (1850).—BREWER, Pr. Bost. Soc. Nat. Hist. VII, p. 307 (1860) (Cuba).

Spinites pallidas Cab. J. f. O. 1856, p. 7 (Cuba).
Spizella socialis Gundl. Repert. Fisico-Nat. Cuba, I, p. 284 (1866); *ib.* J. f. O.
　　1874, p. 121 (Cuba).—Cory, List Bds. W. I. p. 13 (1885).

Accidental in Cuba.

Genus **Sicalis** Boie.

Sicalis Boie, 1828.

Sicalis flaveola (Linn.).

Fringilla flaveola Linn. Syst. Nat. I, p. 321 (1766).
Fringilla flava Müll. Syst. Nat. Suppl. p. 164 (1766).
Emberiza brasiliensis Gmel. Syst. Nat. I, p. 872 (1788).—Lafr. et D'Orb.
　　Syn. Av. 1837, p. 73.
Fringilla brasiliensis Spix, Av. Bras. I, p. 47, pl. 61.—Max. Beitr. III, p.
　　614 (1831).
Sicalis brasiliensis Tsch. Faun. Per. p. 215 (1844).—Cab. in Schomb.
　　Guian. III, p. 679 (1848).—Burm. Syst. Ueb. III, p. 253.—Scl. P.
　　Z. S. 1861, p. 74.—Albrecht, J. f. O. 1862, p. 197.—Taylor, Ibis,
　　1864, p. 83.—Wyatt, Ibis, 1871, p. 328.
Crithagra brasiliensis Gosse, Bds. Jam. p. 245 (1847).—Bp. Consp. I,
　　p. 521 (1820).—March, Pr. Acad. Nat. Sci. Phila. 1863, p. 298.
Sycalis auripectus Bp. Compt. Rend. 37, p. 917.—Scl. P. Z. S. 1855, p.
　　159; *ib.* Cat. Am. Bds. p. 126 (1862).
Sicalis flava Gray, Handl. Bds. II, p. 84 (1870).—A. & E. Newton,
　　Handb. Jamaica, p. 117 (1881).
Sycalis flaveola Pelz. Orn. Bras. p. 231 (1871).—Scl. Ibis, 1872, p. 41.—
　　Scl. & Salv. Nom. Avium Neotr. p. 34 (1873).—Cory, List. Bds.
　　W. I. p. 13 (1885).

Sp. Char.　*Male:*—Above olive green, slightly marked on the back with
　　pale streaks; forehead orange; underparts yellow; wings and tail
　　pale brown, the outer webs of primaries and outer and inner webs of
　　secondaries edged with yellow; tail-feathers edged with yellow.
　　　　Female:—Resembles the male, but much paler; underparts being
　　grayish, tinged with yellow, and under surface becoming white on
　　the belly; orange on the forehead showing very slightly if at all.
　　　　Length (skin), 5.50; wing, 2.75; tail, 2; tarsus, .75.

Habitat.　Jamaica.

The specimens described are from Brazil, as I possess none
from Jamaica.

Carduelis elegans (Linn.) is recorded from Cuba by Dr.
Gundlach (Repert. Fisico-Nat. Cuba, I, p. 396, 1866). It was
probably an escaped cage bird.

GENUS Habropyga CABAN.

Habropyga CABANIS, Wiegm. Archiv, 1847. XIII. p. 331; Mus. Hein. I,
p. 169.

Habropyga melpoda (VIEILL.).

Fringilla melpoda VIEILL. Nouv. Dict. XII, p. 177 (1817).
Habropyga melpoda GUNDL. Anal. Soc. Esp. Hist. Nat. VII, p. 206 (1878)
(Porto Rico).

SP. CHAR. *Male:*—Top of head drab gray; sides of the head and cheeks,
including the eye, reddish orange; back and wings light brown;
lower rump and upper tail-coverts crimson; throat dull white, shading into gray on the sides of the neck, separating the orange of the
cheeks from the brown of the upper parts; rest of underparts dull
grayish, showing a purplish tinge on the flanks; a spot of pale reddish orange near the vent; wings and tail brown.

Female:—Top of head drab gray, paler than in the male; rest of
upper parts like the male; no orange on side of the head; underparts
showing a pale orange yellow tinge.

Length (skin), 4.10; wing, 1.85; tail, 1.75; tarsus, .56.

An introduced species, common in Porto Rico.

FAMILY PLOCEIDÆ.

GENUS Spermestes SWAINS.

Spermestes SWAINSON, Class. Birds. p. 280 (1837).

Spermestes cucullatus (SWAINS.).

Loxia cucullata SWAINS. Zool.
Illust. 1820, pl. 7.
Spermestes cucullata SWAINS.
REICH. Singv. I. 12, pp. 114, 115.
—Br. Consp. I. p. 454 (1850).—
SUNDEV. Oefv. K. Vet. Akad.
Förh. 1869. p. 597.—GRAY, Handl.
Bds. II, p. 54 (1870).
Loxia (Spermestes) cucullata BRY-
ANT, Pr. Bost. Soc. Nat. Hist. X.
p. 254 (1866).

Spermestes cucullatus GUNDL. Anal. Soc. Esp. Hist. Nat. VII, p. 206
(1878).—CORY, List Bds. W. I. p. 13 (1885).

Sp. Char. *Male:*—Top of head dark green, rest of upper surface pale
brown; feathers on the rump alternately banded with dark brown
and white; throat and breast dark green, showing purple in some
lights; belly white, having a patch of metallic green on the sides;
flanks and thighs banded with white and brown; wings pale brown,
a patch of metallic green on the coverts; tail dark brown.

Female:—Above pale brown, darkest on the head; underparts
pale rufous; wings and tail brown.

Length (skin), 4; wing, 2; tail, 1; tarsus, .40.

An introduced species, resident in Porto Rico.

Family ICTERIDÆ.

Genus Icterus Briss.

Icterus Brisson, Orn. II, p. 85 (1760).

Icterus bonana (Linn.).

Oriolus bonana Linn. Syst. Nat. I, p. 162 (1766).
Icterus bonana Daud. Tr. d'Orn. II, p. 332 (1800).—Scl. & Salv. Nom.
    Avium Neotr. p. 36 (1873).—Lawr. Pr. U. S. Nat. Mus. I, p. 355
    (1878).—Scl. Ibis, 1883, p. 358.—Cory, List Bds. W. I. p. 13 (1885).
Pendulinus bonana Vieill. Nouv. Dict. V, p. 316.—Bp. Consp. I, p. 432
    (1850).—Cassin, Pr. Acad. Nat. Sci. Phila. 1867, p. 54.
Xanthornus bonana Cab. Mus. Hein. p. 183.
Icterus bonanæ Scl. Cat. Am. Bds. p. 131 (1862).

Sp. Char. *Male:*—Head, throat, and upper breast dark reddish brown;
upper half of back, wings and tail black; lower half of back, including rump, dull orange; lesser wing-coverts dull brownish orange;
under wing-coverts orange, somewhat brighter than the color of the
belly.

Length (skin), 8.25; wing, 3.50; tail, 3.15; tarsus, .78.

Habitat. Martinique.

Icterus hypomelas (Bonap.).

Icterus dominicensis Vigors, Zool. Journ. 1828, p. 441.—"Albrecht. J. f. O.
    1861, p. 212."
Icterus virescens Vigors, Zool. Journ. 1828, p. 441.
Psarocolius melanopis Wagl. Isis, 1829, p. 759.
Xanthornus dominicensis D'Orb. in La Sagra's Hist. Nat. Cuba, Ois. p.
    115 (1840).—Gundl. Journ. Bost. Soc. Nat. Hist. VI, p. 318 (1852).
    —Cab. J. f. O. 1856, p. 10.—Brewer, Pr. Bost. Soc. Nat. Hist. VII,
    p. 307 (1860).

Pendulinus hypomelas Br. Consp. I. p. 433 (1850).—CASSIN, Pr. Acad.
  Nat. Sci. Phila. 1867, p. 59.
Xanthornus hypomelas GUNDL. Repert. Fisico-Nat. Cuba, I. p. 287 (1866) ;
  ib. J. f. O. 1874, p. 128.
Melanopsar hypomelas GRAY, Handl. Bds. II, p. 32 (1870).
Icterus hypomelas SCL. & SALV. Nom Avium Neotr. p. 36 (1873).—SCL.
  Ibis, 1883, p. 360.—CORY, List Bds. W. I. p. 13 (1885).
Icterus dominicensis var. *hypomelas* BD. BWR. & RIDGW. Hist. N. Am.
  Bds. II, p. 182 (1874).

SP. CHAR. *Male:*—Entire plumage glossy black ; lower half of back,
    thighs and under tail-coverts, and a tinge on the lower belly, bright
    yellow ; wing-coverts, carpus, and under wing-coverts bright yel-
    low ; quills and tail brownish black.
      The young birds of both sexes go through the varied plumages of
    the young of *Icterus dominicensis.*
      Length (skin), 7.75 ; wing, 3.62 ; tail, 3.70 ; tarsus, .90 ; bill, .60.
  HABITAT. Cuba.

Icterus dominicensis (LINN.).

Oriolus dominicensis LINN. Syst. Nat. I, p. 163 (1766).
Icterus dominicensis DAUD. Tr. d'Orn. II. p. 335 (1800).—SALLÉ, P. Z. S.
  1857. p. 232.—BRYANT, Pr. Bost. Soc. Nat. Hist. XI, p. 94 (1866).—
  SCL. & SALV. Nom. Avium Neotr. p. 36 (1873).—CORY, Bull. Nutt.
  Orn. Club, VI. p. 152 (1881) ; *ib.* Bds. Haiti & San Domingo, p. 71
  (1885) ; *ib.* List Bds. W. I. p. 13 (1885).—SCL. Ibis, 1883, p. 361.—
  TRISTRAM, Ibis, 1884, p. 168.
Pendulinus flavigaster VIEILL. Nouv. Dict. V, p. 317.
Melanopsar dominicensis GRAY, Handl. Bds. II, p. 32 (1870).
Icterus dominicensis var. *dominicensis* BD. BWR. & RIDGW. Hist. N. Am.
  Bds. II, p. 182 (1874).

SP. CHAR. *Male:*—General plumage black ; upper wing-coverts, edge of
    carpus, under wing-coverts, lower half of back, rump, flanks, crissum,
    and under tail-coverts bright yellow ; outer surface of wings black ;
    under surface of wings showing the inner webs of the feathers pale,
    becoming dull white at the base ; bill and legs black.
      Young:—Throat, cheeks, and a narrow superciliary stripe black ;
    crown, sides of the head and breast showing a brownish tinge ; rest
    of underparts greenish yellow ; back ashy green, becoming decidedly
    greenish on the rump ; tail olive, brightest on the edges of the feath-
    ers ; primaries and secondaries brown with pale edgings.
      The sexes are apparently similar.
      Length, 7.10 ; wing, 3.60 ; tail, 3.50 ; tarsus, .80 ; bill, .70.
  HABITAT. Haiti and San Domingo.

Icterus portoricensis (BRYANT).

Icterus dominicensis TAYLOR, Ibis, 1864. p. 167.
Icterus dominicensis var. *portoricensis* BRYANT, Pr. Bost. Soc. Nat. Hist.
    X, p. 254 (1866).—SUNDEV. Oefv. K. Vet. Akad. Forh. 1869, p. 597.
    —BD. BWR. & RIDGW. Hist. N. Am. Bds. II. p. 182 (1874).
Pendulinus portoricensis CASSIN, Pr. Acad. Nat. Sci. Phila. 1867, p. 58.
Icterus xanthomus SUNDEV. Oefv. K. Vet. Akad. Förh. 1869, p. 598.
Melanopsar portoricensis GRAY, Handl. Bds. II, p. 32 (1870).
Icterus portoricensis SCL. & SALV. Nom. Avium Neotr. p. 36 (1873).—
    SCL. Ibis, 1883, p. 361.—CORY, List Bds. W. I. p. 13 (1885).
Xanthornus portoricensis GUNDL. Anal. Soc. Esp. Hist. VII, p. 210
    (1878).

SP. CHAR. *Male:*—General plumage black; abdomen, crissum, rump, and
    under wing-coverts yellow; wing-coverts bright yellow, forming a
    broad shoulder patch; some of the feathers on the belly faintly
    edged with yellowish; primaries edged with dull white on the basal
    portion of the inner webs; wings and tail dark brown.
        Immature males and females resemble those of *I. dominicensis*, and
    go through the same varied stages of plumage. A greenish brown
    specimen in my collection is labelled "adult female," but this is
    probably incorrect.
        Length (skin), 7.50; wing, 3.50; tail, 3.25; tarsus, .95; bill, .67.

HABITAT.   Porto Rico.

Icterus laudabilis SCL.

Icterus laudabilis SCL. P. Z. S. 1871, p. 270.—SCL. & SALV. Nom. Avium
    Neotr. p. 36 (1873).—ALLEN. Bull. Nutt. Orn. Club, V, p. 166
    (1880).—SCL. Ibis, 1883, p. 361.—CORY, List Bds. W. I. p. 13 (1885).

SP. CHAR. *Male:*—General plumage glossy black; rump, thighs, lower
    belly and lower half of back yellow, with a tinge of orange; under
    wing-coverts, carpus, and lesser-coverts pale yellow; greater wing-
    coverts edged with white on the inner webs; lower mandible bluish
    at the base; upper mandible and legs brownish black.
        Length (skin), 7.50; wing, 3.85; tail, 3.75; tarsus, .90.

HABITAT.   Santa Lucia.

Icterus cucullatus SWAINS.

Icterus cucullatus SWAINS. Phil. Mag. 1827. p. 436.—GUNDL. Repert.
    Fisico-Nat. Cuba, I, p. 286 (1866); *ib.* J. f. O. 1874, p. 127 (Cuba).
    —CORY, List Bds. W. I. p. 13 (1885).
Hyphantes costotoll GUNDL. J. f. O. 1856. p. 11; *ib.* 1861. p. 413 (Cuba).
Hyphantes bullockii BREWER, Pr. Bost. Soc. Nat. Hist. VII, p. 307 (1860)
    (Cuba).

Recorded from Cuba.

Icterus leucopteryx (WAGL.).

Psarocolius leucopteryx WAGL. Syst. Av. Sp. 16.
Icterus personatus TEMM. Pl. Col. sub tab. p. 482 (1820-39).—BP. Consp
I. p. 435 (1850).
Icterus leucopteryx GOSSE. Bds. Jam. p. 226 (1847).—BP. Consp. I, p.
436 (1850).—SCL. Cat. Am. Bds. p. 34 (1862).—ALBRECHT, J. f. O.
1862, p. 197.—MARCH, Pr. Acad. Nat. Sci. Phila. 1863, p. 299 —SCL.
& SALV. Nom. Avium Neotr. p. 36 (1873).—A. & E. NEWTON,
Handb. Jamaica, p. 104 (1881).—SCL. Ibis, 1883, p. 374.—CORY,
List Bds. W. I. p. 13 (1885).
Pendulinus leucopteryx CASSIN, Pr. Acad. Nat. Sci. Phila. 1867, p. 59.
Melanopsar leucopteryx GRAY, Handl. Bds. II, p. 32 (1870).

SP. CHAR. *Male:*—Throat, forehead, and in front of the eyes black; upper
plumage yellowish green; underparts yellow; wings and tail black;
wing-coverts pure white; showing a broad patch of white on the
wing.
  Female:—Black markings replaced by brownish black; tail pale
greenish yellow instead of black; otherwise resembles the male.
  Length (skin), 7.75; wing, 4.20; tail, 3.25; tarsus, .90; bill, .78.
HABITAT. Jamaica.

Icterus spurius (LINN.).

Oriolus spurius LINN. Syst. Nat. I, p. 162 (1766).
Icterus spurius GUNDL. Repert. Fisico-Nat. Cuba, I, p. 286 (1866);
  ib. J. f. O. 1874, p. 127 (Cuba).—CORY, List Bds. W. I. p. 13 (1885).

Accidental in Cuba.

Icterus oberi LAWR.

Icterus oberi LAWR. Pr. U. S. Nat. Mus. III, p. 351 (1880).—GRISDALE,
Ibis, 1882, p. 487. pl. XIII.—SCL. Ibis, 1883, p. 362.—CORY, List
Bds. W. I. p. 13 (1885).
  "*Male:*—Head, neck, upper part of breast, back, wings and tail
black; lower part of breast, abdomen, under tail-coverts and rump
light brownish chestnut, with the concealed bases of the feathers of
a clear light yellow; the thighs are yellow with a wash of chestnut;
edge of wing and under wing-coverts yellow; bill black, with the
sides of the under mandible bluish for half its length from the base;
tarsi and toes black.
  "Length (skin), 8½ inches; wing, 3⅞; tail, 4; tarsus, 1; bill, ⅞.
  "The female has the upper plumage of a dull greenish olive, with
a yellowish tinge, the front and rump inclining more to yellow; the

tail-feathers are yellowish green ; quills brownish black ; the primaries and secondaries are edged narrowly with dull yellowish gray ;
tertials are margined with fulvous ; wing-coverts dark brown, margined with fulvous ; edge of wing yellow ; the under plumage
is of a rather dull dark yellow ; the breast and under tail-coverts
are of a deeper or warmer color ; the sides are greenish olive ;
bill and legs as in the male.

"The young male resembles the female in plumage, but has the
back somewhat darker." (LAWR. l. c., orig. descr.)

HABITAT. Montserrat.

Icterus icterus (LINN.).

Oriolus icterus LINN. Syst. Nat. I, p. 161 (1766).
Icterus vulgaris DAUD. Tr. d'Orn. II, p. 340 (1800).—BP. Consp. I, p.
434 (1850).—BAIRD, Bds. N. Am. p. 542 (1858).—SCL. Cat. Am.
Bds. p. 133 (1862).—CASSIN, Pr. Acad. Nat. Sci. Phila. 1867. p. 46.
—FINSCH, P. Z. S. 1870, p. 578.—SCL. & SALV. Nom. Avium Neotr.
p. 36 (1873).—GUNDL. Anal. Soc. Esp. Hist. Nat. VII. p. 209 (1878).
—SALV. & GODM. Ibis, 1879, p. 200.—A. & E. NEWTON. Handb.
Jamaica, p. 104 (1881).—SCL. Ibis, 1883, p. 369.—CORY. List Bds.
W. I. p. 13 (1885).
Icterus longirostris VIEILL. Nouv. Dict. 34, p. 547.--BP. Consp. I. p. 435
(1850).—CASSIN, Pr. Acad. Nat. Sci. Phila. 1867, p. 46.

SP. CHAR. *Male:*—Entire head, throat, and a broad patch on the back
black ; wings and tail very dark brown, almost black, the former
having some of the coverts white. forming a wing-band ; outer webs
of secondaries edged with white ; rest of plumage bright orange yellow ; the orange of the rump and nape being separated by the black
back-patch before mentioned ; feathers of the throat narrow and
sharply pointed ; bill black, the base of lower mandible bluish
white.

The sexes are described as similar.

Length (skin), 10 ; wing, 4.50 ; tail, 4 ; tarsus, 1.25 ; bill, 1.15.

Porto Rico, Jamaica, and St. Thomas (introduced).

Icterus galbula (LINN.).

Coracias galbula LINN. Syst. Nat. 10th ed. (1758).
Icterus baltimore LEMB. Aves Cuba, p. 63 (1850) —GUNDL. Repert. Fisico-
Nat. Cuba. I, p 286 (1866) ; *ib* J. f. O. 1871. p. 127 (Cuba).
Hyphantes baltimore CAB. J. f. O. 1856, p. 10 (Cuba).
Yphantes baltimore BREWER, Pr. Bost. Soc. Nat. Hist. VII. p. 307 (1860)
(Cuba).
Icterus galbula COUES, Bull. Nutt. Orn. Club, V, p. 98 (1880).—CORY,
List Bds. W. I. p. 13 (1885).

Accidental in Cuba.

GENUS **Dolichonyx** SWAINS.

Dolichonyx SWAINSON, Zool. Journ. III. p. 351 (1827).

Dolichonyx oryzivorus (LINN.).

Emberiza oryzivora LINN. Syst. Nat. I, p. 311 (1766).
Dolichonyx oryzivorus GOSSE, Bds. Jam. p. 229 (1847).—CAB. J. f. O.
    1856, p. 11 (Cuba).—BREWER. Pr. Bost. Soc. Nat. Hist. VII, p. 307
    (1860) (Cuba).—SCL. P. Z. S. 1861, p. 74 (Jamaica).—ALBRECHT,
    J. f. O. 1862, p. 197 (Jamaica).—GUNDL. Repert. Fisico-Nat. Cuba,
    I, p. 287 (1866); *ib.* J. f. O. 1874, p. 129 (Cuba).—CORY, Bds. Baha-
    ma I. p. 97 (1880).—A. & E. NEWTON, Handb. Jamaica, p. 104
    (1881).—CORY, List Bds. W. I. p. 14 (1885).
Dolichonix oryzivora LEMB. Aves Cuba, p. 57 (1850).
Dolichonyx orizivora BRYANT, Pr. Bost. Soc. Nat. Hist. VII. p. 119 (1859)
    (Bahamas).
Dolichonyx orizivorus MARCH, Pr. Acad. Nat. Sci. Phila. 1863, p. 291
    (Jamaica).

Recorded from the Bahamas, Cuba, Jamaica, and Grenada.

Molothrus bonariensis (Cab.) is recorded from Bisque, Virgin Islands (*Molothrus sericeus* (Licht.) Newton, Ibis, 1860. p. 308). It is a South American species.

GENUS **Agelaius** VIEILL.

Agelaius VIEILLOT, Analyse, p. 33 (1816).

Agelaius humeralis (VIG.).

Leistes humeralis VIG. Zool. Journ. III, p. 442 (1827).
Icterus humeralis D'ORB. in La Sagra's Hist. Nat. Cuba, Ois. p. 114, pl.
    5 (1840).
Agelaeus humeralis BP. Consp. I, p. 430 (1850).—GUNDL. J. f. O. 1856, p.
    13.—SCL. Cat. Am. Bds. p. 136 (1862).—SCL. & SALV. Nom. Avium
    Neotr. p. 37 (1873).— SCL. Ibis, 1884, p. 11.—CORY, List Bds. W. I.
    p. 14 (1885).
Agelaius humeralis BREWER, Pr. Bost. Soc. Nat. Hist. VII, p, 307 (1860).
    —GUNDL. Repert. Fisico-Nat. Cuba, I, p. 288 (1866); *ib.* J. f. O.
    1874, p. 130.—GRAY, Handl. Bds. II, p. 33 (1870).

SP. CHAR. *Male:*—General plumage uniform lustrous black, showing a
    slight brownish tinge on the thighs and quills when held in the
    light; shoulders and lesser coverts dull orange brown, shading into
    pale buff on the middle coverts; bill and feet black.
       The sexes are similar.
       Length (skin), 7.50; wing, 4; tail, 3.15; tarsus, .95; bill, .60.
HABITAT. Cuba.

Agelaius xanthomus SCL.

Agelaius chrysopterus VIEILL. Nouv. Dict. 34. p. 539.—GUNDL. Anal. Soc.
Esp. Hist. Nat. VII. p. 211 (1878); *ib.* J. f. O. 1878. p. 177.
Icterus xanthomus SCL. Cat. Am. Bds. p. 131 (1862).—TAYLOR. Ibis, 1864,
p. 168.—BRYANT, Pr. Bost. Soc. Nat. Hist. X. p. 254 (1866).
Hyphantes xanthomus CASSIN. Pr. Acad. Nat. Sci. Phila. 1867. p. 63.
Agelaius chrysopterus SUNDEV. Oefv. K. Vet. Akad. Förh. 1869. p. 597.—
SCL. &. SALV. Nom. Avium Neotr. p. 37 (1873).
Agelaius xanthomus SCL. Ibis, 1884, p. 12.—CORY, List Bds. W. I. p. 14
(1885).

SP. CHAR. *Male:*—General plumage lustrous black; shoulders and coverts
bright golden yellow; quills and tail showing a faint brownish tinge,
apparently wanting in some specimens.
The sexes are similar.
Length (skin), 8; wing, 4.50; tail, 3.50; tarsus, .95; bill, .60.
HABITAT. Porto Rico.

Agelaius phœniceus (LINN.).

Oriolus phœniceus LINN. Syst. Nat. I, p. 161 (1766).
Agelaius phœniceus CAB. J. f. O. 1856, p. 11 (Cuba).—CORY, Bds. Baha-
ma I. p. 98 (1880); *ib.* List Bds. W. I. p. 14 (1885).
Agelaius phœniceus BRYANT, Pr. Bost. Soc. Nat. Hist. VII. p. 119 (1859)
(Bahamas).

Not uncommon in the Bahama Islands where it probably breeds.
Cuba?

Agelaius assimilis GUNDL.

Agelaius assimilis GUNDL. in Lemb. Aves Cuba, p. 64 (1850); *ib.* J. f. O.
1856. p. 12; *ib.* 1861, pp. 332, 413; *ib.* 1874, p. 131.—BREWER, Pr.
Bost. Soc. Nat. Hist. VII. p. 307 (1860).—GRAY, Handl. Bds. II, p.
33 (1870).
Agelaius phœniceus var. *assimilis* COUES, Bds. N. W. p. 186 (1874).—BD.
BWR. & RIDGW. Hist. N. Am. Bds. II, p. 159 (1874).
Agelaius assimilis SCL. Ibis, 1884, p. 10.—CORY, List. Bds. W. I. p. 14
(1885).

SP. CHAR. *Male:*—Similar to *Agelaius phœniceus,* but somewhat smaller.
Female:—Entirely black; showing a brownish tinge on the head,
back, and breast.
Length (skin), 7.75; wing, 3.75; tail, 3; tarsus, 1; bill, .75.
HABITAT. Cuba.

GENUS Xanthocephalus BONAP.

Xanthocephalus BONAPARTE, Consp. I, p. 431 (1850).

Xanthocephalus xanthocephalus (BONAP.).

Icterus icterocephalus BP. Am. Orn. I, p. 27 (1825).
Xanthocephalus icterocephalus GUNDL. J. f. O. 1862, p. 178 (Cuba); *ib.*
    Repert. Fisico-Nat. Cuba, I. p. 288 (1866); *ib.* J. f. O. 1874, p. 133
    (Cuba).—CORY, List Bds. W. I. p. 14 (1885).

Dr. Gundlach writes me that he purchased a bird of this species in the market in Havana.

GENUS Sturnella VIEILL.

Sturnella VIEILLOT, Analyse, p. 34 (1816).

Sturnella hippocrepis WAGL.

Sturnella hippocrepis WAGL. Isis, 1832, p. 281.—CAB. J. f. O. 1856, p. 14.
    —LAWR. Ann. N. Y. Lyc. VII, p. 266 (1860).—BREWER, Pr. Bost.
    Soc. Nat. Hist. VII, p. 307 (1860).—ALBRECHT, J. f. O. 1861. p. 206.—
    SCL. Ibis, 1861, p. 179; *ib.* Cat. Am. Bds. p. 139 (1862).—GUNDL. J.
    f. O. 1861, pp. 332, 413; *ib.* 1871, p. 276; *ib.* 1874, p. 133.—CASSIN,
    Pr. Acad. Nat. Sci. Phila. 1866, p. 24.—CORY, List Bds. W. I. p. 14
    (1885).
Sturnella ludoviciana SCL. & SALV. Nom. Avium Neotr. p. 38 (1873)
    (Cuba).
Sturnella ludoviciana var. *hippocrepis* BD. BWR. & RIDGW. Hist. N. Am.
    Bds. II, p. 172 (1874).—SCL. Ibis, 1884, p. 25.

SP. CHAR. *Male:*—Above mottled with buff, black and brown; the feathers of the back being dark brown, bordered and blotched with buff and light brown; a narrow imperfect stripe of whitish on the middle of the crown; a stripe of yellow from the nostril, over the eye, continuing in a stripe of dull buffy white to the sides of the neck; cheeks brownish; throat and middle underparts yellow, interrupted by a band of black on the upper breast; sides of the body, lower belly, and under tail-coverts dull buff, the feathers heavily streaked with dark brown; a patch of yellow on the carpus; legs and feet pale brown.

The sexes are similar.

Length (skin), 8.45; wing, 3.95; tail, 2.75; tarsus, 1.50; bill, 1.10.

Has a general resemblance to Florida specimens of *Sturnella magna,* but differing from it in having the legs and claws larger, and the underparts much more streaked.

HABITAT. Cuba.

Genus **Nesopsar** Scl.

Nesopsar Sclater, Ibis, 1859, p. 457.

Nesopsar nigerrimus (Osburn).

Icterus nigerrimus, Osburn, Zoologist. pp. 6661, 6714 (1859).
Nesopsar nigerrimus Scl. Ibis, 1059, p. 457; *ib.* Cat. Am. Bds. p. 139
    (1862).—Albrecht, J. f. O. 1862, p. 197.—March, Pr. Acad. Nat.
    Sci. Phila. 1863, p. 299.—Gray, Handl. Bds. II. p. 34 (1870).—Scl.
    & Salv. Nom. Avium Neotr. p. 38 (1873).—A. & E. Newton,
    Handl. Bds. p. 103 (1881).—Cory, List Bds. W. I. p. 14 (1885).
Ageloius nigerrimus Cassin, Pr. Acad. Nat. Sci. Phila. 1864, p. 12.—
    Pelz. Ibis, 1873, p. 28.
Agelœus nigerrimus Scl. Ibis, 1884, p. 14.

Sp. Char. *Male:*—General plumage glossy blue black; dull black on the
    belly; wings and tail very dark brown, almost black; under-sur-
    face of wings and tail showing a brownish tinge when held in the
    light; bill black, a faint pale mark at the base of the lower mandible.
    The sexes are similar.
        Length (skin), 6.50; wing, 3.50; tail, 2.50; tarsus, .90; bill, .75.
Habitat. Jamaica.

Genus **Quiscalus** Vieill.

Quiscalus Viellot, Analyse, p. 37 (1816).

Quiscalus fortirostris Lawr.

Quiscalus rectirostris Cassin, Pr. Acad. Nat. Sci. Phila. 1866, p. 409?
Quiscalus fortirostris Lawr. Pr. Acad. Nat. Sci. Phila. 1868, p. 360.—
    Scl. Ibis, 1873, p. 324; *ib.* P. Z. S. 1874, p. 175.—Scl. & Salv.
    Nom. Avium Neotr. p. 38 (1873).—Scl. Ibis, 1884, p. 161.—Cory,
    List Bds. W. I. p. 14 (1885).
Holoquiscalus rectirostris Gray, Handl. Bds. II, p. 38 (1870)?
Holoquiscalus fortirostris Gray, Handl. Bds. II, p. 38 (1870).

Sp. Char. *Male:*—Head and back purplish black; a faint greenish tinge
    on the thighs and under tail-coverts; wings and tail black, showing
    greenish reflections; bill and feet black.
        Female:—Similar to the male, but is somewhat duller in colora-
    tion, and is apparently smaller.
        Length (skin), 8.60; wing, 4.25; tail, 3.75; tarsus, 1.25; bill, 1.
Habitat. Barbadoes.

Quiscalus inflexirostris SWAINS.

Quiscalus inflexirostris SWAINS. An. in Men. p. 309 (1838).—Br. Consp.
I, p. 424 (1850).—CASSIN, Pr. Acad. Nat. Sci. Phila. 1866, p. 407.—
SEMPER, P. Z. S. 1872. p. 651.—SCL. P. Z. S. 1874, p. 175.—BD.
BWR. & RIDGW. Hist. N. Am. Bds. II, p. 214 (1874).—LAWR. Pr.
U. S. Nat. Mus. I, pp. 355, 487 (1878).—SCL. Ibis, 1884, p. 160.—
CORY, List Bds. W. I. p. 14 (1885).
Quiscalus barita TAYLOR, Ibis, 1864, p. 168.
Holoquiscalus inflexirostris GRAY, Handl. Bds. II, p. 38 (1870).
Quiscalus luminosus ALLEN, Bull. Nutt. Orn. Club, V, p. 166 (1880).

SP. CHAR. *Male:*—Entire plumage lustrous black, showing a purplish
    tinge on the head, back, and breast, when held in the light; wings
    and tail bluish black; bill and feet black.
    Female?—Top of head grayish brown, becoming darker brown
    on the back; throat dull white; underparts buffy brown; a streak
    of pale buff from the eye to the nape, bordered below by a narrow
    streak of pale brown; quills and tail dark brown, showing a slight
    tinge of bluish.
    Immature males are intermediate between the female? and adult
    male, being light brown and black in patches.
    Length (skin), 9; wing, 4.75; tail, 4; tarsus, 1.25; bill, 1.

HABITAT. Santa Lucia and Martinique.

Quiscalus brachypterus CASSIN.

Quiscalus brachypterus CASSIN, Pr. Acad. Nat. Sci. Phila. 1866, p. 406.—
SCL. & SALV. Nom. Avium Neotr. p. 38 (1873).—SCL. Ibis, 1884, p.
160.—CORY, List Bds. W. I. p. 14 (1885).
Quiscalus crassirostris GUNDL. J. f. O. 1866, p. 188.—BRYANT, Pr. Bost.
Soc. Nat. Hist. X, p. 255 (1886). SUNDEV. Oefv. K. Vet. Akad.
Förh. 1869, p. 598.
Chalcophanes lugubris SUNDEV. Oefv. K. Vet. Akad. Förh. 1869, p. 598.
Holoquiscalus brachypterus GRAY, Handl. Bds. II, p. 38 (1870).
Chalcophanes brachypterus GUNDL. J. f. O. 1874. p. 312; *ib.* 1878, p. 177;
    ib. Anal. Soc. Esp. Hist. Nat. VII, p. 213 (1878).
Quiscalus baritus var. brachypterus BD. BWR. & RIDGW. Hist. N. Am.
    Bds. II, p. 213 (1874).

SP. CHAR. *Male:*—Entire plumage black, showing a purplish tinge when
    held in the light; the wings and tail have a greenish gloss; inner
    surface of wings showing a brownish tinge when held in the light.
    The sexes appear to be similar.
    Length (skin), 8.50; wing, 4.50; tail, 3.75; tarsus, 1.10; bill, .90.

HABITAT. Porto Rico.

Quiscalus crassirostris Swains.

Gracula barita Linn. Syst. Nat. I, p. 165 (1766) ?
Sturnus jamaicensis Daud. Tr. d'Orn. II, p. 317 (1800) ?
Quiscalus crassirostris Swains. An. in Men. p. 355 (1838).—Gosse,
    Bds. Jam. p. 217 (1847).—Bp. Consp. I, p. 425 (1850).—Albrecht,
    J. f. O. 1862, p. 197.—March, Pr. Acad. Nat. Sci. Phila. 1863,
    p. 298.—Scl. & Salv. Nom Avium Neotr. p. 38 (1873).—A. & E.
    Newton, Handb. Jamaica, p. 103 (1881).—Scl. Ibis, 1884, p.
    159.—Cory, List Bds. W. I. p. 14 (1885).
Quiscalus baritus Cassin, Pr. Acad. Nat. Sci. Phila. 1866, p. 405.
Holoquiscalus baritus Gray, Handl. Bds. II, p. 38 (1870).
Quiscalus baritus var. baritus Bd. Bwr. & Ridgw. Hist. N. Am. Bds
    II, p. 213 (1874).

Sp. Char.   Male:—Upper plumage lustrous blue-black, showing a tinge
    of purple when held in the light; underparts black, brownish black
    on the belly; quills and tail brownish black.
        The sexes are described as similar.
        Length (skin), 10; wing, 5; tail, 4.50; tarsus, 1.25; bill, .90.
Habitat. Jamaica.

Quiscalus luminosus Lawr.

Quiscalus sp. Lawr. Pr. U. S. Nat. Mus. I, p. 191 (1878).
Quiscalus luminosus Lawr. Ann. N. Y. Acad. Sci. I, p. 162 (1878); ib.
    Pr. U. S. Nat. Mus. I, pp. 265, 487 (1878).—Ober, Camps in the
    Caribbees, p. 247 (1880).—Scl. Ibis, 1884, p. 161.—Cory, List Bds.
    W. I. p. 14 (1885).

        "Male:—The general plumage is of a lustrous dark bluish violet;
    the upper and under tail-coverts are dull dark green; tail dark glossy
    green; tertials, outer webs of larger quills, and the middle and larger
    wing-coverts, glossy green like the tail; the inner webs of the
    larger quills are black; smaller wing-coverts the color of the back;
    under wing-coverts black; the bill and feet are black; 'iris yellow.'
        "Female:—Upper plumage of a fine dark brown, light on the crown,
    the feathers of which are margined with dull pale rust color; the tail
    is blackish-brown, with a wash of greenish; quills dark brown; the
    under plumage is dark brownish-ash, lighter on the throat and
    breast, and fuliginous on the flanks, lower part of abdomen, and un-
    der tail-coverts; on the lower part of the neck is a wash of dull rust-
    color; bill and feet black; 'iris yellow.'" (Lawr., orig. descr.)
        Length (skin), 10.50; wing, 4.90; tail, 4.25; tarsus, 1.25; bill, 1.25.
Habitat. Grenada.
        Specimens which I have compared appear to differ very slight-
ly from Santa Lucia specimens of Q. inflexirostris. The color-

ation of Grenada specimens is possibly somewhat brighter, but their specific distinctness is questionable.

Quiscalus guadeloupensis LAWR.

Quiscalus guadeloupensis LAWR. Pr. U. S. Nat. Mus. I, pp. 457, 487 (1878).—SCL. Ibis, 1884, p. 160.—CORY, List Bds. W. I. p. 14 (1885).

"*Male:*—The general plumage is of a deep purplish-violet; the wing-coverts have a decided green lustre; tail black, glossed with green; quills black, with a greenish tinge; bill and feet black." (LAWR., orig. descr.)

Female:—Top of head dull brown, becoming darker brown on the back and wings; throat white; a faint moustache-like streak extending from the lower mandible on the sides of the throat; a dull line of buffy-white passing from the upper mandible through the eye; ear-coverts brownish; breast buffy-white; belly dull white, tinged with brownish-olive on the sides; wings and tail dark brown.

Length (skin), 9.75; wing, 5; tail, 4.50; tarsus, 1.25; bill, 1.

HABITAT. Guadeloupe.

Quiscalus gundlachii CASSIN.

Quiscalus barytus D'ORB. in La Sagra's Hist. Nat. Cuba, Ois. p. 120 (1840).—THIENEM. J. f. O. 1857, p. 145.

Chalcophanes barytus CAB. Mus. Hein. I, p. 197 (1851).—GUNDL. J. f. O. 1856, p. 15.

Calcophanes baritus BREWER, Pr. Bost. Soc. Nat. Hist. VII, p. 307 (1860).

Quiscalus gundlachii CASSIN, Pr. Acad. Nat. Sci. Phila. 1866, p. 406.

Holoquiscalus gundlachii GRAY, Handl. Bds. II, p. 38 (1870).

Chalcophanes gundlachii GUNDL. J. f. O. 1874, p. 135.

Quiscalus baritus var. *gundlachii* BD. BWR. & RIDGW. Hist. N. Am. Bds. II, p. 213 (1874).

Quiscalus gundlachi SCL. Ibis, 1884, p. 159.—CORY, List Bds. W. I. p. 14 (1885).

Sp. CHAR. *Male:*—General plumage black, with purplish and bluish reflections; wings and tail showing greenish or bluish reflections when held in the light.

Length (skin), 11.50; wings, 6; tail, 5; tarsus, 1.55; bill, 1.25.

HABITAT. Cuba.

Quiscalus niger (BODD.).

Oriolus niger BODD. Tabl. Pl. Enl. p. 31 (1783).

Quiscalus barita SALLÉ, P. Z. S. 1857, p. 232.

Quiscalus niger CASSIN, Pr. Acad. Nat. Sci. Phila. 1886, p. 407.—SCL. Ibis, 1884, p. 159.—CORY, Bds. Haiti & San Domingo, p. 73 (1885); *ib.* List Bds. W. I. p. 14 (1885).

Quiscalus ater BRYANT. Pr. Bost. Soc. Nat. Hist. XI, p. 94 (1866).—CORY,
    Bull. Nutt. Orn. Club, VI, p. 153 (1881).—TRISTRAM, Ibis, 1884, p.
    168.
Holoquiscalus niger GRAY, Handl. Bds. II, p. 38 (1870).
Quiscalus baritus var. *niger* BD. BWR. & RIDGW. Hist. N. Am. Bds. II, p.
    213 (1874).

SP. CHAR. *Male:*—General plumage lustrous black, showing purple when
    held in the light; wings and tail black with bluish reflections; bill
    and feet black.
        The sexes are similar.
        Length, 10.25; wing, 5.40; tail, 5; tarsus, 1.30; bill, 1.10.
    HABITAT.   Haiti and San Domingo.

Quiscalus atroviolaceus D'ORB.

Quiscalus atroviolaceus D'ORB. in La Sagra's Hist. Nat. Cuba Ois. p.
    121 (1840).—CORY, List. Bds. W. I. p. 14 (1885).
Scaphidurus atroviolaceus GRAY & MITCH. Gen. Bds. II, p. 341.—Br.
    Consp. I, p. 426 (1850).
Scolecophagus atroviolaceus CAB. Mus. Hein. I. p. 196 (1851); *ib.* J. f. O.
    1856, p. 15.—BREWER, Pr. Bost. Soc. Nat. Hist. VII. p. 307 (1860).—
    CASSIN. Pr. Acad. Nat. Sci. Phila. 1866, p. 415.—GUNDL. J. f. O.
    1874. p. 134.
? *Chalcophanes quiscalus* GUNDL. J. f. O. 1856, p. 16; *ib.* 1871, p. 288.
Dives atroviolaceus GRAY, Handl. Bds. II, p. 39 (1870).—SCL. Ibis, 1884,
    p. 152.

SP. CHAR. *Male:*—Entire plumage lustrous black; purplish on the head,
    back, and breast; bluish on the wings and tail; thighs slightly
    brownish in some specimens; bill and feet black.
        Length (skin), 9.25; wing, 5; tail, 4; tarsus, 1.05; bill, .75.
    HABITAT.   Cuba.

FAMILY CORVIDÆ.

GENUS Corvus LINN.

Corvus LINNÆUS, Syst. Nat. I, p. 155 (1766).

Corvus leucognaphalus DAUD.

Corvus leucognaphalus DAUD. Tr. d'Orn. II, p. 231 (1800).—SALLÉ, P.
    Z. S. 1857. p. 232.—TAYLOR, Ibis, 1864, p. 168.—BRYANT, Pr. Bost.
    Soc. Nat. Hist. XI. p. 94 (1866).—SUNDEV. Oefv. K. Vet. Akad.
    Förh. 1869, p. 598.—BD. BWR. & RIDGW. Hist. N. Am. Bds.

II, p. 234 (1874).—SCL. & SALV. Nom. Avium Neotr. p. 40 (1873).—
GUNDL. Anal. Soc. Esp. Hist. Nat. VII. p. 214 (1878).—CORY, Bull.
Nutt. Orn. Club, VI, p. 153 (1881): *ib.* Bds. Haiti & San Domin-
go, p. 74 (1885): *ib.* List Bds. W. I. p. 14 (1885).
Frugilegus leucognaphalus GRAY, Handl. Bds. II, p. 13 (1870).
Microcorax leucognaphalus SHARPE, Cat. Bds. Brit. Mus. III, p. 49 (1877).

SP. CHAR. *Male:*—General plumage black, with faint bluish and purple
reflections in the light: feathers of the throat having the ends separated in hair-like filaments; basal portion of the body-feathers
white; bill and legs black.
The sexes are similar.
Length, 18; wing, 12; tail, 8; tarsus, 2.20; bill, 2.20.

HABITAT. San Domingo and Porto Rico.

Specimens from Porto Rico differ somewhat from those
from San Domingo, but I am of the opinion it would not
be wise to separate them specifically. It is possible they represent good geographical races. The San Domingo bird is
blacker, and shows brighter bluish and purplish reflections when
held in the light. If it should be thought best to separate them,
I would propose the name *dominicensis* for the San Domingo
form.

Corvus jamaicensis GMEL.

Corvus jamaicensis GMEL. Syst. Nat. I, p. 367 (1788).—GOSSE, Bds. Jam.
p. 209 (1847).—DENNY, P. Z. S. 1847, p. 38.—ALBRECHT, J. f. O.
1862, p. 202.—SCL. Cat. Am. Bds. p. 146 (1862).—MARCH, Pr.
Acad. Nat. Sci. Phila. 1863, p. 300.—SCL. & SALV. Nom. Avium
Neotr. p. 49 (1873).—BD. BWR. & RIDGW. Hist. N. Am. Bds. II, p.
234 (1874).—A. & E. NEWTON, Handb. Jamaica, p. 103 (1881).—
CORY, List Bds. W. I. p. 14 (1885).
Frugilegus jamaicensis GRAY, Handl. Bds. II, p. 13 (1870).
Microcorax jamaicensis SHARPE, Cat. Bds. Brit. Mus. III, p. 48 (1877).

SP. CHAR. *Male:*—General color very dark brown, blackish brown on
head and throat; basal portions of most of the feathers gray; wings
and tail dark brown, showing slight purplish reflections when held
in the light; bill and legs black.
The sexes are similar.
Length (skin), 15.75; wing, 8.80; tail, 6.20; tarsus, 1.90; bill,
1.85.

HABITAT. Jamaica.

Corvus solitarius WÜRT.

Corvus solitarius WÜRT. Naumannia, II, p. 55.—Br. Compt. Rend. 37, p.
    829.—CORY, Bds. Haiti & San Domingo, p. 75 (1885); *ib.* List Bds.
    W. I. p. 14 (1885).
Corvus palmarum "WÜRT. Reis. p. 73."
Corvus jamaicensis SALLÉ, P. Z. S. 1857, p. 232.—BRYANT, Pr. Bost. Soc.
    Nat. Hist. XI, p. 94 (1886).
Frugilegus solitarius GRAY, Handb. Bds. II, p. 13 (1870).
Microcorax solitarius SHARPE, Cat. Bds. Brit. Mus. III, p. 49 (1877).

SP. CHAR.   *Male:*—General plumage black, with a purple gloss to the
        feathers; coverts and primaries black; tail black, outer surface
        showing a tinge of purple; bill and legs black.
        The sexes are similar.
        Length, 15; wing, 10; tail, 6; tarsus, 1.75; bill, 1.75.
    HABITAT.   Haiti and San Domingo.

Corvus nasicus TEMM.

Corvus nasicus TEMM. Pl. Col. II, p. 413 (1820-39).—BREWER, Pr. Bost.
    Soc. Nat. Hist. VII, p. 307 (1860).—GUNDL. Repert. Fisico-Nat.
    Cuba, I, p. 290 (1865); *ib.* J. f. O. 1874, p. 137.—BD. BWR. &
    RIDGW. Hist. N. Am. Bds. II, p. 234 (1874).—CORY. List Bds. W.
    I. p. 14 (1885).
Corvus americanus LEMB. Aves Cuba, p. 65 (1850)
Corvus jamaicensis CAB. J. f. O. 1856, p. 16.—THIENEM. J. f. O. 1857,
    p. 152.—GUNDL. J. f. O. 1859, p. 296; *ib.* 1861, p. 414.
Frugilegus nasicus GRAY, Handl. Bds. II, p. 13 (1870).
Microcorax nasicus SHARPE, Cat. Bds. Brit. Mus. III, p. 49 (1877).

SP. CHAR.   *Male:*—General plumage glossy black, showing a purplish
        tinge; basal portion of feathers grayish; nasal bristles short.
        The sexes are similar.
        Length (skin), 17; wing, 10.75; tail, 7.60; tarsus, 1.90; bill, 2.35.
    HABITAT.   Cuba.

Corvus minutus GUNDL.

Corvus minutus GUNDL. Journ. Bost. Soc. Nat. Hist. VI, p. 315 (1852).—
    CAB. J. f. O. 1856, p. 97.—BREWER. Pr. Bost. Soc. Nat. Hist. VII,
    p. 307 (1860).—GUNDL. Repert. Fisico-Nat. Cuba. I. p. 290 (1866);
    ib. J. f. O. 1874, p. 139.—ALLEN, Bull. Mus. Comp. Zool. II. p. 297.
    —BD. BWR. & RIDGW. Hist. N. Am. Bds. II, p. 234 (1874) —CORY,
    List Bds. W. I. p. 14 (1885).
Frugilegus minutus GRAY. Handl. Bds. II. p. 13 (1870).
Colœus minutus SHARPE, Cat. Bds. Brit. Mus. III, p. 49 (1877).

Sp. Char. *Male:*—Entire plumage glossy black, showing purple reflections on the back and wings; underparts glossy black, the purple reflections slightly perceptible when held in the light; basal portion of feathers on the body smoky gray; bill and feet black.

Length (skin), 15; wing, 10.35; tail, 6; tarsus, 2; bill, 1.80.

Habitat. Cuba.

"*Cyanocorax pileatus,*" recorded from Jamaica, "probably a caged bird escaped," is a South American species. First given by Gosse (Bds. Jam. p. 308, 1847), and cited by later authors.

Corvus ossifragus has been recorded from Cuba, but I find no record of its actual capture. Dr. Gundlach writes me that the bird does not occur there.

Family TYRANNIDÆ.

Genus Elainia Sundev.

Elainia Sundevall. Ornithol. System, p. 89 (1836).

Elainia martinica (Linn.).

Muscicapa martinica Linn. Syst. Nat. I, p. 325 (1766).—Gmel. Syst. Nat. I, p. 930 (1788).—Lath. Ind. Orn. II, p. 483 (1790).
? *Muscicapa albicapilla* Vieill. Ois. Am. Sept. I, p. 66 (1807).
Myiobius martinicus Gray, Gen. Bds. I, p. 249, Sp. 27 (1846).
Tyrannula martinica Bp. Consp. I, p. 190 (1850).—Cassin, Pr. Acad. Nat. Sci. Phila. 1860, p. 375.
Elainea riisii Scl. P. Z. S. 1860, p. 313.—Newton, Ibis, 1860, p. 307.— Gray. Handl. Bds. I, p. 352 (1869).—Sundev. Oefv. K. Vet. Akad. Förh. 1869, p. 584.—Scl. P. Z. S. 1870, p. 834.
Elainea martinica Taylor, Ibis, 1864, p. 169.—Gray, Handl. Bds. I, p. 352 (1869).—Scl. P. Z. S. 1871, p. 271; *ib.* 1874. p. 175.—Pelz. Ibis, 1873. p. 113.—Scl. & Salv. Nom. Avium Neotr. p. 48 (1873). —Lawr. Pr. U. S. Nat. Mus. I. pp. 59. 191, 270, 257, 438, 487 (1878). Allen, Bull. Nutt. Orn. Club, V, p. 166 (1880).—Lister. Ibis, 1880, p. 41.—Grisdale, Ibis, 1882, p. 489.—Cory, List Bds. W. I. p. 15 (1885).

Sp. Char. *Male:*—General plumage above brownish olive; feathers on the head having the basal portions white, forming a concealed white patch on the crown: throat gray; breast gray, slightly tinged with olive; sides. flanks, and crissum pale olive, mixed with whitish on the middle of the belly; under wing-coverts pale yellowish; wings and tail dark brown, some of the primaries delicately edged with yellowish white on the outer web, more broadly so on the basal portion of the inner webs, secondaries tipped with the same color; wing-coverts tipped with yellowish white, forming two imperfect wing-bands.

The sexes are similar.

Length (skin), 6.50; wing, 3.25; tail, 3; tarsus, .87; bill, .35.

HABITAT. Lesser Antilles.

Elainia fallax SCL.

Elainea fallax SCL. P. Z. S. 1861, p. 76; *ib.* Cat. Am. Bds. p. 217 (1862). —ALBRECHT, J. f. O. 1862, p. 199—GRAY, Handl. Bds. I, p. 352 (1869).—SCL. & SALV. Nom. Avium Neotr. p. 48 (1873).—A. & E. NEWTON, Handb. Jamaica, p. 107 (1881).—CORY, List Bds. W. I. p. 15 (1885).

SP. CHAR.—"Dusky olive green; wings and tail fuscous, edged externally with olivaceus. The coverts and the secondaries widely margined externally with greenish white; pileum subcrested, interiorly white; beneath yellowish; the throat washed with olivaceus; bill dusky horn color, whitish at base; feet black.

"Length, 5.2; wing, 2.7; tail, 2.5; tarsus, .75." (SCLATER, orig. descr. transl.)

HABITAT. Jamaica.

Elainia cotta GOSSE.

Elainea cotta GOSSE, Ann. N. H. 2d ser. III, p. 257 (1849).—ALBRECHT, J. f. O. 1852, p. 198.—SCL. Cat. Am. Bds. p. 218 (1862).—MARCH. Pr. Acad. Nat. Sci. Phila. 1863. p. 289.—GRAY, Handl. Bds. I, p. 352 (1869).—SCL. & SALV. Nom. Avium Neotr. p. 48 (1873).— A. & E. NEWTON, Handb. Jamaica, p. 107 (1881).—CORY, List Bds. W. I. p. 15 (1885).

SP. CHAR. *Male:*—General plumage above grayish olive; top of head brown; a partially concealed patch of bright yellow on the crown; throat grayish; underparts pale yellowish white; pale yellow on the belly; wings and tail olive brown; secondaries narrowly edged with yellowish.

The sexes are similar.

Length (skin), 5.25; wing, 2.65; tail, 2.40; tarsus, .55; bill, .38.

HABITAT. Jamaica.

GENUS **Pitangus** SWAINS.

Pitangus SWAINSON, Zool. Journ. III, p. 165 (1828).

Pitangus caudifasciatus (D'ORB.).

Tyrannus caudifasciatus D'ORB. in La Sagra's Hist. Nat. Cuba, Ois. p. 82
(1840).—GOSSE, Bds. Jam. p. 177 (1847).—GUNDL. Journ. Bost. Soc.
Nat. Hist. VI, p. 318 (1852).—CAB. J. f. O. 1855. p. 478.—BREWER,
Pr. Bost. Soc. Nat. Hist. VII. p. 307 (1860).—MARCH, Pr. Acad.
Nat. Sci. Phila. 1863, p. 288.—GUNDL. J. f. O. 1872. p. 424.

Pitangus caudifasciatus SCL. P. Z. S. 1861, p. 76: *ib.* Cat. Am. Bds. p.
222 (1862).—ALBRECHT, J. f. O. 1862, p. 199.—GRAY, Handl. Bds. 1,
p. 357 (1869).—SCL. & SALV. Nom. Avium Neotr. p. 50 (1873).—
A. & E. NEWTON, Handb. Jamaica. p. 107 (1881).—CORY, List Bds.
W. I. p. 15 (1885).

SP. CHAR. *Male:*—Top and sides of the head, including the ear-coverts,
dark brown; a concealed patch of bright orange yellow; back
brownish gray; some specimens showing a slight tinge of rufous on
the upper tail-coverts; under wing-coverts very pale yellow; underparts white, slightly tinged with ash on the breast and sides; under
tail-coverts white; tail-feathers having the basal half of the inner
webs of all except the two central feathers pale yellowish white, and
all the feathers narrowly tipped with white; outer tail-feather edged
with white on the outer web.

The sexes are similar.

Length (skin), 8; wing. 4; tail, 3.45; tarsus, .88; bill, .90.

HABITAT. Cuba and Jamaica.

Pitangus taylori SCL.

Pitangus taylori SCL. Ibis, 1864. p. 169.—GRAY, Handl. Bds. I, p. 357
(1869).—SCL. & SALV. Nom. Avium Neotr. p. 50 (1873).—CORY,
List Bds. W. I. p. 15 (1885).

Tyrannus (Pitangus) taylori BRYANT, Pr. Bost. Soc. Nat. Hist. X, p. 249
(1866).

Tyrannus taylori SUNDEV. Oefv. K. Vet. Akad. Förh. 1869, p. 599.—GUNDL.
Anal. Soc. Esp. Hist. Nat. VII, p. 193 (1878).

SP. CHAR. *Male:*—Upper surface dark olive brown, darkest on the head;
concealed portions of the feathers on the crown yellow in front,
succeeded by white, forming a crown patch half yellow and half
white, variable in different specimens; underparts grayish; throat
white; quills and tail dark brown; the primaries showing a faint
rufous edging on the basal portion of the outer web, generally lacking on the first, second, and third; inner webs of primaries and secondaries edged with pale yellowish white; under tail-coverts white;
tail olive brown; the outer web of the outer primary showing a faint
edging of dull white.

The sexes are similar.

Length (skin), 8.50; wing. 4.25; tail, 3.25; tarsus, .75; bill, 90.

HABITAT. Porto Rico.

Pitangus bahamensis Bryant.

Tyrannus candifasciatus Bryant, Pr. Bost. Soc. Nat. Hist. VII, p. 108
    (1859).
Pitangus bahamensis Bryant, Pr. Bost. Soc. Nat. Hist. IX, p. 279 (1864).
    Gray, Handl. Bds. I, p. 357 (1869).—Cory, Bds. Bahama I. p. 102
    (1880); *ib.* List Bds. W. I. p. 15 (1885).

Sp. Char. *Male:*—Above gray, with a tinge of olive on the back; top of
    the head, including the eyes, dark slate color, concealing a patch of
    bright orange yellow upon the crown; underparts ashy white, shad-
    ing into an olive tinge upon the flanks, and pale yellow upon the ab-
    domen and crissum; wings dark brown, edged with yellowish white,
    the coverts with pale brown; under wing-coverts pale yellow; tail
    dark brown; outer webs of first two and tips of rest brownish white;
    upper tail-coverts edged with rufous; bill and feet black.
        The female is similar to the male.
        Length, 8.10; wing, 4.20; tail, 3.50; tarsus, .80; bill, .96.
Habitat.   Bahamas.

Pitangus gabbii Lawr.

Pitangus gabbii Lawr. Ann. N. Y. Lyc. Nat. Hist. XI, p. 288 (1876).—
    Cory, Bull. Nutt. Orn. Club, VI, p. 133 (1881); *ib.* Bds. Haiti &
    San Domingo, p. 76 (1885); *ib.* List Bds. W. I. p. 15 (1885).

Sp. Char. *Male:*—Top of the head and cheeks dark brown, the feathers
    concealing a patch of bright orange yellow; back brown, lighter
    than the head, and becoming still lighter towards the rump: wings
    and tail brown; the primaries heavily edged with rufous on the outer
    edge; some of the secondaries showing pale white on the edges;
    wing-coverts and tail-feathers showing rufous edgings; inner webs of
    primaries and secondaries, and some of the under wing-coverts
    edged with yellowish white, giving the under surface of the closed
    wing a pale yellowish white color; entire under surface white; bill
    and legs black.
        The sexes are similar.
        Length, 7.50; wing, 4; tail, 3.35; tarsus, .85; bill, .85.
Habitat.   Haiti and San Domingo.

Genus Lawrencia Ridgw.

Lawrencia Ridgway, Auk, III, p. 282 (1886).

Lawrencia nanus Lawr.

Empidonax nanus Lawr. Ibis, 1875, p. 386.—Cory, Bds. Haiti & San Domingo,
    p. 82 (1885); *ib.* List Bds. W. I. p. 15 (1885).
Lawrencia nanus Ridgw. Auk, III, p. 282 (1886).

Sp. Char.—"Above dull greenish olive, darker on the crown, and brighter on the rump; tail dark brown, the outer web of the lateral feather pale fulvous; smaller wing-coverts the color of the back; the middle and larger coverts are brownish black, ending with white, forming two bars across the wings; the quill-feathers are dark brown, the third and fourth primaries are narrowly edged with greyish white, the inner quills, just perceptibly edged with light rufous; under lining of wings very pale yellow; throat greyish white; breast, abdomen, and under tail-coverts pale whitish fulvous; thighs light brown; upper mandible brown, the under, whitish horn color, dusky on the sides; tarsi and toes brownish black.

"The first primary is abnormally short, measuring but 1.5-16 inches; third quill longest; tail emarginate, length, 4⅞ inches; wing, 2.3-16; tail, 2; bill, ⅞; tarsus, 11-16." (Lawr. l. c., orig. descr.)

Habitat. San Domingo.

Genus Empidonax Caban.

Empidonax Cabanis, Journ. für Ornith. III, p. 280 (1855).

Empidonax acadicus (Gmel.).

Muscicapa acadica Gmel. Syst. Nat. I, p. 947 (1788).
Muscicapa pusilla Lemb. Aves Cuba, p. 40 (1850).
Myiarchus pusilla Brewer, Pr. Bost. Soc. Nat. Hist. VII, p. 307 (1860) (Cuba)?
Empidonax acadicus Gundl. Repert. Fisico-Nat. Cuba, I, p. 240 (1865); *ib.* J. f. O. 1872, p. 427 (Cuba).—Cory, List Bds. W. I. p. 15 (1885).

Recorded from Cuba.

Genus Contopus Caban.

Contopus Cabanis, "Journ. für Ornith. III, p. 479 (Nov. 1855).

Contopus pallidus (Gosse).

Myiobius pallidus Gosse, Bds. Jam. p. 166 (1847).
Blacicus pallidus Scl. P. Z. S. 1861, p. 77.—Albrecht, J. f. O. 1862, p. 199.—Gray, Handl. Bds. I, p. 363 (1869).
Contopus pallidus Scl. Cat. Am. Bds. p. 231 (1862).—March, Pr. Acad. Nat. Sci. Phila. 1863, p. 290.—Scl. & Salv. Nom. Avium Neotr. p. 52 (1873).—A. & E. Newton, Handb. Jamaica, p. 107 (1881).—Cory, List Bds. W. I. p. 15 (1885).
Contopus caribæus var. *pallidus* Bd. Bwr. & Ridgw. Hist. N. Am. Bds. II, p. 351 (1874).

Sp. Char. *Male:*—General plumage above olive brown, darkest on the head; rump tinged with rufous; wing-coverts edged with rufous, forming two wing-bands; underparts dull grayish olive, slightly

tinged with rufous; throat pale; wings and tail dark brown; upper
mandible dark brown; lower mandible pale.

The sexes are similar.

Length (skin). 5; wing, 2.62; tail, 2.20; tarsus, .60; bill, .45.

HABITAT. Jamaica.

Contopus latirostris (VERR.).

Myiobius latirostris VERR. Nouv. Arch. Mus. Bull. II. p. 22 (1866).—
SCL. P. Z. S. 1871, p. 271.

Contopus latirostris SCL. & SALV. Nom. Avium Neotr. p. 52 (1873).—
ALLEN, Bull. Nutt. Orn. Club, V, p. 166 (1880).—CORY. List Bds.
W. I. p. 15 (1885).

SP. CHAR. *Male:*—General plumage above dark olive; chestnut on the
rump and upper tail-coverts; underparts rufous chestnut, palest
on the throat; wings and tail dark brown, the latter faintly
tipped with buffy white; upper mandible brown; lower mandible
yellowish.

The sexes are similar.

Length (skin) 5.25; wing. 2.30; tail, 2.45; tarsus, .38; bill, .40.

HABITAT. Santa Lucia.

Contopus bahamensis (BRYANT).

Empidonax bahamensis BRYANT, Pr. Bost. Soc. Nat. Hist. VII, p. 109
(1859).—GRAY, Handl. Bds. I, p. 361 (1869).

Contopus caribæus var. *bahamensis* BD. BWR. & RIDGW. Hist. N. Am.
Bds. II, p. 352 (1874).

Contopus bahamensis CORY, Bds. Bahama I. p. 101 (1880); *ib.* List Bds.
W. I. p. 15 (1885).

SP. CHAR. *Male* (winter):—Above brownish olive, becoming darker
upon the crown; a nearly complete circle of white around the eye,
broken above; lores ashy; below pale yellow, with a faint tinge of
olive; wings dark brown; under coverts pale orange yellow;
coverts, secondaries, and tertiaries brownish white, the coverts
forming two distinct bands upon the wing; tail dark brown, lighter
on the outer feathers; legs and upper mandible black; lower mandible pale, becoming darker at tip. One specimen taken had the
yellow of the breast much brighter and deeper, the crissum much
brighter, the olive markings heavier, and the under wing-coverts
pinkish.

Female resembles the male

Length, 5.35; wing, 2.80; tail, 2.60; tarsus, .58; bill, .60.

HABITAT. Bahamas.

Contopus hispaniolensis (BRYANT).

Tyrannula carribæa var. *hispaniolensis* BRYANT, Pr. Bost. Soc. Nat. Hist.
XI, p. 91 (1866).
Blacicus carribæus var. *hispaniolensis* GRAY, Handl. Bds. I. p. 363 (1869).
Contopus carribæus var. *hispaniolensis* ED. BWR. & RIDGW. Hist. N. Am.
Bds. II, p. 351 (1874).
Contopus frazeri CORY, Bull. Nutt. Orn. Club, VIII, p. 94 (1883).
Sayornis dominicensis CORY, Bull. Nutt. Orn. Club. VIII. p. 95 (1883).
Contopus hispaniolensis CORY, Bds. Haiti & San Domingo, p. 81 (1885):
ib. List Bds. W. I. p. 15 (1885).

SP. CHAR. *Male:*—General plumage grayish olive; feathers of the crown
dark brown, edged with olive; throat ashy, becoming olive on the
sides of the breast, and yellowish brown on the abdomen and crissum; wing-coverts pale at the tips, forming two very dull wing-
bands; secondaries very narrowly edged with pale brownish white;
tail brown; under wing-coverts pale yellowish brown.
The sexes are similar.
Length, 5.50; wing, 3; tail, 2.70; tarsus, .58; bill, .52.

HABITAT. Haiti and San Domingo.

Contopus virens (LINN.).

Muscicapa virens LINN. Syst. Nat. I, p. 327 (1766). — D'ORB. in La
Sagra's Hist. Nat. Cuba. Ois. p. 86 (1840).
Myiarchus virens BREWER, Pr. Bost. Soc. Nat. Hist. VII, p. 307 (1860)
(Cuba).
Contopus virens GUNDL. Repert. Fisico-Nat. Cuba, I, p. 239 (1865); *ib.*
J. f. O. 1872, p. 424 (Cuba).—CORY, List Bds. W. I. p. 15 (1885).

Accidental in Cuba.

GENUS Sayornis BONAP.

Sayornis BONAPARTE, "Coll. Delattre, p. 87 (1854)."

Sayornis phœbe (LATH.).

Muscicapa phœbe LATH. Ind. Orn. II, p. 489 (1790).
Muscicapa fusca GMEL. Syst. Nat. I, p. 931 (1788).—LEMB. Aves Cuba.
p. 41 (1850).
Muscicapa lembeyei GUNDL. Journ. Bost. Soc. Nat. Hist. VI, p. 314
(1852) (Cuba).
Aulanax fuscus GUNDL. J. f. O. 1856, p. 1 (Cuba).
Myiarchus lembeyii BREWER, Pr. Bost Soc. Nat. Hist. VII, p. 306 (1860)
(Cuba).

Aulonax lembeyei GUNDL. Repert. Fisico-Nat. Cuba. I, p. 240 (1865); *ib.*
J. f. O. 1872, p. 427 (Cuba).
Sayornis fusca CORY, List Bds. W. I. p. 15 (1885).
Recorded from Cuba.

GENUS Myiarchus CABAN.

Myiarchus CABANIS, "Fauna Peruana, 1844-46, p. 152."

Myiarchus validus CAB.

Tyrannus crinitus GOSSE, Bds. Jam. p. 186 (1847).
Tyrannula gossii BP. Consp. I, p. 186 (1850).—KAUP. P. Z. S. 1851, p. 53.
Myionax validus CAB. et HEIN. Mus. Hein. II, p. 78 (1859).
Myiarchus validus CAB. Orn. Not. II, p. 351.—SCL. P. Z. S. 1861, p. 76.
   —ALBRECHT, J. f. O. 1862, p. 199.—MARCH, Pr. Acad. Nat. Sci. Phila.
   1863. p. 288.—SCL. & SALV. Nom. Avium Neotr. p. 52 (1873).—
   BD. BWR. & RIDGW. Hist. N. Am. Bds. II, p. 331 (1874).—A. & E.
   NEWTON, Handb. Jamaica, p. 107 (1881).—CORY, List Bds. W. I.
   p. 15 (1885).

SP. CHAR. *Male:*—General plumage above dark olive; brownish olive on
   the head; upper tail-coverts rufous; throat gray; belly dull yellow;
   wings dark brown, the feathers heavily bordered with rufous chestnut; wing-coverts edged with dull rufous; outer tail-feathers pale
   rufous, second feather having the inner web rufous, the outer web
   brown, the brown gradually widening on the third and fourth
   feathers, the central feathers being pale olive brown, faintly edged
   with rufous.

    The sexes are similar.

    Length (skin), 8; wing, 4; tail, 4; tarsus, 1; bill, .75.

HABITAT. Jamaica.

Myiarchus stolidus (GOSSE).

Myiobius stolidus GOSSE, Bds. Jam. p. 168 (1847).
Tyrannula stolida KAUP. P. Z. S. 1851, p. 51.
Myiarchus stolidus CAB. J. f. O. 1855, p 479.—SCL. P. Z. S. 1861, p. 77;
   ib. Cat. Am. Bds. p. 234 (1862).—ALBRECHT, J. f. O. 1862, p. 199.
   —MARCH, Pr. Acad. Nat. Sci. Phila. 1863, p. 288.—SCL. & SALV.
   Nom. Avium Neotr. p. 52 (1873).—A. & E. NEWTON, Handb.
   Jamaica, p. 107 (1881).—CORY, List Bds. W. I. p. 15 (1885).
Kaupornis stolidus GRAY, Handl. Bds. I. p. 358 (1869).

SP. CHAR. *Male:*—Head dark brown, shading into grayish olive on the
   back; throat grayish; belly and under tail-coverts dull yellow;
   wings brown; primaries showing a very slight tinge of rufous on

the edges of the basal half of the outer webs; secondaries edged with dull white; under wing-coverts pale yellow; tail brown, the second feather slightly edged with rufous on the inner web, heavier on the third and fourth.

The sexes are similar.

Length (skin), 7.25; wing, 3.50; tail, 3.25; tarsus. 85; bill, .75.

HABITAT. Jamaica.

This species differs from *M. sagræ* by the brighter yellow of the belly and in being very slightly larger.

Myiarchus sagræ (GUNDL.).

Tyrannus phœbe D'ORB. in La Sagra's Hist. Nat. Cuba. Ois. p. 84 (1840).
Muscicapa sagræ, GUNDL. Journ. Bost. Soc. Nat. Hist. VI, p. 313 (1852).
Myiarchus stolidus, BREWER, Pr. Bost. Soc. Nat. Hist. VII, p. 307 (1860).
—GUNDL. Repert. Fisico-Nat. Cuba, I, p. 239 (1865).
Tyrannula (Myiarchus) stolida (var. *lucaysiensis*) BRYANT, Pr. Bost. Soc. Nat. Hist. XI, p. 66 (1866).
Myiarchus phœbe GRAY, Handl. Bds. I, p. 358 (1869).—SCL. & SALV. Nom. Avium Neotr. p. 52 (1873).—CORY, List Bds. W. I. p. 15 (1885).
Myiarchus stolidus var. *phœbe* COUES, Pr. Acad. Nat. Sci. Phila. 1872, p. 78.—BD. BWR. & RIDGW. Hist. N. Am. Bds. II, p. 332 (1874).
Myiarchus sagræ GUNDL. J. f. O. 1872, p. 424.
Myiarchus stolidus var. *lucaysiensis* CORY, Bds. Bahama I. p. 100 (1880).

SP. CHAR. *Male:*—Above brownish olive, becoming darker upon the head, and shading into rufous on the rump; underparts ashy white, shading into yellowish upon the abdomen and crissum; wings dark brown, the coverts tipped and edged with dull white, forming two wing-bands; the basal half of the outer webs of the primaries, except the first two, edged with rufous; some of the secondaries edged with white; under wing-coverts pale yellowish white; tail dark brown, the feathers bordered with rufous upon the inner webs, very faintly upon the two central ones; legs and bill black.

Length (skin), 7.20; wing, 3.25; tail, 3.20; tarsus, .75; bill 65.

HABITAT. Cuba and Bahamas.

This species differs from *M. dominicensis* in lacking the bright rufous edging on the primaries. Both *M. stolidus* and *M. dominicensis* have the belly yellow instead of dull white, as in *M. sagræ*.

Myiarchus antillarum (BRYANT).

Tyrannus (Myiarchus) antillarum BRYANT, Pr. Bost. Soc. Nat. Hist. X, p. 249 (1866).

Myiarchus antillarum SUNDEV. Oefv. K. Vet. Akad. Förh. 1869. p. 599?—
    GRAY. Handl. Bds. I, p. 364 (1869).—SCL. & SALV. Nom. Avium
    Neotr. p. 52 (1873).—GUNDL. Anal. Soc. Esp. Hist. Nat. VII. p. 194
    (1878).—CORY. List Bds. W. I. p. 15 (1885).
Myiarchus stolidus var. *antillarum* COUES, Pr. Acad. Nat. Sci. Phila.
    1872. p. 79.—BD. BWR. & RIDGW. Hist. N. Am. Bds. II, p. 332
    (1874).

SP. CHAR.   *Male:*—Top of head olive brown, shading into grayish olive
    on the back ; rump slightly tinged with rufous ; throat and breast ashy
    gray ; belly white ; a faint tinge of yellowish white on the crissum ;
    quills and tail dark brown, showing a slight rufous edging ; secon-
    daries edged with dull white ; a narrow mark of pale rufous tipping
    the inner webs of some of the tail-feathers.
        The sexes are similar.
        Length (skin), 7 ; wing, 3 ; tail, 3 ; tarsus, .85 ; bill, .65.
HABITAT.   Porto Rico.

This species differs from *M. dominicensis* by the absence of
the rufous tail-markings and in the belly being white. It is
nearest allied to *M. sagræ*, and closely resembles that species,
but is easily distinguished from it by the absence of the broad
rufous edging on the inner webs of the tail-feathers. *M. sagræ*
also shows a tinge of yellow on the belly, which is faint or
wanting in *M. antillarum.*

Myiarchus oberi LAWR.

Myiarchus erythrocercus SCL. P. Z. S. 1871. p. 271.
Myiarchus oberi LAWR. Ann. N. Y. Acad. Sci. I. p. 48 (1878) ; *ib.* Pr. U.
    S. Nat. Mus. I, pp. 59, 191, 217, 487 (1878).—ALLEN, Bull. Nutt.
    Orn. Club, V, p. 166 (1880).—SCL. Ibis, 1880, p. 74.—CORY, List
    Bds. W. I. p. 15 (1885).

SP. CHAR.   *Male:*—"Pileum, nape, and sides of the head dark umber-
    brown upper plumage dark olive-brown, upper tail-coverts edged with
    dull ferruginous ; two middle tail-feathers blackish brown, the other
    feathers are colored the same, except on the outer two-thirds of the
    inner webs, where they are bright ferruginous ; outer web of lateral
    feather and ends of the others, ash color ; quills brownish black, the
    primaries narrowly edged with dark ferruginous ; the outer secon-
    daries are margined with very pale rufous, and the other secondaries
    with pale yellowish white ; wing-coverts dark brown, ending with
    pale ashy tinged with rufous ; under wing-coverts pale, dull yellow ;
    inner margins of quills light salmon-color ; lores, throat, upper part

of breast and sides clear bluish-gray, lower part of breast, abdomen and under tail-coverts pale yellow; bill and feet black.

"Length, 8¼ in.; wing, 3¾; tail, 3⅝; tarsus, ⅞; bill from front, 12-16 "The female does not differ in plumage from the male." (LAWR. l. c., orig. descr.)

HABITAT.   St. Vincent, Dominica, Santa Lucia, and Grenada.

Myiarchus sclateri LAWR.

Myiarchus sclateri LAWR. Pr. U. S. Nat. Mus. I, p. 357 (1878).—CORY, List Bds. W. I. p. 15 (1885).

SP. CHAR.   *Male:—*"The upper plumage is deep dark olive, the head above blackish brown.   Unfortunately, the only feathers left in the tail are the outer four on one side; the outermost two are dark brown and without rufous edgings on the inner webs; the other two feathers are brownish-black, with their inner webs edged with light rufous for about one-quarter their width; quills dark brown, their inner webs bordered with pale salmon-color; wing-coverts edged with dull white; under wing-coverts light ash, with just a tinge of yellow; throat and breast of a clear cinereous gray; abdomen and under tail-coverts dull pale yellow; sides cinereous; bill and feet black.

"Length (fresh), 7½ in.; wing, 3¾; tail, 3½; tarsus, 1; middle toe and claw, 15-16; hind toe to end of claw, ⅝." (LAWR. l. c., orig. descr.)

HABITAT.   Martinique.

Myiarchus dominicensis (BRYANT).

Tyrannula stolida var. *dominicensis* BRYANT, Pr. Bost. Soc. Nat. Hist. XI, p. 90 (1866).
Myiarchus stolidus var. *dominicensis* GRAY, Handl. Bds. I, p. 358 (1881).
Myiarchus stolidus CORY, Bull. Nutt. Orn. Club, VI, p. 153 (1881).
Myiarchus ruficaudatus CORY, Bull. Nutt. Orn. Club, VIII, p. 95 (1883).
Myiarchus dominicensis CORY, Bds. Haiti & San Domingo. p. 79 (1885); *ib.* List Bds. W. I. p. 15 (1885).

SP. CHAR.   *Male:—*Crown dark olive brown, becoming lighter on the back, and showing a more decided grayish tinge; breast ashy; belly, crissum, and under wing-coverts pale yellow; wing-coverts edged with brownish white, forming two dull wing-bands; tertials broadly edged with yellowish white; primaries, except the first, narrowly edged with rufous on the outer web, showing a broader and much paler edging of the same color on the inner webs of the same feathers; two central tail-feathers dark brown, all the rest having more than half of the inner web rufous to the tip; bill and feet black.

The sexes are similar.

Length, 6.50; wing, 3.10; tail, 3; tarsus, .85; bill, .75.

HABITAT.   Haiti and San Domingo.

Myiarchus crinitus (LINN.).

Muscicapa crinita LINN. Syst. Nat. I, p. 325 (1766).
Myiarchus crinitus CAB. J. f. O. 1855, p. 479 (Cuba).—GUNDL. Repert.
     Fisico-Nat. Cuba, I. p. 239 (1865); *ib.* J. f. O. 1872, p. 426 (Cuba).
     —COUES, Pr. Acad. Nat. Sci. Phila. 1872, p. 63 (Cuba).
Tyrannus crinitus BREWER, Pr. Bost. Soc. Nat. Hist. VII, p. 307 (1860)
     (Cuba).

Accidental to Cuba.

GENUS Blacicus CABAN.

Blacicus CABANIS, J. f. O. 1885, p. 480.

Blacicus barbirostris (SWAINS.).

Tyrannula barbirostris SWAINS. Phil. Mag. 1827, p. 367.
Myiobius tristis GOSSE, Bds. Jam. p. 167 (1847).—ALBRECHT, J. f. O. 1862,
     p. 199.
Blacicus tristis CAB. J. f. O. 1855, p. 480.—SCL. Cat. Am. Bds. p. 324
     (1862).—MARCH, Pr. Acad. Nat. Sci. Phila. 1863, p. 290.—GRAY,
     Handl. Bds. I, p. 363 (1869).
Blacicus barbirostris, SCL. P. Z. S. 1871, p. 85.—SCL. & SALV. Nom.
     Avium Neotr. p. 53 (1873).—A. & E. NEWTON, Handb. Jamaica, p.
     108 (1881).—CORY, List Bds. W. I. p. 15 (1885).

SP. CHAR. *Male:*—Top of the head dark brown; pale brownish olive on
     the back; a faint tinge of rufous on the upper tail-coverts; throat
     gray; rest of underparts pale yellow; wings and tail dark brown; the
     secondaries and tail-feathers with pale edgings.

     The sexes are similar.

     Length (skin), 5.75; wing, 2.75; tail, 2.50; tarsus, .75; bill, .55.

HABITAT. Jamaica.

Blacicus caribæus (D'ORB.).

Muscipeta caribæa D'ORB. in La Sagra's Hist. Nat. Cuba, Ois. p. 92
     (1840).—GUNDL. Journ. Bost. Soc. Nat. Hist. VI, p. 316 (1852).
Blacicus caribæus BREWER, Pr. Bost. Soc. Nat. Hist. VII, p. 307 (1860).—
     GUNDL. Repert. Fisico-Nat. Cuba, I, p. 240 (1865); *ib.* J. f. O. 1872,
     p. 426.—GRAY, Handl. Bds. I, p. 363 (1869).—SCL. & SALV. Nom.
     Avium Neotr. p. 53 (1873).—CORY, List Bds. W. I. p. 15 (1885).

SP. CHAR. *Male:*—Top of head brownish olive, the feathers showing a
     narrow streak of dark brown on the shafts, rest of upper plumage
     dull olive; throat gray, with a faint tinge of yellowish near the

breast; belly dull orange rufous, shading into olive on the sides and flanks; quills and tail dark brown; the secondaries with pale edges. The sexes are similar.

Length (skin), 6; wing, 2.75; tail, 2.25; tarsus, .60; bill, .60.

HABITAT. Cuba.

Blacicus brunneicapillus LAWR.

Blacicus brunneicapillus LAWR. Ann. N. Y. Acad. Sci. I, p. 161 (1878); *ib.* Pr. U. S. Nat. Mus. I, pp. 59, 487 (1878).—CORY, List Bds. W. I. p. 15 (1885).

SP. CHAR. *Male:*—Top of head dark brown; back dull brown, tinged with olive; throat grayish, tinged with rufous on the breast, and becoming pale rufus brown on the belly and under tail tail-coverts; wings and tail brown.

The sexes are similar.

Length (skin), 5.25; wing, 2.50; tail, 2.45; tarsus, .60; bill, .48,

HABITAT. Dominica.

Blacicus blancoi GUNDL.

Blacicus blancoi GUNDL. J. f. O. 1874, p. 311; *ib.* Anal. Soc. Esp. Hist. Nat. VII, p. 195 (1878).—CORY, List Bds. W. I. p. 15 (1885).

SP. CHAR. *Male:*—Top of the head olive brown, becoming grayish olive on the back; throat dull white; breast and underparts pale rufous; under wing-coverts rufous; wings and tail pale brown; shafts of the tail-feathers reddish brown; upper mandible dark brown; lower mandible pale.

The sexes are similar.

Length (skin), 5.50; wing, 3; tail, 2.50; tarsus, .50; bill, .53.

HABITAT. Porto Rico.

GENUS **Tyrannus** CUVIER.

Tyrannus "CUVIER Leç. d'Anat. Comp. 1799-1800, tabl. ii."

Tyrannus rostratus SCL.

Tyrannus rostratus SCL. Ibis, 1864, p. 87; *ib.* P. Z. S. 1871, p. 272.— SCL. & SALV. Nom. Avium Neotr. p. 53 (1873).—LAWR. Pr. U. S. Nat. Mus. I, pp. 60, 191, 234, 240, 271, 358 (1878). — RIDGW. Smiths. Misc. Coll. XIX, p. 470 (1880).—ALLEN, Bull. Nutt. Orn. Club, V, p. 196 (1880).—CORY, List Bds. W. I. p. 16 (1885).

Sp. Char. *Male:*—Bill large and heavy; upper plumage slaty-gray, tinged with brownish on the back; a concealed patch of scarlet-orange on the head; ear-coverts dark; throat dull white, grayish on the breast and sides of the body; belly and under tail-coverts dull white, faintly tinged with yellow; under wing-coverts pale yellowish-white; quills and tail brown; secondaries edged with dull white.

The sexes are similar.

Length (skin), 9; wing, 4.50; tail, 4; tarsus, .72; bill, 1.05; width of bill at base, .60.

HABITAT. Lesser Antilles.

Tyrannus magnirostris D'Orb.

? *Tyrannus matutinus* VIEILL. Enc. Méth. 1823, p. 850.
Tyrannus magnirostris D'ORB. in La Sagra's Hist. Nat. Cuba, Ois. p. 80 (1840).—SCL. Cat. Am. Bds. p. 236 (1862).—BRYANT, Pr. Bost. Soc. Nat. Hist. XI, p. 66 (1866).—SCL. & SALV. Nom. Avium Neotr. p. 53 (1873).—RIDGW. Smiths. Misc. Coll. XIX, p. 464 (1880).—CORY, Bds. Bahama I. p. 99 (1880); *ib.* List Bds. W. I. p. 16 (1885).
Melittarchus magnirostris CAB. J. f. O. 1855, p. 447.—BREWER, Pr. Bost. Soc. Nat. Hist. VII, p. 307 (1860).—GUNDL. Repert. Fisico-Nat. Cuba I, p. 238 (1865); *ib.* J. f. O. 1872, p. 421.

Sp. Char. *Male:*—Larger than *T. rostratus.* Head dark brown; ear-coverts and cheeks blackish brown; a concealed patch of bright orange on the crown; back slaty-brown; entire underparts white, showing a faint tinge of yellowish in some specimens; under wing-coverts pale yellow; quills and tail brown; some of the prima-ries and all of the secondaries edged with dull white; wing-coverts edged with dull white.

The sexes are similar.

Length (skin), 10; wing, 5; tail, 4; tarsus, .85; bill, 1.25; width of bill at base, .65.

HABITAT. Cuba. Inagua?

Tyrannus melancholicus VIEILL.

Tyrannus melancholicus VIEILL. Nouv. Dict. XXXV, p. 48, D'ORB. Voy. Ois, p. 311.—SCL. Cat. Am. Bds. p. 235 (1862)—SCL. & SALV. Nom. Avium Neotr. p. 53 (1873).—BD. BWR. & RIDGW. Hist. N. Am. Bds. II, p. 315 (1874).—LAWR. Pr. U. S. Nat. Mus. I, pp. 271, 487 (1878).—RIDGW. Smiths. Misc. Coll. XIX, p. 473 (1880).—CORY, List Bds. W. I. p. 16 (1885).
Muscicapa despotes LICHT. Doubl p. 55 (1823).
Muscicapa furcata SPIX, Av. Bras. II, p. 15 (1825).

Tyrannus crudelis SWAINS. Quart. Journ. Sc. XX, p. 275 (1826).
Tyrannus furcatus MAX. Beitr. III, p. 884 (1831).
Tyrannus albogularis BURM. Syst. Ueb. II, p. 465.
Laphyctes melancholicus CAB. & HEIN. Mus. Hein. II, p. 76 (1859).

SP. CHAR. *Male:*—Top of head gray; a concealed patch of reddish
orange; back dull grayish-olive; throat grayish white, shading into
yellowish-olive on the breast, and having the entire rest of underparts bright yellow; under wing-coverts pale yellow; quills and tail
brown, showing dirty white edgings on some of the coverts and
secondaries.

Length (skin), 8; wing, 4.30; tail, 3.50; tarsus, .75; bill, .75.

HABITAT. Grenada.

Tyrannus dominicensis (GMEL.).

Tyrannus dominicensis BRISS. Orn. II, p. 394, pl. 38, fig. 2 (1760).—RICH.
List 1837.—GOSSE, Bds. Jam. p. 169 (1847).—BAIRD, Cat. N.
Am. Bds. No. 125 (1869).—NEWTON, Ibis, 1859, p. 146.—CASSIN,
Pr. Acad. Nat. Sci. Phila. 1860, p. 375.—ALBRECHT. J. f. O. 1862,
p. 199.—SUNDEV. Oefv. K. Vet. Akad. Förh. 1869, pp. 584. 599.—
BD. BWR. & RIDGW. Hist. N. Ann. Bds. II, p. 319 (1874). —
ALLEN, Bull. Mus. Comp. Zool. II, p. 300 (1881).—RIDGW. Pr. U.
S. Nat. Mus. VII, p. 172 (1884).—CORY. Bull. Nutt. Orn. Club. VII,
p. 153 (1881); *ib.* Bds. Haiti & San Domingo, p. 77 (1885); *ib.*
List Bds. W. I. p. 16 (1885).
Lanius tyrannus var. β. *dominicensis* "GMEL. Syst. Nat. I. p. 202 (1788)."
Tyrannus griseus VEILL. Ois. Am. Sept. I. p. 76, pl. 46 (1807).—SWAINS.
Quart. Journ. Sci. XX. p. 276 (1826).—GRAY, Gen. Bds. I. p. 247
(1844).—BP. Consp. I. p. 192 (1850).—SCL. Cat. Am. Bds. p. 236
(1862).—MARCH. Pr. Acad. Nat. Sci. Phila. 1863, p. 287.—TAYLOR,
Ibis, 1864, p. 169.—LAWR. Ann. Lyc. N. Y. VIII. p. 99 (1864).—
BRYANT, Pr. Bost. Soc. Nat. Hist. XI, p. 90 (1866).—CORY, Bds.
Bahama I. p. 99 (1880); *ib.* Bull. Nutt. Orn. Club. VI, p. 153
(1881).—A. & E. NEWTON, Handb. Jamaica, p. 108 (1881).
Tyrannus matutinus "VIEILL. Enc. Méth. p. 850 (1823)."—D'ORB. in La
Sagra's Hist. Nat. Cuba, Ois. p. 83 (1840).—GRAY, Gen. Bds. I, p.
247 (1844).—SALLÉ, P. Z. S. 1857, p. 232.
Muscicapa dominicensis AUD. Orn. Biog. II, p. 392, pl. 46 (1834); *ib.* Bds.
Am. I. p. 201 (1840).
Tyrannus tiriri "TEMM. Tabl. Méth. p. 24 (1836)."
Tyrannulus dominicensis JARD. Contr. Orn. p. 67 (1850).
Melittarchus dominicensis CAB. J. f. O. 1855, p. 478; *ib.* Mus. Hein. II. p.
80 (1859).—BREWER, Pr. Bost. Soc. Nat. Hist. VII, p. 307 (1860).
Melittarchus griseus GUNDL. J. f. O. 1872, p. 422; *ib.* Anal. Soc. Esp.
Hist. Nat. VII, p. 192 (1878).

Sp. Char.   *Male:*—Above grayish-ash, darkest on the head; a dull black patch behind the eye; underparts whitish, ashy on the sides of the breast; wings brown, secondaries and coverts edged with dull white; under wing-coverts pale yellow; tail brown, feathers faintly tipped and edged with dull white; upper tail-coverts edged with pale rufous.

The sexes are similar.

Length, 8.50; wing, 4.40; tail, 4.10; tarsus, .75; bill, .90.

HABITAT.   Bahamas, Cuba, Haiti, San Domingo, Jamaica, Porto Rico, St. Thomas, St. Croix, St. Bartholomew, and Sombrero.

Tyrannus tyrannus (LINN.).

Lanius tyrannus LINN. Syst. Nat. p. 94 (1758).
Lanius tyrannus var. *carolinensis* et *ludovicianus* GMEL. Syst. Nat. I, p. 302 (1788).
Tyrannus intrepidus? SALLÉ. P. Z. S. 1857. p. 232 (San Domingo).— BREWER, Pr. Bost. Soc. Nat. Hist. VII, p. 307 (1860) (Cuba); *ib.* BRYANT, XI, p. 90 (1867) (San Domingo).
Tyrannus pipiri GUNDL. J. f. O. 1872, p. 423 (Cuba)?
Tyrannus carolinensis CORY, List Bds. W. I. p. 16 (1885).

Accidental in Cuba.   Porto Rico?   San Domingo?

Tyrannus sulphurascens Herz. P. V. Württemberg is an undetermined species mentioned by ·Cabanis (J. f. O. 1857, p. 241). It was originally described as occurring in Cuba and Haiti. Gundlach, in writing of this species (J. f. O. 1871, p. 268), thinks there has been a mistake in the locality.

Milvulus tyrannus is recorded from Grenada by Mr. Wells in his list of the birds of that island.*

· FAMILY COTINGIDÆ.

GENUS Hadrostomus CAB.

Hadrostomus CABANIS, Mus. Hein. II, p. 85 (1859).

Hadrostomus niger (GMEL.).

Lanius niger GMEL. Syst. Nat. I, p. 301 (1788).
Tityra leuconotus GRAY, Gen. Bds. I, pl. 63 (1844).—GOSSE, Bds. Jam. p. 187 (1847).

---

* Wells, List Bds. Grenada, p. 4 (1886).

Pachyrhynchus atterrimus Lafr. Rev. Zool. 1846, p. 320.
Pachyrhamphus nigrescens Cab. Orn. Not. 1, p. 241.—Bp. Consp. I, p 180 (1850).
Pachyrhamphus niger Scl. P. Z. S. 1857, p. 72.
Platypsaris nigra Scl. P. Z. S. 1861, p. 77.—Albrecht, J. f. O. 1862, p. 207.
Hadrostomus niger Cab. & Heine. Mus. Hein. II, p. 85 (1859).— Scl. Cat. Am. Bds. p. 239 (1862).—March, Pr. Acad. Nat. Sci. Phila. 1863, p. 290.—Scl. & Salv. Nom. Avium Neotr. p. 56 (1873).—A. & E. Newton, Handb. Jamaica, p. 108 (1881).—Cory, List Bds. W. I. p. 16 (1885).

Sp. Char. *Male:*—Top of the head black, shading into dark brown, with a blackish gloss on the back; throat, breast and belly smoke-color; a faint tinge of rufous on the flanks; wings and tail dark brown, almost black; tertials and some of the wing-coverts heavily marked with white, forming a partially concealed white patch at the junction of the wing and back.

Female:—Top of the head dark brown; a malar stripe of light brown; throat brownish white, rest of underparts dull white; crissum brownish olive; back and rump dark slate-color; wings and tail brown, pale rufous on the inner webs of the primaries; outer webs of secondaries, and some of the inner primaries, showing dull rufous brown.

Length (skin), 7.20; wing, 4; tail, 3.50; tarsus, .85; bill, .55.

Habitat. Jamaica.

Family CAPRIMULGIDÆ.

Genus Nyctibius Vieill.

Nyctibius Vieillot, Analyse, p. 38 (1806).

Nyctibius jamaicensis (Gmel.).

Caprimulgus jamaicensis Gmel. Syst. Nat. I, p. 1029 (1788).—Denny, P. Z. S. 1847, p. 38.
Nyctibius jamaicensis Gosse, Bds. Jam. p. 41 (1847).—Bp. Consp. I, p. 58 (1850).—Scl. P. Z. S. 1861, p. 77; *ib.* Cat. Am. Bds. p. 278 (1862). —Albrecht, J. f. O. 1862, p. 199.—March, Pr. Acad. Nat. Sci. Phila. 1863, p. 286.—Gray, Handl. Bds. I, p. 56 (1869).—Scl. & Salv. Nom. Avium Neotr. p. 95 (1873).—A. & E. Newton, Handb. Jamaica, p. 108 (1881).—Cory, List Bds. W. I. p. 16 (1885).

Sp. Char. *Male:*—General plumage grayish, heavily marked, streaked, and blotched with brown and white; some of the feathers sparsely tinged with pale rufous on the back, wing-coverts, and underparts; throat dull white, the shafts of the feathers brown, giving the throat the appearance of being streaked with narrow lines of brown; these lines are broader on the shafts of the feathers on the belly, many of the feathers being tipped with brown, and showing the pale rufous edging before mentioned; wings and tail brown, imperfectly banded with pale markings; upper surface of tail showing imperfect white bands; under surface of tail thickly mottled with dull white; under surface of wings brown, with white dots.

Length (skin). 16; wing, 12; tail, 8.50; culmen, 1.

HABITAT.   Jamaica.

Nyctibius pallidus GOSSE.

Nyctibius pallidus GOSSE, Bds. Jam. p. 49 (1847).—BP. Consp. I, p. 58 (1850).—ALBRECHT, J. f. O, 1862, p. 199.—MARCH, Pr. Acad. Nat. Sci. Phila. 1863, p. 286.—SCL. P. Z. S. 1866, p. 129 (?).—GRAY. Handl. Bds. I, p. 56 (1869).—A. & E. NEWTON, Handb. Jamaica, p. 108 (1881).—CORY, List Bds. W. I. p. 16 (1885).

"Length 11 inches, expanse 22, rictus 1⅜. beak from feathers to tip ⅞, flexture, 6, tail 3¾.

"The nostrils prominent, tubulated, and covered with a membrane; from the nostrils runs a deep groove or furrow towards the tip. The beak was bent like the end of an Owl's, and when closed was longer than the under mandible; the latter was of a subulated form, shorter and bending in a contrary direction to the upper one: it was broader than the upper; its margins were inverted, and received the upper one exactly, when closed. There were no bristles on the angle of the mouth. The tibiæ (tarsi?) or shank-bones are shortened into a heel, so that the measure of what is usually called the leg, from the bend of the knee to the first joint of the middle toe, is only 2-8 of an inch. The length of that part which ought to be called the leg, (tibia?) is 1½ inch, and the bone of the thigh 1 inch. Toes four, three before, one behind; covered with ash-coloured scales, very flat beneath, and all connected by narrow membrane. Claws brown, strong, gently curved, and compressed; middle claw thinned to an edge on the inner side, but not serrate. Tail of ten feathers, equal, broad, rounded, barred with blackish and grey, and these bars again marked with less black bars. Wing quills coloured chiefly like the tail, but deeper; secondaries edged with clay-colour; winglet and long coverts immediately beneath it, black, with a few whitish bars; greater coverts black, edged with clay-colour; the next row of coverts whitish, with black shafts; the next row black, making a large triangular black spot in the expanded wing. Eyes very large, irides bright yellow. Head, neck, and throat, white, with black shafts; above each eye some black and white streaked feathers in an erect position, forming two small roundish rings. On the breast, clay-coloured feathers with black shafts and black spots. Sides, belly, and vent, white with black shafts. A line of black feathers down the middle of the back; rump ashy, with narrow black shafts. On shoulders a mixture of ash and clay-colour, with black shafts. Plumage very loose. Weight, 3 oz., 7 sc." (GOSSE, from Robinson's MSS., Bds. Jam. pp. 49, 50. 1847.)

HABITAT. Jamaica.

This is a very doubtful species, not generally recognized by authors. Probably the same as *N. jamaicensis.*

GENUS Chordeiles SWAINS.

Chordeiles SWAINSON, Fauna Bor. Amer. II. p. 496 (1831).

Chordeiles minor CAB.

Chordeiles minor CAB. J. f. O. 1856. p. 5.—SCL. Cat. Am. Bds. p. 279 (1862).—ALBRECHT, J. f. O. 1862. p. 199.—MARCH, Pr. Acad. Nat. Sci. Phila. 1863, p. 286.—SCL. & SALV. Nom. Avium Neotr. p. 96 (1873).—GUNDL. J. f. O. 1874, p. 117; *ib.* Anal. Soc. Esp. Hist. Nat. VII, p. 202 (1878).—A. & E. NEWTON, Handb. Jamaica, p. 109 (1881).—CORY, Bds. Bahama I. p. 106 (1880); *ib.* Bds. Haiti & San Domingo, p. 85 (1885); *ib.* List Bds. W. I. p. 16 (1885).

Chordeiles gundlachii LAWR. Ann. Lyc. N. Y. VI, p. 165 (1856).

Chordeiles popetue BRYANT, Pr. Bost. Soc. Nat. Hist. VII, p. 108 (1859). —GUNDL. Repert. Fisico-Nat. Cuba, I, p. 282 (1865); *ib.* J. f. O. 1874, p. 117.

Chordeiles gundlachi BREWER, Pr. Bost. Soc. Nat. Hist. VII, p. 306 (1860).

Chordeiles popetue var. *minor* BD. BWR. & RIDGW. Hist. N. Am. Bds. II, p. 400 (1874).

SP. CHAR. *Male:*—Above dark brown, variegated with white and tawny; underparts tawny, banded with brown; throat tawny, becoming whitish on the breast; a white line from sides of throat to chin; first two primaries with a spot on the inner web, and the second and third with a band of white; edge of carpus white.

The female differs from the male by having the sides of the throat rufous instead of white.

Length, 8.25; wing, 7; tail, 4; tarsus, .50; bill, .20.

HABITAT. Antilles.

Chordeiles virginianus (BRISS.).

Caprimulgus virginianus BRISSON, Orn. p. 477 (1760). — GMEL. Syst. Nat. I, ii, 1028 (1788).

Chordeiles virginianus Gosse. Bds. Jam. p. 33 (1847).—Lemb. Aves Cuba,
    p. 51 (1850) (?).—Sundev. Oefv. K. Vet. Akad. For. 1869, p. 600 (?),
    —Cory. List Bds. W. I. p. 16 (1885).
Chordeiles popetue var. *popetue* Bo. Bwa. & Ridgw. Hist. N. Am. Bds. II.
    p. 401 (1874) (Greater Antilles).

I have never seen a specimen of *C. virginianus* from the West
Indies; several authors have recorded it, but it is possible that
they may have mistaken *C. minor* for this species.[*]

Genus **Antrostomus** Gould.

Antrostomus "Gould, Icones Avium, 1838."

Antrostomus rufus (Bodd.).

Caprimulgus rufus "Bodd. et Gmel. ex Pl. Enl. p. 735."
Antrostomus rufus Cassin, Pr. Acad. Nat. Sci. Phila. V, p. 183. —
    Scl. P. Z. S., 1866, p. 136.—Gray, Handl. Bds. I, p. 59 (1869).—
    Scl. & Salv. Nom. Avium Neotr. p. 96 (1873).—Cory, List Bds.
    W. I. p. 16 (1885).
Antrostomus rutilus Burm. Syst. Ueb. II, p. 385.—Allen, Bull. Nutt.
    Orn. Club, V, p. 169 (1880).

Sp. Char.—Upper surface mottled and varied with brown and black; the
    terminal portions of the feathers on the head with broad patches of
    black in the centre of the feathers; underparts darkening on the
    breast, but becoming heavily tinged with rufous on the abdomen and
    crissum; primaries broadly blotched with light rufous, heaviest on
    the outer webs; a large blotch of white on the terminal portion of
    the inner web of the outer tail-feather, showing upon both webs of
    the second and third feathers; central tail-feathers dark brown,
    heavily mottled with rufous; feet black. In general appearance the
    bird is smaller and much darker than *A. carolinensis.*
        Length, 10.50; wing, 7.50; tail, 5; tarsus, .60.

Recorded from Santa Lucia, W. I.

Antrostomus carolinensis (Gmel.).

Caprimulgus carolinensis Gmel. Syst. Nat. I, p. 1028 (1788).—D'Orb. in
    La Sagra's Hist. Nat. Cuba, Ois. p. 96 (1840).
Antrostomus carolinensis Gundl. Repert. Fisico-Nat. Cuba, I, p. 283
    (1865); *ib.* J. f. O. 1874, p. 120 (Cuba).—Scl. P. Z. S. 1866, p. 136
    (Jamaica).—Gundl. Anal. Soc. Esp. Hist. Nat. VII, p. 201 (1878)
    (Porto Rico).—A. & E. Newton, Handb. Jamaica. p. 107 (1881).—

---

[*] Dr. Gundlach writes me *C. virginianus* is not uncommon in Cuba in winter.

CORY, Bds. Bahama I. p. 104 (1880); *ib.* Bull. Nutt. Orn. Club. VI.
p. 153 (1881) (Haiti); *ib.* List Bds. W. I. p. 16 1885).
Androstomus carolinensis CORY, Bds. Haiti & San Domingo, p. 84 (1885).

Recorded from Cuba, Porto Rico, Jamaica, Haiti, San
Domingo, and Bahamas.

Antrostomus cubanensis LAWR.

Caprimulgus vociferus D'ORB. in La Sagra's Hist. Nat. Cuba, Ois. p. 98
(1840).—LEMB. Aves Cuba, p. 130 (1850).
Antrostomus vociferus GUNDL. J. f. O. 1856. p. 6.—BREWER, Pr. Bost.
Soc. Nat. Hist. VII. p. 306 (1860).—SCL. & SALV. Nom. Avium
Neotr. p. 96 (1873).
Antrostomus cubanensis LAWR. Ann. Lyc. Nat. Hist. N. Y. 1862. p. 260.
—GUNDL. Repert. Fisico-Nat. Cuba, I. p. 283 (1865); *ib.* J. f. O.
1874. p. 120.—GRAY. Handl. Bds. I, p. 59 (1869).—CORY, List Bds.
W. I. p. 16 (1885).
Antrostomus macromystax var. *cubanensis* BD. BWR. & RIDGW. Hist. N.
Am. Bds. II, p. 409 (1874).

"*Adult male:*—Upper plumage dark ash, minutely mottled with
dull rufous and grey, the feathers conspicuously marked with longitudinal stripes of black in their centres; a line extends from the
bill over the eye and along the crown of greyish white, tinged with
pale rufous and intermixed with black; the tertiaries ochraceous-
white, beautifully variegated with black, and having near the
end of each feather an irregular patch of velvety black; wing-
coverts the same color as the back, some of them marked near
their ends with ochraceous spots; primaries dark reddish-brown
sprinkled with dull rufous and grey at their ends, and having
bright rufous spots arranged regularly on their outer webs, there
are spots also on their inner webs, more obscure in color and assuming a mottled form; secondaries dark brown, mottled with grey
on the outer webs, and tinged with rufous on the inner; tail very
full, of a fine deep brown, the two central tail-feathers closely
banded with curving bars of mottled grey and pale rufous, the next
feather on each side, with the bars dull rufous, and rather narrowly
tipped with ochraceous-white, less in extent on the inner web, the
three outer feathers are irregularly barred with dull rufous mottling
for their basal half, their ends for about an inch creamy-white,
with ochraceous edges; throat dark brown, minutely freckled with
rufous, the neck immediately below this color crossed with a band
of pale rufous; a line of pale rufous-white or ochraceous spots extend along below the under mandible, and down the side of the
neck, a few spots of the same in a line below the eye; on the side
of the neck enclosed by these spots and the band across the throat,

is a triangular blackish-brown patch, speckled with rufous; sides of
the head brown, freckled with minute rufous spots; feathers of the
breast and abdomen ochraceous white, more or less tinged with
rufous, and having their centres dark brown, and their sides and
ends barred and mottled with the same color; the exposed ends of
the feathers being but little mottled give quite a light appearance
to the under plumage; lower part of the abdomen and under tail-
coverts dull pale rufous, the feathers of the latter with dark markings along their shafts; sides under the wings dull rufous narrowly
barred with dark brown; under wing-coverts brown mottled with
rufous; tarsi clothed in front with rufous brown feathers; the bill
is light brown, black at the point, and having very strong bristles,
some of which are nearly two inches in length, and furnished with
lateral filaments; feet brown. Length about 11½ inches; wing 7½;
tail 5¾; tarsus ½." (LAWR. l. c. orig. descr.)

HABITAT. Cuba.

GENUS Stenopsis CASSIN.

Stenopsis CASSIN, Pr. Acad. Nat. Sci. Phila. 1851. p. 179.

Stenopsis cayennensis (GMEL.).

Caprimulgus cayennensis GMEL. Syst. Nat. I, p. 1031 (1788).—CAB. in
    Schomb. Guiana, III, p. 710 (1848).
Caprimulgus cayanus LATH. Ind. Orn. II, p. 587 (1790).
Caprimulgus leopetes JARD. & SELBY, Ill. Orn. II, pl. 87.
Caprimulgus odontopterou LESS. Rev. Zool. 1839, p. 105.
Antrostomus cayennensis BP. Consp. I. 61 (1850).
Stenopsis cayennensis CASSIN. Pr. Acad. Nat. Sci. Phila. 1851, p. 179.—
    CAB. & HEIN. Mus. Hein. III, p. 91.—SCL. Cat. Am. Bds. p. 280
    (1862).—GRAY. Handl. Bds. I, p. 59 (1869).—SCL. & SALV. Nom
    Avium Neotr. p. 96 (1873) (Martinique).—CORY, List Bds. W. I. p.
    16 (1885).

SP. CHAR. *Male:*—Upper surface a mixture of gray, rufous, dark brown,
    and white, the feathers mottled and edged with the different colors;
    the two central tail-feathers gray, curiously marked with dark brown,
    rest of the tail feathers white, edged with brown on the outer webs,
    and banded near the centre, the brown lacking on the outer web of
    the outer feather; under surface of tail-feathers white, showing a
    band of brown across the centre; throat and abdomen white; breast
    heavily mottled with rufous; wings dark brown, the coverts mottled
    with rufous and blotches of white; a heavy band of white crossing
    the middle of the primaries.
        Female.—Entirely lacks the white markings on the wings and
    tail; the general plumage is dull brown, variously marked with

brown and rufous; the under surface being dull rufous, narrowly banded with brown; wings and tail brown, marked with rufous.

Length (skin), 8.75; wing, 5.50; tail, 4.50; tarsus, .60.

It is claimed that this species occurs in the Lesser Antilles. A specimen in my collection is labelled "Trinidad," and Messrs. Sclater and Salvin give it from Martinique.

Genus **Siphonorhis** Scl.

Siphonorhis Sclater, P. Z. S. 1861, p. 77.

Siphonorhis americanus (Linn.).

'*Caprimulgus jamaicensis* Briss. Av. II, p. 480."
Caprimulgus americanus Linn. Syst. Nat. I, p. 346 (1766).
Chordeiles americanus Br. Consp. I. p. 63 (1850).
Siphonorhis americanus Scl. P. Z. S. 1861, p. 77; *ib.* Cat. Am. Bds. p. 282 (1862).—Albrecht, J. f. O. 1862, p. 199.—March, Pr. Acad. Nat. Sci. Phila. 1863, p. 286.—Gray. Handl. Bds. I. p. 60 (1869).—Scl. & Salv. Nom. Avium Neotr. p. 97 (1873).—A. & E. Newton, Handb. Jamaica, p. 109 (1881).
Siphonorhis americana Cory, List Bds. W. I. p. 16 (1885).

Sp. Char. *Male:*—General plumage above rufous brown, mottled and streaked with gray, dull white, bright rufous, and dark brown; a patch of dull white on the throat; breast rufous, delicately dotted and lined with brown; feathers of the underparts broadly tipped with dull white; tail dull rufous, streaked and marked with brown, showing a sub-terminal band of brown, the feathers tipped with white; primaries dark brown, broadly dotted with rufous on the outer webs, showing various markings of rufous on the inner webs.

Length (skin), 9; wing, 5; tail, 4.75; tarsus, .90.

Habitat. Jamaica.

Family CYPSELIDÆ.

Genus **Cypselus** Illig.

Cypselus Illiger, Prodr. Syst. Mamm. et Avium, p. 229 (1811).

Cypselus phœnicobius (Gosse).

Tachornis phœnicobia Gosse, Bds. Jam. p. 58 (1847).—Gundl. J. f. O. 1856, p. 5.—Albrecht, J. f. O. 1862, p. 194.—March, Pr. Acad. Nat. Sci. Phila. 1863, p. 287.—A. & E. Newton, Handb. Jamaica, p. 108 (1881).

Cypselus iradii LEMB. Aves Cuba, p. 50 (1850).
Cypselus phœnicobia BR. Consp. I, p. 66 (1850).
Cypselus cayenuensis SALLÉ. P. Z. S. 1857, p. 232.
Tachornis gradii BREWER. Pr. Bost. Soc. Nat. Hist. VII, p. 306 (1860).
Cypselus phœnicobius SCL. P. Z. S. 1865, p. 634.—SCL. & SALV. Nom.
    Avium Neotr. p. 94 (1873).—CORY. Bull. Nutt. Orn. Club, VI, p.
    153 (1881); *ib.* Bds. Haiti & San Domingo, p. 87 (1885); *ib.* List
    Bds. W. I. p. 17 (1885).
Tachornis iradii GUNDL. Repert. Fisico-Nat. Cuba, I, p. 282 (1866); *ib.*
    J. f. O. 1874, p. 116.
Cypselus cayanensis BRYANT, Pr. Bost. Soc. Nat. Hist. XI, p. 95 (1866).
Tachornis phœnicobius GRAY, Handl. Bds. I, p. 64 (1869).

SP. CHAR.   *Male:*—General plumage dull greenish black; throat, rump,
    abdomen, and a narrow line in the centre of the belly white; bill
    and feet black.
        The sexes are similar.
        Length, 3.75; wing, 3.70; tail, 1.75; tarsus, .20; bill, .15.
    HABITAT.   Cuba, Jamaica, Haiti, and San Domingo.

GENUS Cypseloides STREUBEL.

Cypseloides STREUBEL, Isis, 1848, p. 360.

Cypseloides niger (GMEL.).

Hirundo niger GMEL. Syst. Nat. I, p. 1025 (1788).
Cypselus niger GOSSE, Bds. Jam. p. 63 (1847).—GUNDL. & LAWR. Ann.
    N. Y. Lyc. VI, p. 268 (1858).—BREWER, Pr. Bost. Soc. Nat. Hist.
    VII, p. 306 (1860).—ALBRECHT, J. f. O. 1861, p. 207; *ib.* 1862, p.
    194. — MARCH, Pr. Acad. Nat. Sci. Phila. 1863, p. 287.—A. & E.
    NEWTON, Handb. Jamaica, p, 108 (1881).
Cypselus nigra BP. Consp. I, p. 66 (1850).
Cypselus borealis KENN. Pr. Acad. Nat. Sci. Phila. 1857, p. 202.—SCL.
    P. Z. S. 1865, p. 615.
Nephocætes niger BAIRD, CASS. & LAWR. Bds. N. Am. p. 142 (1858).—
    GRAY, Handl. Bds. I, p. 68 (1869).—GUNDL. J. f. O. 1874, p. 115;
    ib. Anal. Soc. Esp. Hist. Nat. VII, p. 200 (1878).
Cypseloides niger SCL. P. Z. S. 1865, p. 615.—SCL. & SALV. Nom. Avium
    Neotr. p. 95 (1873).—LAWR. Pr. U. S. Nat. Mus. I, pp. 459, 487
    (1878).—SCL. Ibis, 1880, p. 74.
Nechopætes niger GUNDL. Repert. Fisico-Nat. Cuba, I, p. 281 (1866.)
Nephœcetes niger COOPER, Orn. Cal. I, p. 349 (1870).—HD. BWR. &
    RIDGW. Hist. N. Am. Bds. II, p. 429 (1874).—CORY, Bds. Haiti &
    San Domingo, p. 88 (1885); *ib.* List Birds W. I. p. 17 (1885).

SP. CHAR.   *Male:*—Entire plumage dark brown, showing slight greenish reflections when held to the light; forehead slightly washed with white; a dark spot in front of the eye; bill and feet black.

The sexes are similar.

Length, 6; wing. 6; tail, 2.50; tarsus, .40; bill, .20.

HABITAT. San Domingo, Jamaica, Cuba, Porto Rico, and Guadeloupe.

GENUS **Chætura** STEPH.

Chætura STEPHENS, Shaw's Gen. Zool. Birds, XIII, pt. ii, p. 76 (1825).

Chætura dominicana LAWR.

Chætura poliura LAWR. Pr. U. S. Nat. Mus. I, p. 62 (1878).
Chætura dominicana LAWR. Ann. N. Y. Acad. Sci. I, p. 255 (1878); *ib.*
Pr. U. S. Nat. Mus. I, p. 487 (1878).—SCL. Ibis, 1880, p. 75.—CORY,
List Bds. W. I. p. 17 (1885).

SP. CHAR.—Entire upper surface dark brown, almost black, showing a faint olive tinge to the feathers when held in the light; under portions dark smoky brown, palest on the throat; wings and tail dark brown; rump lighter than than the back; bill and feet black.

Length, 4; wing. 3.80; tail, 2.

HABITAT. Dominica.

GENUS **Hemiprocne** NITZSCH.

Hemiprocne NITZSCH, Pterylogr. p. 123 (1840).

Hemiprocne zonaris (SHAW).

Hirundo zonaris SHAW, in Mill. Cim. Phys. pl. 55.
Hirundo albicollis VIEILL. Nouv. Dict. XIV, p. 524.
Cypselus collaris TEMM. Pl. Col. p. 195 (1820-39).—MAX. Beitr. III. p. 344 (1831).—BREWER, Pr. Bost. Soc. Nat. Hist. VII, p. 306 (1860).—ALBRECHT. J. f. O. 1861, p. 206.—GUNDL., J. f. O. 1874, p. 114.
Hemiprocne collaris NITZSCH, Pterylogr. p. 123 (1840).

Pallene collaris BOIE, Isis, 1844, p. 168.

Acanthylis collaris GOSSE. Bds. Jam. p. 51 (1847).—GRAY. List Sp. Fiss-
p. 15.—BR. CONSP. I. p. 64 (1850).—BURM. Syst. Ueb. II, p. 364.

Hemiprocne torquata STREUBEL, Isis, 1848. p. 362.

Acanthylis albicollis SCL. P. Z. S. 1854. p. 10.

Hemiprocne zonaris SCL. & SALV. Ibis. 1860, p. 37.—CAB. & HEIN. Mus.
Hein. III, p. 84 (1860).—SCL. & SALV. Nom. Avium Neotr. p. 95
(1873).—CORY, List Bds. W. I. p. 17 (1885).

Chætura zonaris SCL. P. Z. S. 1861, p. 79; *ib.* Cat. Am. Bds. p. 282
(1862); *ib.* P. Z. S. 1865, p. 609.—ALBRECHT. J. f. O. 1862, p. 201.

Nephocætes collaris GUNDL. J. f. O. 1862, p. 177 (?); *ib.* Contrib. Orn.
Cuba, p. 83 (1876).

Chætura collaris MARCH, Pr. Acad. Nat. Sci. Phila. 1863, p. 286.

Acanthyllis zonaris A. & E. NEWTON, Handb. Jamaica. p. 108 (1881).

SP. CHAR.   *Male:*—Very large.   Entire plumage brownish-black, deepest
on the back, and showing a tinge of bluish when held in the light,
lightest on the throat and primaries; an unbroken collar of white
passes around the neck.

The female seems to be similar, but some specimens show more
white where the collar touches the breast.

Length (skin), 7.50; wing. 8; tail, 2.75.

HABITAT.   Jamaica and Cuba.   San Domingo?

FAMILY TROCHILIDÆ.

GENUS Glaucis BOIE.

Glaucis BOIE, Isis, 1831, p. 545.

Glaucis hirsuta (GMEL.).

Trochilus hirsutus GMEL. Syst. Nat. I, p. 490 (1788).

Trochilus brasiliensis LATH. Ind. Orn. I, p. 308 (1790).

Trochilus ferrugineus WEID, Beitr. IV, p. 120, Sp. 21.

Trochilus mazeppa LESS. Troch. p. 18, pl. 3 (1831).

Trochilus superciliosus LESS. Colib. t. 6 (1831) ♀ ?

Glaucis hirsuta BOIE, Isis, 1831, p. 545.—REICH. Aufz. Colib. p. 15 (1853).
—BP. Rev. Mag. Zool. 1854, p. 249.—CAB. & HEIN. Mus. Hein, III,
p. 4 (1860).—GOULD, Mon. Troch. I, pl. 5 (1861).—SALV. &
ELLIOT, Ibis, 1873, p. 276.—MULS. Hist. Nat. Ois. Mouch. I, p. 39
—ELLIOT, Mon. Troch. p. 6 (1878).—CORY, List Bds. W. I. p. 17
(1885).

Trochilus dominicus LICHT. (*nec* Linn.) Doubbt. p. 10, Sp. 110.

Polytmus hirsutus GRAY, Gen. Bds. I, p. 108 (1844).

Glaucis mazeppa REICH. Aufz. Colib. p. 15 (1853).—BY. Rev. Mag. Zool.
　　1854, p. 249.—GOULD. Mon. Troch. I, pl. 6 (1861).
Glaucis melanura GOULD, P. Z. S. 1860, p. 364; *ib.* Mon. Troch. I, pl. 9
　　(1861).
Glaucis lanceolata GOULD, Mon. Troch. I, pl. 8 (1861).
Glaucis ænea LAWR. Pr. Acad. Nat. Sci. Phila. 1867, p. 232.
Glaucis hirsutus SCL. & SALV. Nom. Avium Neotr. p. 78 (1873).—LAWR.
　　Pr. U. S. Nat. Mus. I, pp. 271, 487 (1878).

SP. CHAR.　*Male:*—Bill stout, long and curved; the upper mandible dark,
　　the lower mandible light; top of head dull brown; back green,
　　the feathers delicately edged with rufous; tail bronze green on the
　　central feathers, the rest rufous, showing a sub-terminal bar of
　　greenish brown, and all the feathers tipped with white; underparts
　　dull rufous; the throat showing greenish feathers in places.
　　　Female:—Similar to the male, but lacks the mottling on the
　　throat, the entire surface being rufous.
　　　Length (skin), 4.75; wing, 2; tail, 1.75; bill, 1.25.
　HABITAT.　Grenada.

GENUS **Lampornis** SWAINS.

Lampornis SWAINSON, Zool. Journ. III, p. 358 (1827).

Lampornis dominicus (LINN.).

Trochilus dominicus LINN. Syst. Nat. I, p. 191 (1766).—GMEL. Syst. Nat.
　　I, p. 489 (1788).—LATH. Ind. Orn. I, p. 309 (1790).
Trochilus margaritaceus GMEL. Syst. Nat. I, p. 490 (1788).
Trochilus auraleutus VIEILL. Ois. Dor. pl. XII (1802).—SHAW, Gen.
　　Zool. VIII, p. 306 (1811).
Polytmus margaritaceus GRAY, Gen. Bds. I. p. 108 (1844).
Lampornis margaritaceus BP. Consp. I, p. 72 (1850).
Eulampis auruleutis BP. Consp. I, p. 71 (1850); *ib.* Rev. Mag. Zool.
　　1854, p. 250.
Margarochrysis aurulenta REICH. Aufz. Colib. p. 11 (1853).
Hypophania dominica REICH. Aufz. Colib. p. 11 (1853).
Lampornis aurulenta SALLÉ, P. Z. S. 1857, p. 233.
Lampornis aurulentus CASSIN, Pr. Acad. Nat. Sci. Phila. 1860, p. 377.—
　　GOULD, Mon. Troch. II, pl. 79 (1861).—MULS. Hist. Nat. Ois.
　　Mouch. I, p. 152.—GUNDL. Anal. Soc. Esp. Hist. Nat. VII, p. 223
　　(1878).—CORY, Bull. Nutt. Orn. Club, VI, p. 153 (1881).—TRIS-
　　TRAM, Ibis, 1884, p. 168.
Lampornis virginalis GOULD, Mon. Troch. II, pl. 80 (1861).
Tro hilus (Lampornis) aurulentus BRYANT, Pr. Bost. Soc. Nat. Hist.
　　XI, p. 95 (1866).—SUNDEV. Oefv. K. Vet. Akad. För. 1869, p.
　　600.

Lampornis dominicus ELLIOT, Ibis, 1872, p. 349; *ib.* Mon Troch. p. 41
(1878). — CORY. Bds. Haiti & San Domingo, p. 90 (1885); *ib.*
List Bds. W. I. p. 17 (1885).

SP. CHAR.   *Male:*— Entire upper parts yellowish green; throat bright
golden green; breast and belly purplish black; flanks green, showing a spot of white; under tail-coverts dark purple; wings purplish
brown; outer tail-feathers violet-purple, bordered with steel blue;
median feathers bronze green; bill and feet black.

Female:—Underparts dull gray, whitening on the throat; tail
tipped with white; rest as in the male.

Immature specimens have the underparts dull brownish white,
with a line of metallic green passing down the middle of the throat,
continuing in a line of black down the middle of the breast and
abdomen to the vent.

Length, 4.90; wing, 2.60; tail, 1.85; bill, .93.

HABITAT.   Haiti, San Domingo, Porto Rico, and St. Thomas?

Lampornis viridis (VIEILL.).

Trochilus viridis AUD. & VIEILL. Ois. Dor. I, p. 34 (1802).—SUNDEV.
Oefv. K. Vet. Akad. För. 1869, p. 600.
Lampornis viridis BP. Consp. I, p. 71 (1850).—GOULD, Mon. Troch. II,
pl. 78 (1861).—SCL. & SALV. Nom. Avium Neotr. p. 81 (1873).—
GUNDL. Anal. Soc. Esp. Hist. Nat. VII. p. 222 (1878).—CORY, List
Bds. W. I. p. 17 (1885).
Chalybura viridis REICH. Aufz. Colib. p. 10 (1853).
Agyrtria viridis REICH. Troch. Enum. p. 7 (1855).

SP. CHAR.   *Male:*—General plumage bright green, showing a bluish
tinge on the under surface when held in the light; tail steel blue;
wings dark brown; bill black.

Female:— Upper surface bright golden green; head brownish;
underparts dull ashy-white, tinged with green on the sides and
flanks; central tail-feathers bronze green, rest of tail-feathers showing dark blue on their inner webs, and golden brown on the outer,
all of the feathers narrowly tipped with white.

Length (skin), 4; wing, 2.50; tail. 1.75.

HABITAT.   Porto Rico.

Lampornis mango (LINN.).

Trochilus mango LINN. Syst. Nat. I, p. 191 (1766).—GMEL. Syst. Nat. I,
p. 491 (1788).
Trochilus porphyrurus SHAW, Nat. Misc. IX, p. 333.
Polytmus porphyrurus GRAY, Gen. Bds. I, p. 108 (1844).

Lampornis mango GOSSE, Bds. Jam. p. 88 (1847).—BP. Consp. I, p. 72
(1850).—GOULD, Mon. Troch. II, pl. 74 (1861).—MARCH, Pr. Acad.
Nat. Sci. Phila. 1863, p. 284.—ELLIOT, Ibis, 1872, p. 350.—SCL. &
SALV. Nom. Avium Neotr. p. 81 (1873).—ELLIOT, Mon. Troch. p.
39 (1878).—A. & E. NEWTON, Handb. Jamaica, p. 108 (1881).—
CORY, List Bds. W. I. p. 17 (1885).

Floresia porphyrura REICH. Aufz. Colib. p. 11 (1853).

Lampornis floresi BP. Rev. Mag. Zool. 1854, p. 250.

Anthracothorax porphyrurus REICH. Troch. Enum. p. 8 (1855).

Lampornis porrhyrura CAB. & HEIN. Mus. Hein. III, p. 19 (1860).—SCL.
P. Z. S. 1861, p. 79.— ALDRECHT, J. f. O. 1862, p. 201.

Lampornis porphyrurns GOULD, Mon. Troch. II, pl. 81 (1861).—MULS.
Hist. Nat. Ois. Mouch. I, p. 163.

Eudoxa porphyrura HEINE, J. f. O. 1863, p. 179.

SP. CHAR. *Male:*—Upper parts olive green; a golden brownish tinge
on the back, showing a gloss of purple on the nape; a broad band
of metallic purple passes from the bill on each side of the neck;
throat and upper breast greenish black; dull black on the belly; tail-
feathers purple, edged with steel blue, two central feathers dull
black.

The female of this species is described as having the chin green
and the throat greenish purple.

Length, 4.85; wing, 2.70; tail, 1.90; bill, .88.

HABITAT. Jamaica.

GENUS **Eulampis** BOIE.

Eulampis BOIE, Isis, 1831, p. 547.

Eulampis jugularis (LINN.).

Trochilus jugularis LINN. Syst. Nat. I, p. 190 (1766).—LATH. Ind. Orn.
I, p. 305 (1790).

Trochilus auratus GMEL. Syst. Nat. I, p. 487 (1788).

Trochilus violaceus GMEL. Syst. Nat. I, p. 488 (1788).

Trochilus venustissimus GMEL. Syst. Nat. I, p. 490 (1788).

Trochilus cyanomelas GMEL. Syst. Nat. I, p. 498 (1788).

Trochilus cyaneus LATH. Ind. Orn. I, p. 309 (1790).

Trochilus granatinus LATH. Ind. Orn. I, p. 305 (1790).

Trochilus bancrofti LATH. Ind. Orn. I, p. 317 (1790).

Sauimanga prasinoptère VIEILL. Ois. Dor. II, p. 65 (1802).

"*Certhia prasinoptera* SPARR. Mus. Carls. t. 81"?

Trochilus (Eulampis) auratus LESS. Syn. Genr. Troch. p. 7 (1831).

Polytmus jugularis GRAY, Gen. Bds. I, p. 108 (1844).

Topaza violacea GRAY, Gen. Bds. I, p. 110 (1844).

Eulampis jugularis BP. Consp. I, p. 72 (1850).—REICH. Aufz. Colib. p.

11 (1853).— Cab. & Hein. Mus. Hein. III, p. 17 (1860).—Gould.
Mon. Troch. II, pl. 82 (1861).—Taylor, Ibis, 1864, p. 169.—Scl.
P. Z. S. 1871, p. 272.—Elliot, Ibis, 1872, p. 352.—Muls. Hist. Nat.
Ois. Mouch. II, p. 131.—Scl. & Salv. Nom. Avium Neotr. p. 81
(1873).—Elliot, Mon. Troch. p. 43 (1878).—Lawr. Pr. U. S. Nat.
Mus. I, pp. 60, 192, 358, 458, 487 (1878).—Allen, Bull. Nutt. Orn.
Club, V, p. 167 (1880). — Lister, Ibis, 1880, p. 42. — Grisdale,
Ibis, 1882, p. 486.—Cory, List Bds. W. I. p. 17 (1885).

Sp. Char. *Male:*—Upper surface velvet black; wings metallic green;
upper and under tail-coverts bright metallic bluish green; entire
throat including the chin beautiful purple, dull golden in some
lights; tail bluish green; bill black.

    Length, 4.50; wing, 3.05; tail, 1.65; bill .90.

Habitat.   Lesser Antilles.

Eulampis holosericeus (Linn.).

Trochilus holosericeus Linn. Syst. Nat. I, p. 191 (1766).— Lath. Ind.
   Orn. I, p. 305 (1790).—Less. Colib. p. 76 (1831).—Sundev. Oefv.
   K. Vet. Akad. För. 1869, p. 585.
Polytmus holosericeus Gray. Gen. Bds. I, 108 (1844).
Eulampis holosericeus Br. Consp. I, p. 72 (1850).—Cassin, Pr. Acad.
   Nat. Sci. Phila. 1860, p. 377.—Gould, Mon. Troch. II, pl. 83 (1861).
   —Taylor, Ibis, 1864, p. 170.—Scl. P. Z. S. 1871, p. 272; ib. 1874,
   p. 175.—Elliot, Ibis, 1872. p. 352.—Scl. & Salv. Nom. Avium
   Neotr. p. 81 (1873).—Muls. Hist. Nat. Ois. Mouch. I, p. 134.—
   Lawr. Pr. U. S. Nat. Mus. I, pp. 60, 192, 234, 272, 358, 458, 487
   (1878).—Elliot, Mon. Troch. p. 42 (1878).—Allen, Bull. Nutt.
   Orn. Club, V, p. 167 (1880).—Lister, Ibis, 1880, p. 42.—Grisdale,
   Ibis, 1882, p. 486.—Cory, List Bds. W. I. p. 17 (1885).
Sericotes chlorolæmus Reich. Aufz. Colib. p. 11 (1853).
Sericotes holosericeus Reich. Aufz. Colib. p. 11 (1853).
Eulampis chlorolæmus Br. Rev. Mag. Zool. 1854, p 250.—Gould, Mon.
   Troch. II, pl. 84 (1861).
Anthracothorax (Sericotes) holosericeus Reich. Troch. Enum. p. 9 (1855).
Anthracothorax chlorolæmus Reich. Troch. Enum. p. 9 (1855).
Lampornis holosericeus Cab. & Hein. Mus. Hein. III. p. 19 (1860).—
   Gundl. Anal. Soc. Esp. Hist. Nat. VII. p. 224 (1878).
L. mpornis chlorolæmus Cab. & Hein. Mus. Hein. III, p. 19 (1860).
Eulampis longirostris Gould, Intr. Troch. octavo ed. p. 69 (1861).
Trochilus (Lampornis) holosericeus Sundev. Oefv. K. Vet. Akad. För
   1869, p. 600.

Sp. Char. *Male:*—Upper plumage dark green, with a slight golden
tinge on the back; throat bright green, ending with a patch of blue

on the breast; belly greenish black; upper and under tail-coverts
bright bluish-green; tail dark blue, showing slight purple reflections.

The sexes are described as similar.

Length, 4.30; wing, 3; tail, 1.50; bill, .85.

HABITAT. Lesser Antilles.

GENUS Aithurus CAB. & HEIN.

Aithurus CAB. & HEIN. Mus. Hein. III, p. 50 (1860).

Aithurus polytmus (LINN.).

Trochilus polytmus LINN. Syst. Nat. I, p. 189 (1766).—GMEL. Syst. Nat. I,
  p. 486 (1788).—LATH. Ind. Orn. I, p. 302 (1790).—GOSSE, Bds. Jam.
  p. 97 (1847).—GOULD, Mon. Troch. IV, pl. 98 (1861).
Ornismya cephalatra LESS. Ois. Mouch. p. 78 (1829).
Trochilus maria HILL, Ann. Mag. Nat. Hist. III, p. 258 (1849).—GOSSE,
  Ill. Bds. Jam. pl. 22.
Polytmus cephalatra BP. Consp. I, p. 72 (1850).
Polytmus cephalater BP. Rev. Mag. Zool. 1854, p. 254.—SCL. P. Z. S.
  1861, p. 79.—ALBRECHT, J. f. O. 1862, p. 201.
Aithurus polytmus CAB. &. HEIN. Mus. Hein. III, p. 50 (1860).—GOULD,
  Intr. Troch. octavo ed. p. 75 (1861).—MARCH, Pr. Acad. Nat. Sci.
  Phila. 1863, p. 284.—GRAY, Handl. Bds. I, p. 134 (1869).—SCL. &
  SALV. Nom. Avium Neotr. p. 82 (1873).—ELLIOT, Mon. Troch. p.
  96 (1878).—A. & E. NEWTON, Handb. Jamaica, p. 108 (1881).—
  CORY, List Bds. W. I. p. 17 (1885).
Aithurus fuliginosus MARCH, Pr. Acad. Nat. Sci. Phila. 1863, p. 285.—
  GRAY, Handl. Bds. I, p. 134 (1869).

SP. CHAR. *Male:*—Top of head, with elongated feathers, velvety black;
back dark green; throat and underparts bright green; wings
brown, with a tinge of purple; tail black, the two long tail-feathers
showing peculiar uneven edging of the webs; bill dull red.

Female:—Upper parts green, brownish on the head; underparts
white, tinged with green on the sides and flanks; middle tail-feathers green, bluish near the tip, outer tail-feathers heavily tipped with
white, wanting on the two middle feathers.

Length, 8.50; wing, 2.50; tail, 5.50; bill, .80.

HABITAT. Jamaica.

GENUS Thalurania GOULD.

Thalurania GOULD, P. Z. S. 1848, p. 13.

Thalurania bicolor (GMEL.).

Trochilus bicolor GMEL. Syst. Nat. I. p. 496 (1788).—VIEILL. Ois. Dor. p. 75 (1802).

Ornismya wagleri LESS. Hist. Ois. Mouch. p. 203 (1829).

Hylocharis wagleri GRAY, Gen. Bds. I. p. 114 (1844).

Thalurania wagleri BP. Consp. I, p. 77 (1850).—REICH. Aufz. Colib. p. 7 (1853).—CAB. & HEIN. Mus. Hein. III, p. 24 (1860).—GOULD, Mon. Troch. II, pl. 109 (1861).—SCL. & SALV. Nom. Avium Neotr. p. 83 (1873).—SALV. & ELLIOT, Ibis, 1873, p. 360.—LAWR. Ann. N. Y. Acad. Sci. I, p. 46 (1878); *ib.* Pr. U. S. Nat. Mus. I, pp. 61, 487 (1878).

Thalurania bicolor GRAY, Handl. Bds. I, p. 130 (1869).—ELLIOT, Mon. Troch. p. 102 (1878).—CORY, List Bds. W. I. p. 17 (1885).

SP. CHAR. *Male:*—Entire head and throat deep blue, but very slightly metallic; back dark green; breast and abdomen metallic golden-green; tail-coverts greenish-blue; tail steel blue; wings brown; upper mandible black; under mandible flesh-color, tipped with black.

Female:—Upper surface green, showing slight bronze reflections; underparts dull white, marked with green on the flanks and sides; outer tail-feathers tipped with white; rest of tail-feathers green, broadly marked with blue on the terminal portion.

Length, 3.80; wing, 2.35; tail, 1.70; bill, .60.

HABITAT. Dominica.

GENUS Trochilus LINN.

Trochilus LINNÆUS, Syst. Nat. I, p. 189 (1766).

Trochilus colubris LINN.

Trochilus colubris LINN. Syst. Nat. I, p. 191 (1766).—CAB. J. f. O. 1856, p. 98 (Cuba).—GUNDL. Repert. Fisico-Nat. Cuba, I, p. 291 (1866); *ib.* J. f. O. 1874, p. 141 (Cuba); *ib.* 1878, p. 159 (Porto Rico); *ib.* Anal. Soc. Esp. Hist. Nat. VII, p. 221 (1878) (Porto Rico). — ELLIOT, Mon. Troch. p. 105 (1878) (Bahamas).—CORY, List Bds. W. I. p. 18 (1885).

Orthorhynchus colubris D'ORB. in La Sagra's Hist. Nat. Cuba, Ois. p. 126 (1840).

Melisuga colubris BREWER, Pr. Bost. Soc. Nat. Hist. VII, p. 306 (1860) (Cuba).

Recorded from Bahamas, Cuba, and Porto Rico.

GENUS Mellisuga BRISS.

Mellisuga BRISSON, Orn. III, p. 695 (1760).

Mellisuga minima (LINN.).

Trochilus minimus LINN. Syst. Nat. I, p. 193 (1766).—GMEL. Syst. Nat. I,
    p. 500 (1788).—LATH. Ind. Orn. I, p. 320 (1790).
Trochilus minutulus VIEILL. Ois. Am. Sept. II, p. 73 (1807).
Trochilus vieilloti SHAW, Gen. Zool. VIII, p. 347 (1812).
Ornismya minima LESS. Ois. Mouch. pl. 79 (1829).
Hylocharis nigra GRAY. Gen. Bds. I, p. 114 (1844).
Mellisuga humilis GOSSE, Bds. Jam. p. 127 (1847).—MARCH, Pr. Acad.
    Nat. Sci. Phila. 1863. p. 285.
Trochilus catherinæ SALLÉ, Rev. Zool. 1849, p. 498.
Hylocharis niger BP. Consp. I, p. 81 (1850).
Mellisuga minima BP. Consp. I, p. 81 (1850).—REICH. Aufz. Colib. p. 6
    (1853).—SALLÉ, P. Z. S. 1857, p. 233.—GOULD, Mon. Troch. III, p.
    133.—MARCH. Pr. Acad. Nat. Sci. Phila. 1863, p. 285.—MULS. Hist.
    Nat. Ois. Mouch. IV, p. 82 (1877).—ELLIOT, Mon. Troch. p. 103
    (1878).—A. & E. NEWTON, Handb. Jamaica, p. 108 (1881).—CORY,
    Bull. Nutt. Orn. Club, VI, p. 153 (1881); *ib.* Bds. Haiti & San
    Domingo, p. 92 (1885); *ib.* List Bds. W. I. p. 18 (1885).
Mellisuga humila ALBRECHT, J. f. O. 1862. p. 201.
Trochilus (Mellisuga) minimus BRYANT, Pr. Bost. Soc. Nat. Hist. XI, p.
    95 (1866).

SP. CHAR. *Male:*—Above bright green; flanks green; throat dull white,
spotted with brown, heaviest on the lower part; underparts white;
under tail-coverts tipped with green; tail black; bill and feet black.
    Female:—Resembles the male, but lacks the spots on the throat;
lateral tail-feathers tipped with white.
    Length, 2.70; wing, 1.50; tail, .60; bill, .45.

HABITAT. Jamaica, Haiti, and San Domingo.

GENUS Calypte GOULD.

Calypte GOULD, Intr. Troch. octavo ed. p. 87 (1861).

Calypte helenæ (GUNDL.)

Orthorhynchus helenæ GUNDL. in LEMB. Aves Cuba, p. 70 (1850).—BREWER, Pr.
    Bost. Soc. Nat. Hist. VII, p. 306 (1860).
Orthorhynchus boothi CAB. J. f. O. 1856, p. 99.—GUNDL. J. f. O. 1859, p.
    347.
Calypte helenæ GOULD, Mon. Troch. III, pl. 136 (1861).—GRAY, Handl.
    Bds. I, p. 145 (1869).—ELLIOT, Ibis, 1872, p. 354.—GUNDL. J. f. O.
    1874. p. 144.—MULS. Hist. Nat. Ois. Mouch. IV, p. 77 (1877).—
    ELLIOT, Mon. Troch. p. 108 (1878).—CORY, List Bds. W. I. p. 18
    (1885).

Sp. Char.   *Male:*—Head, throat, and elongated feathers of the neck
metallic red, almost pink in some lights; upper parts greenish-blue,
becoming steel-blue on the tail; breast grayish-white; belly and
flanks greenish; wings purplish-brown.

> *Female:*—Head dull brown; back green, shading into blue on the
> lower part; underparts grayish-white; tail bluish-green, outer
> feathers tipped with white.

> Length (skin), 2.50; wing, 1.25; tail, .80; bill, .45.

Habitat.   Cuba.

Genus Doricha Reich.

Doricha Reich. Aufz. der Colib. p. 12 (1853).

Doricha evelynæ (Bourc.).

Calothorax evelynæ Gray, Gen. Bds. I, p. 110 (1844).—Reich. Aufz.
    Colib. p. 13 (1853).—Gould, Mon. Troch. III, pl. 156 (1861).
Trochilus evelynæ Bourc. P. Z. S. 1847, p. 44.
Callothorax evillina Bp. Rev. Mag. Zool. 1854, p. 257.
Lucifer evelina Reich. Troch. Enum. p. 10 (1855).
Trochilus bahamensis Bryant, Pr. Bost. Soc. Nat. Hist. VII, p. 106
    (1859).
Doricha evelynæ Gould, Intr. Troch. octavo ed. p. 95 (1861).—Elliot,
    Ibis, 1872, p. 353.—Muls. Hist. Nat. Ois. Mouch. IV, p. 38 (1877).—
    Elliot, Mon. Troch. p. 125 (1878).—Cory, Bds. Bahama I. p. 108
    (1880); *ib.* List Bds. W. I. p. 18 (1885).

Sp. Char.   *Male:*—Above green, showing slight golden reflections on
the back, with the tips of the feathers, in some specimens, bluish;
head darker; throat beautiful purple-violet, below which is a band
of white; underparts green, mixed with rufous, shading into white
on the flanks; crissum pale rufous white; wings brownish purple;
tail appearing black, very dark purple in some lights; outer feathers
with faint terminal spot of rufous, second with inner web, and third
with inner and basal half of outer web cinnamon; bill and feet
black.

> *Female:*—Purple gorget wanting and replaced by dull white, with
> a slight tinge of rufous; upper parts paler than in the male; sides
> cinnamon, becoming brightest under the wings; central feathers of
> the tail bright green, the rest cinnamon; an oblique purplish band
> on the tips of the fourth feathers.

> Length, 3.40; wing, 1.70; tail, 1.40; tarsus, .15; bill, .70.

Habitat.   Bahamas.

Doricha lyrura Gould.

Doricha lyrura Gould, Ann. Mag. Nat. Hist. 4th ser. IV, pp. 111, 112
    (1869).—Elliot, Ibis, 1872, p. 354.—Scl. & Salv. Nom. Avium

Neotr. p. 85 (1873). — MULS. Hist. Nat. Ois. Mouch. IV, p. 41 (1877).—ELLIOT, Mon. Troch. p. 126 (1878).—CORY, Bds. Bahama I. p. 110 (1880); ib. List Bds. W. I. p. 18 (1885).

SP. CHAR. *Male:*—General appearance the same as *D. evelynæ*, but differs from it by showing the beautiful purple-violet on the forehead as well as on the throat, and also having a much longer tail, formed somewhat in the shape of a lyre, from which this bird has derived its name. The throat of *D. lyrura* shows bright blue in some lights on the lower part, while that of *D. evelynæ* is almost entirely purple-violet, showing the bluish tinge very slightly if at all.

Female:—Upper parts brownish-green; throat and upper breast dull white; rest of underparts pale rufous brown; central tail-feathers green, rest of tail-feathers pale rufous, showing a black band in the centre.

Length, 3.64; wing, 1.60; tail, 1.58; tarsus, .13; bill, .60.

HABITAT. Inagua and Long Island.

GENUS Bellona MULS. & VERR.

Bellona MULS. & VERR. Class. Troch. p. 75 (1865).

Bellona cristata (LINN.).

Trochilus cristatus LINN. Syst. Nat. I, p. 192 (1766).—GMEL. Syst. Nat. I, p. 498 (1788).—LATH. Ind. Orn. I, p. 317 (1790).—SCHOMB. Hist. Barb. p. 681.

Trochilus puniceus GMEL. Syst. Nat. I, p. 497 (1788).

Trochilus pileatus LATH. Ind. Orn. I, p. 318 (1790).

Ornismya cristata LESS. Troch. p. 20 (1831).

Mellisuga cristata GRAY, Gen. Bds. I, p. 113 (1844).

Orthorhynchus cristatus BP. Consp. I, p. 83 (1850).—REICH. Aufz. Colib. p. 11 (1853).—CAB. & HEIN. Mus. Hein. III. p. 61 (1860).—GOULD, Mon. Troch. IV, pl. 205 (1861).—ELLIOT, Ibis, 1872, p. 355.—SCL. P. Z. S. 1874, p. 175.—LAWR. Pr. U. S. Nat. Mus. I, pp. 272, 487 (1878).

Orthorhynchus ornatus GOULD, Mon. Troch. IV, pl. 206 (1861).—SCL. P. Z. S. 1871, p. 272.—ELLIOT. Ibis, 1872, p. 355.—LISTER, Ibis, 1880, p. 42.

Trochilus exilis SUNDEV. Oefv. K. Vet. Akad. För. 1869, p. 584.

Bellona cristata MULS. Hist. Nat. Ois. Mouch. III, p. 193 (1876). — ELLIOT, Mon. Troch. p. 178 (1878).—CORY, List Bds. W. I. p. 18 (1885).

Orthorhynchus exilis ALLEN, Bull. Nutt. Orn. Club, V, p. 167 (1880).

SP. CHAR. *Male:*—Forehead and crown bright golden-green, feathers lengthened, forming a crest, the green gradually fading and becoming dark blue on the crest; upper plumage green; throat dull smoke brown; underparts dull black, showing a faint purplish tinge; bill and feet black.

Female:—Upper parts bronze-green: underparts grayish-brown; central tail-feathers bronze-green, rest of tail-feathers brownish, tipped with ash on the outer feathers.

Length, 2.90; wing, 2; tail, 1.35; bill, .40.

HABITAT. Santa Lucia, Barbadoes, St. Vincent, Martinique, and St. Bartholomew.

Bellona exilis (GMEL.).

Trochilus exilis GMEL. Syst. Nat. I, p. 484 (1788).—LATH. Ind. Orn. I, p. 310 (1790).

Trochilus cristatellus LATH. Ind. Orn. Supp. p. 39 (1790).

Mellisuga exilis GRAY, Gen. Bds. I, p. 113 (1844).

Orthorhynchus chlorolophus BP. Consp. I. p. 83 (1850).

Orthorhynchus exilis REICH. Aufz. Colib. p. 11 (1853).—BP. Rev. Mag. Zool. 1854, p. 256.—A. & E. NEWTON, Ibis, 1859, p. 141.—GOULD, Mon. Troch. IV, pl. 207 (1861).—TAYLOR, Ibis, 1864, p. 170.—ELLIOT, Ibis, 1872, p. 355.—LAWR. Pr. U. S. Nat. Mus. I, pp. 234, 458, 487 (1878).—GRISDALE, Ibis, 1882, p. 486.

Trochilus (Orthorhynchus) exilis SUNDEV. Oefv. K. Vet. Akad. För. 1869, p. 600.

Bellona exilis MULS. Hist. Nat. Ois. Mouch. III. p. 196 (1876).—ELLIOT, Mon. Troch. p. 179 (1878).—CORY, List Bds. W. I. p. 18 (1885).

SP. CHAR. *Male:*—Forehead and crest bright golden-green, becoming darker green at the tip; upper parts dark green, tinged with bronze on the upper tail-coverts; throat smoke-brown, becoming dull purplish-black on the belly; tinged with green on the sides and flanks; central tail-feathers dull green, rest of tail-feathers dark purple.

Female:—Upper plumage dark green; underparts smoke-gray, tinged with green on the sides; wing-coverts bronzy-green; wings purplish-brown; two central tail-feathers dull green; outer tail-feathers tipped with gray.

Length, 3.40; wing, 2.05; tail, 1.45; bill, .55.

HABITAT. Porto Rico, St. Thomas, Dominica, St. Croix, Montserrat, Nevis, and Martinique.

GENUS Sporadinus BP.

Sporadinus BONAPARTE, Rev. Mag. Zool. 1854, p. 255.

Sporadinus elegans (VIEILL.).

Trochilus elegans VIEILL. Ois. Dor. I, p. 32 (1802).

Ornismya swainsonii LESS. Ois. Mouch. pp. 17, 197 (1829).

Hylocharis elegans GRAY, Gen. Bds. I. p. 114 (1844).
Lampornis elegans Br. Consp. I. p. 72 (1850).
Riccordia elegans REICH. Aufz. Colib. p. 8 (1853).
Sporadinus elegans Br. Rev. Mag. Zool. 1854, p. 255.—SALLÉ, P. Z. S.
    1857, p. 233.—GOULD, Mon. Troch. V, pl. 347 (1861).—SCL. & SALV.
    Nom. Avium Neotr. p. 94 (1873).—ELLIOT, Mon. Troch. p. 241 (1878).
    —CORY, Bull. Nutt. Orn. Club, IV. p. 153 (1881); *ib.* Bds. Haiti &
    San Domingo, p. 93 (1885); *ib.* List Bds. W. I. p. 18 (1885).
Chlorestes elegans REICH. Troch. Enum. p. 4 (1855).
Sporadicus elegans CAB. & HEIN. Mus. Hein. III, p. 25 (1860).
Trochilus (Sporadinus) elegans BRYANT, Pr. Bost. Soc. Nat. Hist. XI, p.
    95 (1866).

SP. CHAR. *Male:—*Upper parts bronze-green; throat bright metallic
    green; a portion of the breast black; wings brownish-purple; tail
    dark brown, with a bronze lustre on the upper surface; bill flesh
    color, tip black.
        *Female:—*Above bronze-green; top of head grayish; underparts
    brownish-gray; central tail-feathers bronze-green; rest of tail-feath-
    ers gray, with subterminal black bar; some of the feathers glossed
    with green.
        Length, 4; wing, 2.20; tail, 1.70: bill, .70.
    HABITAT.   Haiti and San Domingo.

Sporadinus riccordi (GERV.).

Trochilus riccordi GERV. Rev. Mag. Zool. 1835, pls. 41, 42.—Br. Consp.
    I, p. 81 (1850).
Ornismya parzudaki LESS. Rev. Zool. 1838, p. 315.
Orthorhynchus riccordi D'ORB. in La Sagra's Hist. Nat. Cuba, Ois. p. 128
    (1840).
Hylocharis riccordi GRAY, Gen. Bds. I, p. 114 (1844).
Riccordia raimondi REICH. Aufz. Colib. p. 8 (1853).
Sporadinus riccordi Br. Rev. Mag. Zool. 1854, p. 255.—GOULD, Mon.
    Troch. V, pl. 348 (1861).—MULS. Hist. Nat. Ois. Mouch. II, p. 74
    (1875).—ELLIOT, Mon. Troch. p. 241 (1878).—CORY, List Bds. W.
    I. p. 18 (1885).
Chlorestes raimondi REICH. Troch. Enum. p. 4 (1855)
Chlorestes riccordi GUNDL. J. f. O. 1856, p. 99.—BREWER, Pr. Bost. Soc.
    Nat. Hist. VII, p. 306 (1860).—GUNDL. Repert. Fisico-Nat. Cuba, I,
    p. 291 (1856); *ib.* J. f. O. 1874, p. 142.
Sporadicus riccordi CAB. & HEIN. Mus. Hein. III, p. 25 (1860).
Sporadinus ricordi SCL. & SALV. Nom. Avium Neotr. p. 94 (1873).—
    CORY, Bds. Bahama I. p. 111 (1880).
Sporadinus bracei LAWR. Ann. N. Y. Acad. Sci. I, p. 50 (1877).—CORY,
    Bds. Bahama I. p. 113 (1880).

SP. CHAR. *Male:*—Entire plumage bronzy green, becoming metallic on
the throat; wings purplish brown; four central tail-feathers bronze,
the remainder purplish black, showing bronze on the outer webs;
under tail-coverts white; upper mandible dark brown; lower mandible pale, becoming dark at the tip; tail forked.

   Female:—Resembles the male, except having the crown brownish;
throat and centre of abdomen pale buff; under tail-coverts grayish-
white.

   Length, 3.60; wing, 1.80; tail, 1.50; tarsus, .15; bill, .75.

HABITAT. Cuba and Bahamas.

Sporadinus maugæi (VIEILL.).

Trochilus maugæus VIEILL. Dict. Hist. Nat. VII, p. 568 (1817).
Ornismya maugei LESS. Ois. Mouch. p. 194 (1829).
Thaumatias ourissia BP. Consp. I. p. 73 (1850).
Sporadinus maugei BP. Rev. Mag. Zool. 1854, p. 255.—GOULD, Mon.
   Troch. V, pl. 349 (1861).—SCL. & SALV. Nom. Avium Neotr. p. 91
   (1873).—MULS. Hist. Nat. Ois. Mouch. II, p. 77 (1875).—ELLIOT,
   Mon. Troch. p. 242 (1878).—CORY, List Bds. W. I. p. 18 (1885).
Trochilus maugei SUNDEV. Oefv. K. Vet. Akad. För. 1869. p. 600.
Chlorestes gertrudis GUNDL. J. f. O. 1874, p. 315.
Chlorolampis gertrudis CAB. J. f. O. 1875, p. 223.
Sporadinus (Marsyas) maugei MULS. Cat. Ois. Mouch. p. 13 (1875).
Chlorolampis maugæus GUNDL. Anal. Soc. Esp. Hist. Nat. VII. p. 225
   (1878).

SP. CHAR. *Male:*—Entire plumage bright green, the feathers showing a
golden tinge when held in the light; throat dark blue, golden-green
in some lights; tail dark blue; wings dark brown.

   Female:—Underparts dull white; the central feathers of the tail
green, the rest grayish-green, with a band of blue near the tip;
outer feathers tipped with grayish-white.

   Length (skin), 3.35; wing, 2; tail, 1.25; bill, .55.

HABITAT. Porto Rico.

FAMILY TROGONIDÆ.

GENUS Priotelus GRAY.

Priotelus GRAY, List. Gen. Bds. 1840.

Priotelus temnurus (TEMM.).

Trogon temnurus TEMM Pl. Col.
No. 326 (1820-39). — VIGORS,
Zool. Journ. 1827, p. 443.—GOULD,
Mon. Trog. pt. II (1835). —
D'ORB. in La Sagra's Hist. Nat.
Cuba, Ois. p. 165(1840).—GUNDL.
Journ. Bost. Soc. Nat. Hist. VI, p.
319 (1857).

Priotelus temnurus BP. Consp. I,
p. 150 (1850).— GUNDL. J. f. O.
1856, p. 106.—BREWER, Pr. Bost.
Soc. Nat. Hist. VII, p. 307 (1860).
GRAY, Handl. Bds. I, p. 83. (1869)

—GUNDL. Repert. Fisico-Nat. Cuba, I, p. 298 (1866); *ib.* J. f. O.
1874, p. 165.

Prionteles temnurus SCL. & SALV. Nom. Avium Neotr. p. 103 (1873).
Prionoteles temnurus CORY, List Bds. W. I. p. 18 (1885).

SP. CHAR. *Male:*—Top of the head dark blue, purplish on the crown;
back bright green, showing a tinge of bluish in some lights; rump
showing distinctly bluish, green in some lights; throat white, shading into gray on the breast; belly and under tail-coverts bright red:
tail-feathers square at the tips, the two central feathers green on
the inner webs, bluish on the outer; rest of tail, except the
three outer feathers, blue; outer tail-feather having the terminal
half dull white, grayish on the outer web; basal half of the inner
web of outer tail-feather dark blue; the second and third feathers
nearly the same, but having the white on the outer web more restricted; on the third feather the white appears only in two or three
spots, but the terminal portion of the feather for an inch or more is
entirely dull grayish white; primaries dark brown, the feathers
heavily barred with white on the outer web; some of the coverts
also banded with white; lower mandible reddish; upper mandible
dark brown.

The sexes are similar.

Length, 10: wing, 5; tail, 5; tarsus, .50; bill, .62.

HABITAT. Cuba.

GENUS Temnotrogon BONAP.

Temnotrogon "BONAPARTE, Consp. Voluer. Zygodact. No. 8, p. 14, 1854."

Temnotrogon roseigaster (VIEILL.).

Couroucou à ventre rouge, de Saint Dominique, BUFF. Hist. Nat. Ois. VI,
p. 287 (1779).

"Le Couroucou à caleçon rouge, ou Le Couroucou Damioiseaux, LE VAILL.
-    Nat. Cour. pl. 13. p. 18."
Trogon roseigaster VIEILL. Ency. Méth. III. p. 1358 (1820).—GOULD,
        Mon. Trog. pl. 20 (1838).—Br. Consp. I. p. 149 (1850).—SALLÉ, P.
        Z. S. 1857, p. 235.— BRYANT, Pr. Bost. Soc. Nat. Hist. XI, p. 95
        (1866).
Trogon rhodogaster TEMM. Pl. Col. III (1820–39).
Temnotrogon roseigaster Br. Consp. Voluer. Zygodact. No. 8, p. 14
        (1854).—GRAY, Handl. Bds. I, p. 83 (1869).— CORY, Bds. Haiti
        & San Domingo, p. 95 (1885).— List Bds. W. I. p. 18 (1885)
Temnotrogon rhodogaster SCL. & SALV. Nom. Avium Neotr. p. 103
        (1873).

SP. CHAR.   *Male:*—Top of the head, back, and upper tail-coverts lustrous
            golden green; breast and throat gray, showing a tinge of green
            when held in the light; belly and under tail-coverts bright red;
            primaries and secondaries dark slaty brown, the outer webs barred
            with white; wing-coverts green, narrowly barred with white; under
            surface of tail dark blue, the three outer feathers having the outer
            webs and tips white, but showing a spot of black on the outer web
            near the tip; the inner webs of the two central tail-feathers dull
            greenish, extending nearly to the tip where it is replaced by the
            blue of the outer web; bill yellow; feet brownish.
        The sexes are similar.
        Length, 11; wing, 5.40; tail, 6.40; tarsus, .65; bill, .65.
HABITAT.   San Domingo.

FAMILY CUCULLIDÆ.

GENUS Crotophaga LINN.

Crotophaga LINNÆUS, Syst. Nat. I, p. 154 (1766).

Crotophaga ani LINN.

Crotophaga ani LINN. Syst. Nat. I, p. 154 (1766).—GOSSE, Bds. Jam. p.
        282 (1847).—Br. Consp. I, p. 99 (1850).—SALLE, P. Z. S. 1857, p.
        234.—CASSIN, Pr. Acad. Nat. Sci. Phila. 1860. p. 377.—BREWER, Pr.
        Bost. Soc. Nat. Hist. VII, p. 307 (1860).—ALBRECHT, J. f. O. 1862,
        p. 203.—MARCH, Pr. Acad. Nat. Sci. Phila. 1863, p. 153.—BRYANT,
        Pr. Bost. Soc. Nat. Hist. XI, p. 95 (1866).— SUNDEV. Oefv. K.
        Vet. Akad. För. 1869, p. 600.—SCL. & SALV. Nom. Avium Neotr. p.
        107 (1873).—BD. BWR. & RIDG. Hist. N. Am. Bds. II, p. 488 (1874).
        —GUNDL. J. f. O. 1874, p. 159; *ib.* Anal. Soc. Esp. Hist. Nat. VII, p.
        233 (1878).—LAWR. Pr. U. S. Nat. Mus. I, pp. 193, 273, 487 (1878).

—Allen, Bull. Nutt. Orn. Club, V. p. 169 (1880).—Lister, Ibis,
1880, p. 41.—Cory, Bds Bahama I. p. 118 (1880); *ib.* Bull. Nutt.
Orn. Club. VI, p. 154 (1881); *ib.* Bds. Haiti & San Domingo, p. 100
(1885); *ib.* List Bds. W. I. p. 18 (1885).—Ridgw. Pr. U. S. Nat.
Mus. VII, p. 172 (1884).

Crotophaga minor Less. Tr. Orn. p. 130 (1831).

Crotophaga lævirostris? Bryant, Pr. Bost. Soc. Nat. Hist. VII, p. 105
(1859).

Crotophaga rugirostris Gundl. Repert. Fisico-Nat. Cuba, I, p. 296 (1866).

Sp. Char. *Male:*—Upper mandible much curved; culmen rising above
the head, flattened to a sharp edge; nostrils situated in the middle
of the lower half of the upper mandible; general color black, showing bluish reflections; the feathers of the head, neck, breast, and
upper part of the back with metallic bronze borders; iris brown.

The sexes are similar.

Length, 12.25; wing, 6.20; tail, 7.50; tarsus, 1.50; bill, 1.10.

Habitat. West Indies.

Genus Saurothera Vieill.

Saurothera Vieillot, "Analyse, p. 36, 1816."

Saurothera vetula (Linn.).

Cuculus vetula Linn. Syst. Nat. I, p. 169 (1766).
Saurothera jamaicensis Lafr. Rev. Zool. 1847, p. 354.
Saurothera vetula Gosse, Bds. Jam. p. 273 (1847).—Br. Consp. I, p. 96
(1850).—Albrecht, J. f. O. 1862, 202.—Scl. Cat. Am. Bds. p. 323
(1862).—March, Pr. Acad. Nat. Sci. Phila. 1863, p. 283.—Gray,
Handl. Bds. II, p. 208 (1870).—Scl. & Salv. Nom. Avium Neotr.
p. 107 (1873).—A. & E. Newton, Handb. Jamaica, p. 109 (1881).—
Cory, List Bds. W. I. p. 18 (1885).
Coccygus vetula Schleg. Mus. Pays-Bas, I, p. 39 (1864).

Sp. Char.—Top of head, including the eye, dark olive brown; lighter
brown on the nape; rest of back and wings light gray; throat white;
breast and belly tinged with pale rufous; under surface of wings
chestnut rufous; primaries chestnut rufous, tipped with pale olive;
tail feathers, except central ones, bluish black, tipped with white.

The sexes are similar.

Length, 14.50; wing, 5; tail, 7.50; tarsus, 1; bill, 1.50.

Habitat. Jamaica.

Saurothera dominicensis Lafr.

Saurothera dominicensis Lafr. Rev. Zool. 1847, p. 355 —Salle. P. Z. S.
1857, p. 234.—Bryant, Pr. Bost. Soc. Nat. Hist. XI, p. 95 (1866).—

GRAY, Handl. Bds. II, p. 208 (1870).—SCL. & SALV. Nom. Avium
Neotr. p. 107 (1873).—CORY, Bull. Nutt. Orn. Club, VI, p. 154
(1881); *ib.* Bds. Haiti & San Domingo, p. 98 (1885); *ib.* List Bds.
W. I. p. 18 (1885)—TRISTRAM, Ibis, 1884, p. 168.
Coccygus dominicensis SCHLEG. Mus. Pays-Bas, I, p. 40 (1864).

SP. CHAR. *Male:*—Head, back, breast, and two central tail-feathers slate
color, darkest on the head, where it sometimes shows a faint brownish tinge, and lightest, being almost ashy on the breast; wing-
coverts and tertiaries slaty gray, showing pale greenish reflections
when held in the light; primaries and some of the secondaries bright
rufous brown, the first two edged with dull greenish, and all tipped
with the same color; outer tail-feathers bluish, tipped with white,
becoming dull olive at the base; two central tail-feathers tipped
with black; throat and abdomen pale rufous; a bare space encircling
the eye bright red; bill and legs slaty.

The sexes are similar.

Length, 15.50; wing, 5.50; tail, 9; tarsus, 1.40; bill, 1.60.

HABITAT.   Haiti and San Domingo.

Saurothera vieilloti Br.

Saurothera vetula VIEILL. Nouv. Dict. XXXII, p. 348.—LAFR. Rev. Zool.
1847, p. 357.
Saurothera vieilloti Br. Consp. I, p. 97 (1850).—SCL. Cat. Am. Bds. p.
324 (1862). SUNDEV. Oefv. K. Vet. Akad. För. 1869, p. 599.—
GRAY, Handl. Bds. II, p. 208 (1870).—SCL. & SALV. Nom. Avium
Neotr. p. 107 (1873).—GUNDL. Anal. Soc. Esp. Hist. Nat. VII, p.
230 (1878).—CORY, List Bds. W. I p. 18 (1885).
Saurothera vieilloti var. *rufescens* BRYANT, Pr. Bost. Soc. Nat. Hist. X,
p. 256 (1866).

SP. CHAR.—Entire upper surface pale olive brown; throat dull white,
shading into gray on the breast; belly and under tail-coverts chestnut brown; tail olive, tipped with black, and narrowly edged on the
tip with white; primaries having the outer webs pale olive, and
and heavily marked with rufous on the terminal portions of the
inner webs; secondaries showing distinctly olive green.

The sexes are similar.

Length, 16; wing, 5; tail, 8.50; tarsus, 1.05; bill, 1.40.

HABITAT.   Porto Rico.

Saurothera merlini D'Orb.

Saurothera merlini D'ORB. in La Sagra's Hist. Nat. Cuba, Ois. p. 152
(1840).—BR. Consp. I, p. 97 (1850).—CAB. J. f. O. 1856, p. 104.—
GUNDL. Journ. Bost. Soc Nat. Hist. VI, p. 319 (1857).—BREWER,

Pr. Bost. Soc. Nat. Hist. VII. p. 307 (1860).—GUNDL. Repert. Fisico-Nat. Cuba, I. p. 296 (1866); *ib.* J. f. O. 1874, p. 158.—GRAY, Handl. Bds. II, p. 208 (1870).—SCL. & SALV. Nom. Avium Neotr. p. 107 (1873).—CORY, List Bds. W. I. p. 18 (1885).

Coccygus merlini SCHLEG. Mus. Pays-Bas. I. p. 40 (1864).

SP. CHAR.—Very large. The head and back brown, with a tinge of olive, brightest on the head and rump; throat dull ashy white, shading distinctly ashy on the breast; rest of underparts pale chestnut brown; under surface of wings rufous; pale chestnut on the under coverts; primaries dark chestnut rufous, tipped with olive, showing slight metallic reflections; wing-coverts pale olive; two central tail-feathers dull olive, showing a brownish tinge on basal portions; rest of tail-feathers dull olive, showing a brownish tinge slightly on the inner webs, and having a subterminal band of black, tipped with white.

Length, 21; wing, 7.25; tail, 12; tarsus, 1.50; bill, 2.

HABITAT. Cuba.

Saurothera bahamensis BRYANT.

Saurothera vetula BRYANT, Pr. Bost. Soc. Nat. Hist. VII, p. 106 (1859).

Saurothera bahamensis BRYANT, Pr. Bost. Soc. Nat. Hist. IX, p. 280 (1864).—CORY, Bds. Bahama I. p. 116 (1880); *ib.* List Bds. W. I. p. 18 (1885).

SP. CHAR. *Male:*—Above pale olive, showing slight greenish reflections; throat and breast dull grayish white; belly and crissum tawny; primaries mostly rufous; tail-feathers, except the two central ones, tipped with pale brownish white; legs slaty blue; soles of the feet yellow; eyelids vermillion red; upper mandible brownish, shading into slate color at the base; iris brown.

The sexes are similar.

Length, 18; wings, 6.25; tail, 9.50; tarsus, 1.50; bill, 1.80.

HABITAT. Bahamas.

GENUS Coccyzus VIEILL.

Coccyzus VIEILLOT. Analyse. p. 28, 1816.

Coccyzus americanus (LINN.).

Cuculus americanus LINN. Syst. Nat. I, p. 170 (1766).

Coccyzus carolinensis D'ORB. in La Sagra's Hist. Nat. Cuba, Ois. p. 150 (1840).

Coccyzus americanus GOSSE, Bds. Jam. p. 279 (1847).—A. & E. NEWTON, Ibis. 1859. p. 146 (St. Croix).—BREWER, Pr. Bost. Soc. Nat. Hist.

VII, p. 307 (1860) (Cuba).—ALBRECHT, J. f. O. 1862, p. 202 (Jamaica).—GUNDL. Repert. Fisico-Nat. Cuba, I, p. 295 (1866).—SCL. P. Z. S. 1866, p. 166 (Jamaica).—GUNDL. J. f. O. 1874, p. 156 (Cuba); *ib.* Anal. Soc. Esp. Hist. VII, p. 233 (1878) (Porto Rico).—CORY, Bds. Bahama I. p. 117 (1880).—A. & E. NEWTON, Handb. Jamaica, p. 109 (1881).

Coccyzus dominicus MARCH, Pr. Acad. Nat. Sci. Phila. 1863, p. 154 (Jamaica).

Coccyzus bairdi SCL. P. Z. S. 1864, p. 120 (Jamaica).

Coccyzus americanus CAB. J. f. O. 1856, p. 104 (Cuba).—CORY, List Bds. W. I. p. 19 (1885).

This species occurs in the Bahama Islands, Cuba, Jamaica, and Porto Rico. It has also been recorded from St. Croix.

Coccyzus minor (GMEL.).

Cuculus minor GMEL. Syst. Nat. I, p. 411 (1788).

Cuculus seniculus LATH. Ind. Orn. I, p. 219 (1790).

Coccyzus seniculus VIEILL. Ency. Méth. p. 1346.—GOSSE, Bds. Jam. p. 281 (1847).—BP. Consp. I. p. 111 (1850).—SALLÉ, P. Z. S. 1857, p. 234.—NEWTON, Ibis, 1859, p. 150.—BREWER, Pr. Bost. Soc. Nat. Hist. VII, p. 307 (1860).—CASSIN, Pr. Acad. Nat. Sci. Phila. 1860, p. 377.—CAB. & HEIN. Mus. Hein. IV, p. 78 (1861).—ALBRECHT, J. f. O. 1862, p. 202.—SCL. P. Z. S. 1864, p. 121.—PELZ. Orn. Bras. p. 273 (1871).

Coccyzus helviventris CAB. in Schomb. Guian. III, p. 714 (1848).

Coccygus minor BAIRD, Bds. N. Am. p. 78 (1858).—CAB. J. f. O. 1856, p. 104.—GUNDL. Repert. Fisico-Nat. Cuba, I, p. 295 (1866).—BRYANT, Pr. Bost. Soc. Nat. Hist. XI, p. 96 (1866).—BD. BWR. & RIDGW. Hist. N. Am. Bds. II, p. 482 (1874).—ALLEN, Bull. Nutt. Orn. Club, V, p. 169 (1880); *ib.* CORY, VI, p. 154 (1881); *ib.* Bds. Haiti & San Domingo. p. 101 (1885); *ib.* List Bds. W. I. p. 19 (1885).

Coccyzus nesiotes CAB. & HEIN. Mus. Hein. IV, p. 78 (1861).—TAYLOR, Ibis, 1864, p. 121.

Coccyzus domisicus SCL. Cat. Am. Bds. p. 323 (1862).

Coccygus seniculus SCHLEG. Mus. Pays-Bas. I, p. 38 (1864).

Coccystes seniculus SUNDEV. Oefv. K. Vet. Akad. För. 1869, p. 599.

Coccyzus minor LÉOT. Ois. Trin. p. 353.—BRYANT, Pr. Bost. Soc. Nat. Hist. X, p. 255 (1866).—SCL. P. Z. S. 1870, p. 166.—GUNDL. J. f. O. 1874, p. 157; *ib.* Anal. Soc. Esp. Hist. Nat. VII, p. 231 (1878).—LAWR. Pr. U. S. Nat. Mus. I, pp. 62, 193, 234, 487 (1878).—CORY, Bds. Bahama I. p. 117 (1880).—A. & E. NEWTON, Handb. Jamaica, p. 109 (1881).—RIDGW. Pr. U. S. Nat. Mus. VII, p. 172 (1884).

SP. CHAR. *Male.* Above grayish olive, tinged with ash on the head; underparts yellowish brown, darkest on the thighs, and becoming pale on the throat; a streak of dark brown behind the eye, passing

under it; quills and under wing-coverts yellowish brown; outer tail-
feathers black, tipped with white, and showing slight bronze reflections; the others lighter, except the central ones, tipped with white;
under mandibles yellow, except at the tip.

The sexes are similar.

Length, 11.80; wing, 5.20; tail, 6.50; tarsus, 1.08; bill, .90.

HABITAT. Bahamas and Antilles.

Coccyzus erythrophthalmus (WILS.).

Cuculus erythrophthalmus WILSON, Am. Orn. IV, p. 16 (1811).
Coccyzus erythrophthalmus LEMB. Aves Cuba, p. 73 (1850).—BREWER, Pr.
    Bost. Soc. Nat. Hist. VII. p. 307 (1860) (Cuba).—GUNDL. Repert.
    Fisico-Nat. Cuba. I. p. 295 (1866); *ib.* J. f. O. 1874, p. 157 (Cuba).
Coccyzus erythrophthalmus CAB. J. f. O. 1856, p. 104 (Cuba). — CORY,
    List Bds. W. I. p. 19 (1885).

Accidental in Cuba.

GENUS Hyetornis SCL.

Hyetornis SCLATER, Cat. Am. Bds. p. 321 (1862).

Hyetornis pluvialis (GMEL.).

Cuculus pluvialis GMEL. Syst. Nat. I, p. 411 (1788).
Piaya cinnamomeiventris LAFR. Rev. Zool. 1846, p. 321.—DES MURS.
    Icon. Orn. pl. 65.
Piaya pluvialis GOSSE, Bds. Jam. p. 277 (1847).—BP. Consp. I, p. 111
    (1850).—SCL. P. Z. S. 1861, p. 79.—ALBRECHT. J. f. O. 1862, p. 203.
    —MARCH, Pr. Acad. Nat. Sci. Phila. 1863, p. 283.
Hyetornis pluvialis SCL. Cat. Am. Bds. p. 321 (1862).—SCL. & SALV.
    Nom. Avium Neotr. p, 108 (1873).—A. & E. NEWTON, Handb. Ja-
    maica, p. 109 (1881).—CORY, List Bds. W. I. p. 19 (1885).
Hyetormantis pluvialis CAB. J. f. O. 1862, p. 203.
Coccygus pluvialis SCHLEG. Mus. Pays-Bas, I, p. 39 (1864).
Hyetornis pluvianus GRAY, Handl. Bds. II, p. 212 (1870).

SP. CHAR.—Top of the head smoky brown; rest of the upper surface
    olive; throat dull white showing a tinge of chestnut; rest of under-
    parts dark chestnut brown; under wing-coverts rufous chestnut;
    primaries dull olive brown, with a slight metallic tinge of green on
    the inner webs of the primaries and distinctly perceptible on the sec-
    ondaries; tail dull black, showing a slight purplish tinge; all of the
    tail-feathers tipped with white.

The sexes are similar.

Length, 18; wing, 7; tail 10; tarsus, 1.60; bill, 1.25.

HABITAT. Jamaica.

Bucco cayennensis Sallé (P. Z. S. 1857. p. 234), from San Domingo, it is impossible to identify, as Sallé gives no description, and the genus does not occur in the West Indies.

FAMILY ALCEDINIDÆ.

GENUS Ceryle BOIE.

Ceryle BOIE. Isis, 1828, p. 316.

Ceryle stictipennis LAWR.

Ceryle torquata LAWR. Pr. U. S. Nat. Mus. I, pp. 459, 487 (1878).—CORY, List Bds. W. I. p. 19 (1885).
Ceryle stictipennis LAWR. Pr. U. S. Nat. Mus. VIII, p. 623 (1885).

"*Male.*—The upper plumage is ashy-blue, with a broad, pure white band across the hind neck, connecting with the white of the throat; lores black; a spot of white anterior to the eye, and another of the same color below it; the upper tail-coverts are colored like the back, and are barred on both webs with white; the two middle tail feathers are also colored like the back; the shafts are black, bordered narrowly on each side with deep black; they are conspicuously marked with triangular-shaped white spots on the middle of each web, these are nine in number on each side; the other tail feathers are black, with their outer edges colored like the back, and having pure white spots rounded in form on each web, those on the inner webs are much the largest; all the tail feathers are tipped with white; the quills are black, largely white on their inner webs and marked on the primaries with quadrate white spots on the outer ones, rather far apart; the secondaries have also small white spots on their outer webs, and on the inner webs large round spots, the outer webs are margined with ashy-blue; the tertiaries are largely ashy-blue on their outer webs, spotted and barred with white, the inner webs are brownish-black and marked with large spots of white; the wing-coverts are like the back in color, the larger ones are sparsely marked with small white spots; the scapulars are narrowly barred with white; the under wing-coverts are white, largely intermixed with cinnamon color; the throat is white; the lower part of the neck, the breast, and the abdomen are of a very dark cinnamon color; the lower part of the abdomen, the flanks, crissum and under tail-coverts are white, closely spotted and barred with rather dull ashy-blue; the tarsi and toes are dark brown; the bill is black, with the basal half of the under mandible yellow.

"Length, fresh, 17 inches; wing, $7\frac{3}{4}$; tail, $5\frac{1}{2}$; bill, 3.

"*Habitat.*—Guadeloupe, West Indies. Type in my collection.

"*Female.*—The color above is similar to that of the male, but it has the entire back and wings marked sparsely with small white spots; in the markings on the head, wings, and tail they are much alike; it has the white throat and band on the hind neck as in the male; across the lower part of the throat and upper part of the breast there is a broad band of ashy-blue, minutely freckled with white; this band is bordered narrowly below with white; the entire under plumage besides, and the under wing-coverts are deep cinnamon.

"Length (skin), 17½ inches; wing, 8; tail, 5½; bill, 3.

"The type of the female is in the National Museum, Washington, to which it was sent from Guadeloupe by Mr. L. Guesde." (LAWR., l. c., orig. descr.)

It has always been supposed that *Ceryle torquata* did not differ from the Guadeloupe species, but Mr. Lawrence considers the West Indian bird as new and has described it as above. The greatest difference seems to be the white spotting of the wings and back, and the rufous marking of the under wing-coverts. Were it not for the isolated position where it occurs it would represent a variety of *C. torquata*, but as we have no record of its occurrence elsewhere in the West Indies, and as it is claimed to be resident in Guadeloupe, intergradation seems hardly probable, still the Lesser Antilles are not, as yet, so well known as to justify us in saying *C. torquata* does not occur elsewhere in the West Indies.

Ceryle alcyon (LINN.).

Alcedo alcyon LINN. Syst. Nat. I. p. 180 (1766).—D'ORB. in La Sagra's Hist. Nat. Cuba. Ois. p. 130 (1840).—BRYANT, Pr. Bost. Soc. Nat. Hist. X, p. 255 (1866) (Porto Rico).—SUNDEV. Oefv. K. Vet. Akad. För. 1869, p. 585 (St. Bartholomew); *ib.* p. 600 (Porto Rico). *Ceryle alcyon* GOSSE, Bds. Jam. p. 81 (1847).—SALLÉ, P. Z. S. 1857, p. 233 (San Domingo).—A. & E. NEWTON, Ibis, 1859, p. 67 (St. Croix).—BRYANT, Pr. Bost. Soc. Nat. Hist. VII, p. 108 (1859) (Bahamas).—BREWER, *ib.* p. 306 (1860) (Cuba).—SCL. P. Z. S. 1861, p. 77 (Jamaica).—ALBRECHT, J. f. O. 1862, p. 199 (Jamaica).—GUNDL. Repert. Fisico-Nat. Cuba, I. p. 292 (1866).—LAWR. Pr. U. S. Nat. Mus. I, p. 62 (1878) (Dominica); *ib.* p. 193 (St. Vincent); *ib.* p. 292 (Grenada); *ib.* p. 339 (Martinique); *ib.* p. 459 (Guadeloupe).—GUNDL. Anal. Soc. Esp. Hist. Nat. VII, p. 216 (1878) (Porto Rico).—SCL. P. Z. S. 1879, p. 765 (Montserrat).—ALLEN, Bull.

Nutt. Orn. Club, V. p. 169 (1880) (Santa Lucia).—Cory, Bds. Bahama I. p. 115 (1880).—A. & E. Newton, Handb. Jamaica, p. 119 (1881).—Cory, Bull. Nutt. Orn. Club, VI, p. 153 (1881) (Haiti).—Grisdale, Ibis. 1882, p. 486 (Montserrat).—Cory, Bds. Haiti & San Domingo, p. 103 (1885); *ib.* List Bds. W. I. p. 19 (1885).

Alcedo (Ceryle) alcyon Bryant, Pr. Bost. Soc. Nat. Hist. XI, p. 95 (1866) (San Domingo).

Common throughout the West Indies.

Family TODIDÆ.

Genus Todus Linn.

Todus Brisson, Orn. IV, p. 528 (1760).— Linnæus, Syst. Nat. I, p. 198 (1766).

Todus viridis Linn.

Todus viridis Linn. Syst. Nat, I, p. 178 (1766).—Gosse, Bds. Jam. p. 72 (1847).—Denney. P. Z. S. 1847, p. 38.—Bp. Consp. I, p. 182 (1850). —Albrecht, J. f. O. 1862, p. 199. — Scl. Cat. Am. Bds. p. 263 (1862).—Scl. & Salv. Nom. Avium Neotr. p. 103 (1873).—Sharpe, Ibis, 1874, p. 349.—A. & E. Newton, Handb. Jamaica, p. 109 (1881).—Cory, List Bds. W. I. p. 19 (1885).

Sp. Char.—Entire upper surface bright grass-green; throat bright red, the feathers showing delicate tippings of white when held in the light; a narrow stripe of white on each side of the throat, becoming grayish as it reaches the breast; breast dull white, strongly tinged with green; belly very pale yellow; a patch of pink on the sides of the body; primaries dark brown, narrowly edged with green; the inner primaries and secondaries having nearly the entire outer web green; under surface of tail brownish; upper surface of tail green.

The sexes are similar.

Length, 4.60; wing, 1.80; tail, 1.40; tarsus, .45; bill, .62.

Habitat. Jamaica.

Todus angustirostris Lafr.

Todus angustirostris Lafr. Rev. Zool. 1851, p. 478.—Sallé, P. Z. S. 1857, p. 233—Bryant, Pr. Bost. Soc. Nat. Hist. XI. p. 91 (1866). —Sharpe, Ibis. 1874, p. 352.—Cory, Bds. Haiti & San Domingo, p. 107 (1885); *ib.* List Bds. W. I. p. 19 (1885).

SP. CHAR. *Male:*—Above bright green; throat dark crimson; the feathers slightly touched with white; underparts white; flanks pinkish; under wing- and tail-coverts pale yellow; a line of white extending from the base of the mandible, separating the colors of the head and throat, becoming grayish as it reaches the sides of the neck; entire upper mandible and terminal half of lower mandible dark brown; legs black.

The sexes are similar.

Length, 4; wing, 1.90; tail, 1.50; tarsus, .45; bill, .60; width of bill at middle, .12.

HABITAT. San Domingo.

Todus subulatus GOULD.

Le Todier de St. Domingue, "BUFF. Pl. Enl. p. 585, figs. 1, 2 (1783)."
Todus viridis "VIEILL. Nouv. Dict. XXXIV, p. 184, pl. 29, fig. 4 (1819)."

Todus subulatus GOULD (Fig. sine descr.).—GRAY & MITCH. Gen. Bds. I, p. 63 pl. 22 (1847). —BP. Consp. I, p. 182 (1850).— GRAY, Handl. Bds. I. p. 79 (1869).—SHARPE, Ibis, 1874, p. 351.—TRISTRAM, Ibis, 1884, p. 168.—CORY, Bds. Haiti & San Domingo, p. 105 (1885); *ib.* List Bds. W. I. p. 19 (1885).

Todus dominicensis LAFR. Rev. Zool. 1847. p. 331.—SALLÉ, P. Z. S. 1857, p. 233.—BRYANT, Pr. Bost. Soc. Nat. Hist. XI, p. 91 (1866).—SCL. & SALV. Nom. Avium Neotr. p. 103 (1873).—CORY, Bull. Nutt. Orn. Club, VI, p. 154 (1881).

SP. CHAR. *Male:*—Above bright green; throat crimson red, the feathers faintly tipped with white; underparts dull yellowish, the feathers edged with pale red on the basal portions, the color only slightly showing on the surface, and giving the chest and belly the appearance of being pale yellowish, faintly streaked with red; flanks pale pinkish red; wings dark brown, edged with green; tail green; crissum and under tail-coverts pale yellow; a line of dull white extending from the base of the mandible, separating the green of the head from the red of the throat; upper mandible dark brown; the tip of lower mandible dull brown.

The sexes are similar.

Length, 4.35; wing, 2.05; tail, 1.60; tarsus, .60; bill, .70; width of bill at middle, .20.

HABITAT. Haiti and San Domingo.

Todus pulcherrimus SHARPE.

Todus pulcherrimus SHARPE, Ibis, 1874, p. 353, pl. xiii, f. 3.—CORY.
List Bds. W. I. p. 19 (1885).

"Above bluish-green, rather tinged with olive on the lower back,
the wing-coverts showing a very strongly pronounced blue shade;
quills blackish, bordered narrowly with light green, shading off into
bluish towards the tips of the secondaries; tail dull greenish, with
narrow margins of bluish green; forehead lighter and rather more
olive-green than the back, and tinged with orange near the base of
the beak; lores tinged with orange; sides of face yellowish green;
sides of neck dull rufous; chin white; throat bright carmine, with
silvery white margins to most of the feathers; rest of under surface
with a light crimson blush, varied on the breast with white oval
spots to the feathers, producing an ocellated appearance, the crimson colour brightest on the flanks, shading off into ochraceous buff on
the sides of the vent; on each side of the upper breast a patch of
greenish; under wing-coverts ochraceous buff, the outermost smaller
coverts washed with pale carmine; upper mandible blackish, lower
one yellowish; feet black.

"Total length, 3.5 inches; culmen, 0.85; wing, 1.9; tail, 1.4; tarsus, 0.65.

"HAB. Jamaica?

"This new species comes nearest to *T. subulatus* of S. Domingo, having, like that species, the white ocellations on the
breast: but it differs from that bird and all other members of the
genus by its brilliant coloration below, and by its being bluish
green above.

"The type is in the British Museum." (SHARPE, l. c., orig.
descr.)

Todus hypochondriacus BRYANT.

Todus viridis DESM. H. N. Tang. pl. 67.—VIEILL. Gal. Ois. I, pl. 124
(1825).
Todus mexicanus LESS. Ann. Soc. Nat. XI, p. 167.—LAFR. Rev. Zool.
1847, p. 333.—BP. Consp. I, p. 182 (1850).--BAIRD, Ibis, 1867, p.
260.
Todus hypochondriacus BRYANT, Pr. Bost. Soc. Nat. Hist. X, p. 250
(1866). — SUNDEV. Oefv. K. Vet. Akad. För. 1869, p. 598. — SCL.
& SALV. Nom. Avium Neotr. p. 103 (1873).—SHARPE, Ibis, 1874, p.
354.—GUNDL. Anal. Soc. Esp. Hist. Nat. VII, p. 219 (1878).—CORY,
List Bds. W. I. p. 19 (1885).

SP. CHAR.—Entire upper plumage bright green; throat red, the feathers showing faint edgings of white when held in the light; a narrow stripe of white down the sides of the throat; cheeks green; breast grayish, becoming white on the belly; sides of the body and under tail-coverts yellow; a faint bluish tinge is perceptible on the carpus; under surface of tail dull brown; upper surface green.

Length, 4.75; wing, 1.75; tail, 1.10; tarsus, .50; bill, .60.

HABITAT. Porto Rico.

General appearance of *T. viridis*, but lacking the broad pink patch on the sides of the body. Some specimens show a slight pinkish tinge.

Todus multicolor GOULD.

Todus multicolor GOULD, Icon. Av. pl. 2 (1837).—D'ORB. in La Sagra's Hist. Nat. Cuba. Ois. p. 132 (1840).—BP. Consp. I. p. 182 (1850).— CAB. J. f. O. 1856, p. 101.—BREWER, Pr. Bost. Soc. Nat. Hist. VII. p. 307 (1860).—GUNDL. J. f. O. 1871, p. 288; *ib.* 1874. p. 146.— SCL. & SALV. Nom. Avium Neotr. p. 103 (1873).—SHARPE, Ibis, 1874, p. 352.—CORY, List Bds. W. I. p. 19 (1885).

Todus portoricensis LESS. Ann. Soc. Nat. XI, p. 167 (1838).—GRAY, Gen. Bds. I, p. 63 (1844).—LAFR. Rev. Zool. 1847. p. 332.—LEMB. Aves Cuba, p. 131 (1850).—BREWER, Pr. Bost. Soc. Nat. Hist. VII, p. 307 (1860).

SP. CHAR.—Entire upper plumage grass-green; a tinge of yellow in front of the eye; cheeks pale blue; throat red, showing the faint whitish tippings when held in the light; breast dull white, becoming gray on the belly; sides of the body pale pink; under tail-coverts yellowish green; under surface of tail pale brown, showing a slight bluish reflection; upper surface of tail green.

Length, 3.80; wing, 1.50; tail, 1.25; tarsus, .50; bill, .55.

HABITAT. Cuba.

FAMILY PICIDÆ.

GENUS Picumnus TEMM.

Picumnus TEMMINCK, Nouv. Rec. de Pl. Col. d'Ois. IV, 1820-39.

Picumnus micromegas SUNDEV.

Chloronerpes passerinus SALLÉ, P. Z. S. 1857. p. 234.

Picus (Chloronerpes) passerinus BRYANT, Pr. Bost. Soc. Nat. Hist. XI, p. 96 (1866).

Phœnicomanes iora SHARPE, P. Z. S. 1874, p. 427, is a synonym of an Eastern species. It was supposed to have been sent from Jamaica with a collection of birds from that island.

Picumnus micromegas SUNDEV. Consp. Avi. Picin. p. 95 (1866).—BRYANT,
    Pr. Bost. Soc. Nat. Hist. XI, p. 96 (1866).— TRISTRAM, Ibis, 1884,
    p. 167.
Picumnus lawrencei CORY, Bull. Nutt. Orn. Club. VI. p. 129 (1881); *ib.*
    Bds. Haiti & San Domingo, p. 109 (1885); *ib.* List Bds. W. I. p. 19
    (1885).

SP. CHAR.   *Male:*—Tail soft, composed of ten feathers; general plumage
    above olive green; forehead showing a tinge of black; top of the
    head bright yellow, cut by a band of red, again becoming yellow at
    the base; underparts yellowish, palest on the throat, mottled and
    streaked with brown feathers; on the sides of the neck marked
    with dull white, nearly joining above, forming an imperfect collar;
    wing-coverts and outer webs of primaries and secondaries olive
    green; inner webs brown, becoming pale on the edges; bill, legs,
    and feet dark slate color; iris reddish brown.
        The adult female differs from the male in wanting the red band
    on the head.
        Length, 5.10; wing. 2.75; tail, 1.85; tarsus, .70; bill, .62.

HABITAT.   Haiti and San Domingo.

GENUS Campephilus GRAY.

Campephilus GRAY. List Gen. Bds. p. 54 (1840).

Campephilus principalis bairdi.

Campephilus principalis Cab. J. f. O. 1856, p. 102.— BREWER, Pr. Bost.
    Soc. Nat. Hist. VII, p. 307 (1860).
Campephilus bairdii CASSIN, Pr. Acad. Nat. Sci. Phila. 1863, p. 322.—
    GUNDL. Repert. Fisico-Nat. Cuba, I. p. 293 (1866);—*ib.* J. f. O.
    1874, p. 148.
Campephilus principalis var. *bairdi* BD. BWR. & RIDGW. Hist. N. Am.
    Bds. II. p. 496 (1874).
Campephilis principalis bairdi CORY, List Bds. W. I. p. 19 (1885).

    "Much resembling *C. principalis*, but smaller and with the black
anterior feathers of the crest larger than those succeeding, which
are scarlet. White longitudinal line on the neck reaching quite to
the base of the bill. In *C. principalis* the scarlet plumes of the
crest are the longer, and the line on the neck does not extend to the
base of the bill, both of which characters are very accurately shown
in Audubon's plates B. of Am. pl. 66, and oct. ed. IV. pl. 256. Colors
of all other parts in the present bird are the same as those of *C.
principalis.*
    "Total length about 18½ inches, wing, 9¼, tail, 6½ inches." (CAS-
SIN, l. c., orig. descr.)

HABITAT.   Cuba.

I have now before me eighteen examples of *C. principalis* and two of *C. bairdi*, and have examined three other specimens of the latter bird. In none of the Cuban birds does the white cheek stripe quite reach the bill and in most of them it lacks a quarter of an inch or more. Some examples of *C. principalis* show the white stripe *nearly* as long as the Cuban form. The length of this stripe is quite variable in the series from Florida. The "black anterior feathers of the crest" do not appear to be a constant character, as two examples of *C. principalis* from Florida show it as much as any Cuban specimens which I have seen.

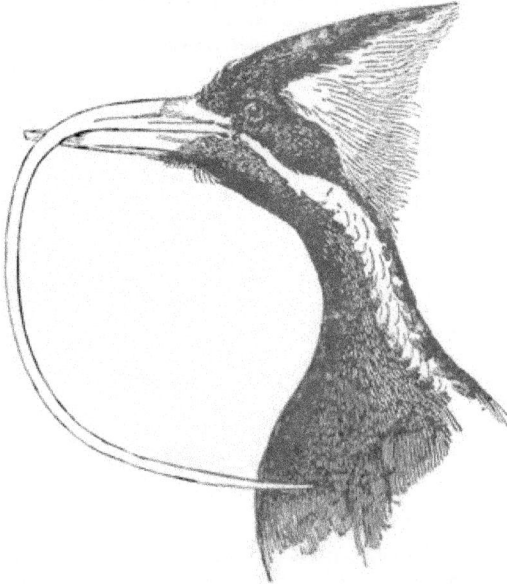

In Dr. Gundlach's private museum at Ingenio Fermina, Cuba, I had the pleasure of seeing a most curious *C. bairdi*, which might well be called an ornithological monstrosity. The bird was perfect in every respect with the exception of the upper mandible, which had grown to the enormous length of nearly twelve inches, curving downward and passing the body on the left side, as shown in the accompanying figure.

Genus Dryobates Boie.

Dryobates Boie, Isis, 1826, p. 977.

Dryobates villosus maynardi.

Picus villosus Bryant, Pr. Bost. Soc. Nat. Hist. VII, p. 106 (1859).—Cory, Bds.
    Bahama I. p. 120 (1880).
Picus insularis Mayn. The Nat. in Florida, I, No. 4 (1885); not of Gould, 1862.
Picus villosus insularis Cory, List Bds. W. I. p. 19 (1885).
Dryobates villosus maynardi Ridgw. Man. N. A. Bds. p. 282 (1887).

> Sp. Char.   *Male:*—Above black, with a white band down the middle of
> the back, finely lined with black; all the quills, middle and larger
> wing-coverts with numerous spots of white; crown black; a patch
> over the eye, and a stripe from the mandible to the nape white; a
> black stripe from the eye, passing through the cheeks, over the
> nape, and joining the black of the back; a scarlet crescent around the
> base of the skull, joining the white superciliary stripe; underparts
> ashy, with the sides mottled and striped with black; two outer tail-
> feathers white, edged and tipped with pale brown; third black, with
> a patch of pale brown upon the outer web, the others black.
>   *Female:*—The scarlet crescent wanting, replaced by white.
>   Length, 7.25; wing, 4.20; tail, 3; tarsus, .70; bill, 1.

Habitat.   Northern Bahama Islands.

This form differs from *P. villosus* in the greater extent of
white in front of the eye, the black streaks on the sides of the
breast, and black shaft-lines on the white feathers of the back.

Genus Sphyrapicus Baird.

Sphyrapicus Baird, Bds. N. A. p. 101 (1858).

Sphyrapicus varius (Linn.).

Picus varius Linn. Syst. Nat. I. p. 176 (1766).—D'Orb. in La Sagra's
    Hist. Nat. Cuba, Ois. p. 141 (1840).—Gosse, Bds. Jam. p. 270
    (1847).—Cab. J. f. O. 1856, p. 102 (Cuba).—Newton, Ibis, 1859.
    p. 308 (St. Croix).—Bryant, Pr. Bost. Soc. Nat. Hist. VII, p. 106
    (1859) (Bahamas): *ib.* Brewer, p. 396 (1860) (Cuba).—Albrecht.
    J. f. O. 1862, p. 202 (Jamaica).—March. Pr. Acad. Nat. Sci. Phila.
    1863, p. 284 (Jamaica).
Sphyrapicus varius Gundl. Repert. Fisico-Nat. Cuba. I, p. 294 (1866):
    ib. J. f. O. 1874, p. 150 (Cuba).—Cory, Bds. Bahama I. p. 121
    (1880).—A. & E. Newton, Handb. Jamaica, p. 109 (1881).—Cory,
    List Bds. W. I. p. 20 (1885).
Picus (Sphyrapicus) varius Bryant. Pr. Bost. Soc. Nat. Hist. XI, p. 65
    (1867) (Bahamas).

Recorded from the Bahama Islands, Jamaica, Cuba, and St. Croix.

GENUS Xiphidiopicus BONAP.

Xiphidiopicus BONAPARTE, "Consp. Vol. Zygodact. p. 11 (1854)."

Xiphidiopicus percussus (TEMM.).

Picus percussus TEMM. Pl. Col. pp. 390, 424 (1820-39). — VIGORS, Zool.
    Journ. III, p. 444 (1827).—D'ORB. in La Sagra's Hist. Nat. Cuba,
    Ois. p. 143 (1840).—THIENE. J. f. O. 1857, p. 153.
Dendrobates percussus GRAY, Gen. Bds. II, p. 437 (1844-49).
Chloropicus percussus MALH. Mém. Acad. Metz. 1848-49, p. 352.
Picus ruppellii WAGL. Syst. Av. sp. 29 (1827).
Chloronerpes percussus Br. Consp. I, p. 118 (1850).—CAB. J. f. O. 1856,
    p. 102.—BREWER, Pr. Bost. Soc. Nat. Hist. VII, p. 307 (1860).—
    GUNDL. Repert. Fisico-Nat. Cuba, I, p. 294 (1866); *ib.* J. f. O.
    1874, p. 151.
Xiphidiopicus percussus Br. Consp. Vol. Zygodact. p. 11 (1854).—SCL.
    Cat. Am. Bds. p. 339 (1862).—GRAY, Handl. Bds. II, p. 199 (1870).
    —SCL. & SALV. Nom. Avium Neotr. p. 100 (1873).—Cory, List Bds.
    W. I. p. 20 (1885).

SP. CHAR.—Forehead dull white, extending in a superciliary stripe down
    the sides of the head to the neck; top of the head bright red; a nar-
    row stripe of smoky black passing from the eye down the sides of
    the neck; a narrow stripe of black on the centre of the throat from the
    chin, succeeded on the lower throat and upper breast by a broad
    patch of red; sides of throat dull white; back and upper surface of
    wings yellowish green; middle portion of breast pale yellow; feath-
    ers on the sides of the body barred with smoky black and white,
    showing a tinge of yellow; primaries dark brown, blotched with dull
    white on the edges of the webs; inner primaries and secondaries
    heavily edged with pale green on the outer web; two central tail-
    feathers dull black, showing a silvery tinge when held in the light;
    under surface of the tail showing feathers with narrow alternate
    bands of gray and pale brown.

The female differs from the male in having the top of the head
    black, the feathers showing narrow shaft lines of white; the red
    restricted to the base of the skull.

Length, 9; wing, 5; tail, 3.50; tarsus, .80; bill, .90.

HABITAT. Cuba.

GENUS Melanerpes SWAINS.

Melanerpes SWAINSON, F. B. A. II, 1831.

Melanerpes portoricensis (DAUD.).

Picus portoricensis DAUD. Ann. du Mus. II, p. 383, pl. 51 (1803).—BRYANT.
Pr. Bost. Soc. Nat. Hist. X, p. 256 (1866). — SUNDEV. Oefv. K.
Vet. Akad. För. 1869, p. 599.
Picus rubidicollis VIEILL. Ois. Am. Sept. II, p. 63, pl. 117 (1807).—TEMM.
Nat. Syst. p. 210 (1807).
Melanerpes rubidicollis GRAY, Gen Bds. II. p. 444 (1844-49).—BP. Consp.
I, p. 115 (1850).
Melanerpes portoricensis A. & E. NEWTON, Ibis, 1859, p. 377.—CASSIN,
Pr. Acad. Nat. Sci. Phila. 1860, p. 377.—SCL. Cat. Am. Birds, p.
34; (1862).—TAYLOR, Ibis, 1864, p. 170.—SCL. & SALV. Nom. Avium
Neotr. p. 100 (1873).—GUNDL. Anal. Soc. Esp. Hist. Nat. VII, p.
227 (1878).—CORY, List Bds. W. I. p. 20 (1865).
Melampicus portoricensis MALH. Mon. Pic. II, p. 205 (1862).
Asyndesmus portoricensis GRAY, Handl. Bds. II. p. 201 (1870).

SP. CHAR. *Male:*—Entire upper surface black with greenish reflections;
forehead dull white, reaching and nearly encircling the eye; throat,
breast, and underparts dull crimson red, mixed with brownish on
the sides of the body; a spot of white near the carpus; rump white;
wings and tail brownish black.

Length, 7.50; wing, 5; tail, 3; tarsus, .90; bill, 1.

HABITAT.   Porto Rico and St. Thomas.

Melanerpes l'herminieri (LESS.).

Picus hermanieri LESS. Traité Orn. p, 228 (1831).
Melampicus hermanieri MALH. Mém. Acad. Metz, 1848-49. p. 365.
Linnæipicus herminieri MALH. N. Class. Pic. p. 53 (1850).—BP. Consp.
Vol Zygodact. sp. 255 (1854).
Melanerpes hermanieri BP. Consp. I, p. 515 (1850).—REICH. Handb. p.
351 (1853).
Linncopicus hermanieri GRAY, Cat. Gen. Bds. p. 93 (1855).
Melanerpes l'herminieri SCL. & SALV. Nom. Avium Neotr. p. 100 (1873).
—LAWR. Pr. U. S. Nat. Mus. I, p 459 (1878).—CORY, List Bds.
W. I. p. 20 (1885).

SP. CHAR. *Male:*—General plumage black, showing a steel blue gloss on
the back when held in the light; feathers on the breast tinged with
dull red; a faint reddish tinge is perceptible on the forehead.

Female:—Similar to the male, perhaps smaller, and lacks the
tinge of red on the forehead, although the reddish tinge on the
forehead of the male is possibly not constant.

Length (skin), 10; wing, 5.50; tail, 4; tarsus 1; bill, 1.25.

HABITAT.   Guadeloupe.

GENUS **Centurus** SWAINS.

Centurus SWAINSON, Class. Birds. II, p. 310. 1837.

Centurus striatus (MÜLL.).

Picus dominicensis, striatus. "BRISS. Orn. IV, p. 65, pl. 4, fig. 2 (1760)
  (♂ ad)."
Picus dominicensis "BRISS. t. c. pl. 3. fig. 2 (♀ ad)."
Le Pic rayé de St. Dominique "BRISS. Orn. IV, p. 65, pl. 4, fig. 1 (1760)
  (♂ ad).
Picus striatus MÜLL. Syst. Nat. Suppl. (1766).—GMEL. Syst. Nat. I, p.
  427 (1788).—CUV. Rég. An. I, p. 451 (1829).—BRYANT, Pr. Bost.
  Soc. Nat. Hist. XI, p. 96 (1866).
Centurus striatus GRAY, Gen. Bds. II, p. 442 (1841-49).—BP. Consp. I, p.
  119 (1850).—SALLÉ, P. Z. S. 1857, p. 234.—SCL. & SALV. Nom.
  Avium Neotr. p. 100 (1873).—RIDGW. Pr. U. S. Nat. Mus. IV, p.
  117 (1881).—CORY, Bull. Nutt. Orn. Club. VI, p. 154 (1881); *ib.*
  Bds. Haiti & San Domingo, p. 111 (1885); *ib.* List Bds. W. I, p. 20
  (1885).—TRISTRAM, Ibis, 1884, p. 168.
Zebrapicus striatus MALH. Mon. Pic. II, p. 231 (1862).

SP. CHAR. *Male:*—Underparts yellowish green, showing a tinge of
brownish olive on the breast, and gray on the throat; forehead and
sides of the head gray; a patch of bright crimson red extending from
the forehead to the nape, covering nearly the entire top of the head;
rump crimson red; back alternately banded with black and yellowish green; upper surface of wings having the appearance of the
back, except that they are alternately banded with black and yellow;
upper surface of tail dark brown; bill and feet dark slate color.

Female:—Top of head black; the nape showing bright crimson
red, which encroaches slightly upon the back of the head; rest as in
the male.

Length, 9; wing, 4.60; tail, 3.60; tarsus, .90; bill, 1.20.

HABITAT. Haiti and San Domingo.

Centurus radiolatus (WAGL.).

Picus varius medius jamaicensis RAY, Syn. Av. p. 181, No. 11.
Picus jamaicensis EDW. Gleanings. Pl. 244 (♂ ad.).
Pic rayé Femelle de la Jamaique BUFF. Pl. Enl. p. 597 (♂ ad.).
Picus carolinus. part. LINN. Syst. Nat. I, p. 175 (1766).
Picus radiolatus WAGL. Syst. Av. No. 59 (1827); *ib.* Isis, 1829, p. 572.
Centurus radiolatus GOSSE, Bds. Jam. p. 271 (1847).—BP. Consp. I, p.
  118 (1850).—REICH. Handb. p. 409 (1854).—SCL. Cat. Am. Bds. p.
  343 (1862).—ALBRECHT. J. f. O. 1862, p. 203.—MARCH, Pr. Acad.
  Nat. Sci. Phila. 1863, p. 284.—SCL. & SALV. Nom. Avium Neotr. p.

100 (1873).—RIDGW. Pr. U. S. Nat. Mus. IV, p. 111 (1881).—A. &
E. NEWTON, Handb. Jamaica, p. 109 (1881).—CORY, List Bds. W. I.
p. 20 (1885).
Zebrapicus radiolatus MALH. Mon. Pic. II. p. 237 (1862).

SP. CHAR. *Male:*—Forehead and throat dull white; crown bright red;
breast smoke gray, becoming yellowish-olive on the underpart ; a
faint reddish tinge near the vent, sometimes lacking; back black,
the feathers delicately barred with dull white; rump black, barred
with white; tail black, the inner webs of the two central feathers
delicately barred with white; wings black, showing numerous cot-
tings of white on the secondaries and the basal portions of the outer
webs of some of the primaries.

Female:—Forehead white; top of the head smoky-brown; a band
of red at the base of the skull; otherwise similar to the male.

Length, 10.50; wing, 5; tail, 3.50; tarsus, .90; bill, 1.25.

HABITAT.   Jamaica.

Centurus superciliaris (TEMM.).

Picus superciliaris TEMM. Pl. Col. IV, p. 433 (1820-39).—CUV. Rég. An.
p. 451 (1829).—WAGL. Isis, 1829, p. 515.—LESS. Traité d'Orn. p. 227
(1831). —THIENE. J. f. O. 1857, p. 153.
Colaptes superciliaris VIG. Zool. Journ. III, p. 445 (1827).—D'ORB. in
La Sagra's Hist. Nat. Cuba, Ois. p. 146 (1840).
Colaptes superciliosus GRAY, Gen. Bds. II, p. 446 (1844-49).
Centurus superciliaris BP. Consp. I, p. 118 (1850).—REICH. Handb. p.
408 (1854).—GUNDL. J. f. O. 1856, p. 103: ib. 1874, p. 152.—BREWER,
Pr. Bost. Soc. Nat. Hist. VII, p. 307 (1860).—SCL. Cat. Am. Bds.
p. 342 (1862).—SCL. & SALV. Nom. Avium Neotr. p. 101 (1873).—
RIDGW. Pr. U. S. Nat. Mus. IV, p. 115 (1881).—CORY. List Bds. W
I. p. 20 (1885).
Zebrapicus superciliaris MALH. Mém. Acad. Metz, 1848-49, p. 361.

SP. CHAR. *Male:*—Forehead white, showing a faint tinge of orange red at
base of the upper mandible; crown deep red, the color extending to
the nape; a patch of black over and back of the eye; rest of head
and throat dull grayish white; breast yellowish olive, becoming
distinctly yellowish on the belly; a patch of deep red on the middle
of the belly; back and wings heavily banded with black and white,
showing a faint yellowish tinge on the back; primaries dark brown,
narrowly tipped with white on the fifth and sixth; the two central
tail-feathers heavily banded with black and white; the outer tail-
feathers banded with black and white near the tip; rest of tail-
feathers faintly tipped with white; bill black; feet black.

Female:—Similar to the male, but having top of the head white succeeded by a band of dull black, which is in turn replaced by red on the base of the skull, extending to the nape.

Length, 11.50; wing, 5.80; tail, 4; tarsus, 1; bill, 1.50.

HABITAT. Cuba.

GENUS Colaptes SWAINS.

Colaptes SWAINSON, Zool. Journ. III, p. 353, Dec. 1827.

Colaptes chrysocaulosus GUNDL.

Colaptes chrysocaulosus GUNDL. Ann. N. Y. Lyc. Nat. Hist. VI, p. 273 (1858).—BREWER, Pr. Bost. Soc, Nat. Hist. VII, p. 307 (1860).—ALBRECHT, J. f. O. 1871, p. 210.—GUNDL. Report. Fisico-Nat. Cuba. I, p. 294 (1866); *ib.* J. f. O. 1874, p. 153.—SCL. & SALV. Nom. Avium Neotr. p. 101 (1873).—CORY, List B I's. W. I. p. 20 (1885).
Colaptes auratus CAB J. f. O 1855, p. 103.—BREWER, Pr. Bost. Soc. Nat. Hist. VII, p. 307 (1860).
Colaptes auratus var. *chrysocaulosus* BD. Bwa. & RIDGW. Hist. N. Am. Bds. II, p. 575 (1874).

SP. CHAR. *Male:*—Top of head gray; a nuchal patch of red; a malar stripe of black; sides of the head and throat pale chocolate; a patch of black on the breast; underparts dull yellowish white, heavily spotted with black; upper surface pale chocolate brown, banded with black; upper surface of primaries brown, having the shafts bright yellow; under surface of wings pale yellow; tail dark brown, some of the feathers showing narrow bands on the edges; under surface of tail yellow, feathers tipped with black; bill black and feet dark brown.

The female resembles the male, but differs from it in lacking the black malar stripe, having the throat and sides of the head entirely pale chocolate brown.

Length, 10.50; wing, 5.40; tail, 4.50; tarsus, .80; bill, 1.20.

HABITAT. Cuba.

Colaptes gundlachi CORY.

Colaptes gundlachi CORY, Auk, III, pp. 498, 502 (1886).

SP. CHAR.—Resembles *Colaptes chrysocaulosus*, but is smaller; the yellow of the quills and tail much brighter; under surface of outer tail feathers distinctly banded; feathers of the rump, white, heavily blotched with black; shafts of the quills and tail very bright yellow; first primary much longer than in the Cuban species.

Length, 8.90; wing, 5.20; tail, 4; tarsus, .85; bill, 1.10.

HABITAT. Grand Cayman.

GENUS Nesoceleus SCL. & SALV.

Nesoceleus SCLATER & SALVIN. App. Nom. Avium Neotr. p. 155 (1873).

Nesoceleus fernandinæ (Vig.).

Colaptes fernandinæ Vig. Zool. Journ. 1827, p. 445.—D'Orb. in La
   Sagra's Hist. Nat. Cuba, Ois. p. 148 (1840).—Cab. J. f. O. 1856, p.
   104.—Brewer, Pr. Bost. Soc. Nat. Hist. VII. p. 307 (1860).—
   Gundl. Repert. Fisico-Nat. Cuba, I, p. 295 (1866); *ib.* J. f. O.
   1874. p. 155.—Gray, Handl. Bds. II, p. 202 (1870).
Picus fernandinæ Wagl. Isis. 1829, p. 517.
Colaptes fernandina Denny, P. Z. S. 1847, p, 39.
Geopicus fernandinæ Malh. Mém. Acad. Metz, 1848-49, p. 359; *ib.* Mon.
   Pic. II, p. 273 (1862).
Colaptes fernandiæ Reich. Handb. Orn. p. 415, No. 975 (1854).
Nesoceleus fernandinæ Scl. & Salv. Nom. Avium Neotr. p. 101 (1873).—
   Cory, List Bds. W. I. p. 20 (1885).

Sp. Char. *Male:*—Top of the head pale yellowish brown, each feather
   having a central stripe of dark brown, giving the head a delicate
   striped appearance; a space around the eye, extending beyond the
   ear-coverts, pale rufous brown; a stripe of black passes from the
   under mandible down the sides of the throat; chin and throat
   black, the feathers edged with yellowish white; rest of plumage
   having the feathers banded with yellow and dark brown, brightest
   on the underparts; under wing-coverts pale yellow, somewhat
   marked with brownish; wings and tail dark brown, thickly banded
   with yellow; under surface of wings and tail showing the shafts of
   the feathers clear pale yellow, the shafts being brown on the
   upper surface.

   Female:—Similar to the male, but having the head and cheeks
   browner, and lacking the black stripe on the side of the throat.

   Length (skin), 12; wing, 6. 25; tail, 5; tarsus, 1; bill, 1.60.

Habitat.   Cuba.

GENUS **Ara** BRISS.

Ara BRISSON, Orn. 1760.

Ara tricolor (BECHST.).

Psittacus tricolor BECHST. Kurze Ueb. p. 64, pl. 1 (1811).
Macrocercus tricolor VIEILL. Nouv. Dict. p. 262.—LESS. Traité d'Orn. p.
    186 (1831).—CAB. J. f. O. 1856, p. 105.—BREWER, Pr. Bost. Soc.
    Nat. Hist. VII. p. 307 (1860).—GUNDL. Repert. Fisico-Nat. Cuba, I,
    p. 297 (1866); *ib.* J. f. O. 1874. p. 163.
Sittace tricolor WAGL. Mon. Psitt. p. 669 (1832).—FINSCH, Die Papag.
    Mon. Bearb. I, p. 409 (1867).
Ara tricolor D'ORB. in La Sagra's Hist. Nat. Cuba, Ois. p. 161 (1840).—
    GRAY, Gen. Bds. II, No. 5 (1844-49).—GOSSE, Bds. Jam. p. 260
    (1847).—ALBRECHT, J. f. O. 1862, p. 202.—SCL. & SALV. Nom.
    Avium Neotr. p. 111 (1873).—A. & E. NEWTON, Handb. Jamaica
    p. 110 (1881).—CORY, List Bds. W. I. p. 20 (1885).
Macrocercus makavouanna STEPH. Gen. Zool. XIV. p. 112.
Macrocercus (Aracanga) tricolor LEMB. Aves Cuba, p. 132 (1850).—BP
    Rev. Mag. Zool. 1854. p. 149.
Arara tricolor BP. Naum. 1856.—SCHLEG. Mus. Pays-Bas, Psitt. p. 4
    (1864).
Primolinus tricolor GRAY, Handl. Bds. II, p. 145 (1870).

SP. CHAR. *Male:*—Forehead red, becoming yellowish red on the top of
    the head, and shading into bright yellow on the nape; feathers of the
    upper back cinnamon red, edged with greenish; lesser wing-coverts
    brown, with reddish edgings; entire underparts scarlet red, showing a
    tinge of orange on the cheeks and throat, some of the feathers show-
    ing yellow on the belly; primaries and secondaries showing bright
    blue on the upper surface; under surface of primaries pale brownish
    red; upper surface of tail-feathers showing the feathers cinnamon red,
    shading into bright blue on the tips, some feathers showing more

blue than cinnamon, while in others the cinnamon predominates;
under surface of tail-feathers cinnamon red, showing bright orange
when held in the light : crissum pale blue ; bare skin around the eye
probably dull white ; bill dark brown, pale at the tip ; legs and feet
brownish ; iris pale yellow.

The sexes are probably similar.

Length (skin), including tail, about 18: wing, 11 ; tail, 10; tarsus,
.85 ; bill, curve, 2, straight line from base to tip, 1.75.

HABITAT. Cuba. Jamaica.

It is not impossible that *Ara militaris* may have occurred in
Cuba and Jamaica, but it is improbable. The bird recorded as
such was perhaps *A. tricolor* wrongly identified ; Gosse remarks, however, that every description he received of the bird
agreed with that of *A. militaris*, "the Great Green Macaw of
Mexico." Dr. Gundlach writes me he believes *A. tricolor* is still
to be found in the swamps of Cuba, and that years ago he killed
a number of birds of this species in the swamps of southern Cuba.
He has several fine specimens in his collection.

GENUS Conurus KUHL.

Conurus KUHL, Consp. Psitt. 4, 1830.

Conurus euops (WAGL.).

Sittace euops WAGL. Mon. Psitt. p. 638 (1832).
Psittacus euops HALM. Orn. Atl. Pap. p. 95 (1836).
Conurus guianensis D'ORB. in La Sagra's Hist. Nat. Cuba, Ois. p. 162
(1840).—LEMB Aves Cuba, p. 132 (1850).

Conurus euops GRAY, Gen. Bds.
II, No. 26 (1844-49).—FINSCH,
Die Papag. Mon. Bearb. I, p.
474 (1867).— SCL. & SALV.
Nom. Avium Neotr. p. 112
(1873). — CORY, List Bds.
W. I. p. 20 (1885).

Evopsitta evops BP. Rev. Mag.
Zool. 1854, p. 151.

Psittacara euops SOUANCÉ, Rev.
Mag. Zool. 1856, p. 69.

Conurus guianensis CAB. J. f O.
1856, p. 106. — BREWER, Pr.
Bost. Soc. Nat. Hist. VII, p.
307 (1860).

Conurus evops GUNDL. Repert. Fisico-Nat. Cuba. I, p. 297 (1866); *ib.*
  J. f. O. 1874, p. 164; *ib.* Anal. Soc. Esp. Hist. Nat. VIII, p. 229
  (1878).
Evopsitta guyanensis GRAY, Handl. Bds. II, 146 (1870).

SP. CHAR. *Male:*—General plumage green, dark above, light beneath; the
  head dotted here and there with small touches of bright red; this
  marking does not seem at all constant; a broad patch of vermillion
  red on the under wing-coverts, extending upon the carpus; shafts of
  wing and tail-feathers brownish black; bare skin around the eye
  bluish white.
  The sexes are apparently similar.
  Length (skin) 9.75; wing. 5.50; tail, 5.50; tarsus, .50.
HABITAT. Cuba.

Conurus xantholæmus SCL.

Conurus xantholæmus SCL. Ann. Nat. Hist. 3d ser. IV, p. 225; *ib.* Cat.
  Am. Bds. p. 348 (1862).—A. & E. NEWTON, Ibis, 1859, p. 374.—
  CASSIN, Pr. Acad. Nat. Sci. Phila. 1860, p. 377.—TAYLOR, Ibis,
  1864, p. 171.—GRAY, Handl. Bds. II, p. 148 (1870).—CAB. J. f. O.
  1879, p. 222.—CORY, List Bds. W. I. p. 20 (1885).
Conurus xanthogenius SCL. & SALV. Nom. Avium Neotr. p. 112 (1873).

SP. CHAR. *Male:*—Forehead, sides of the head and chin bright orange;
  rest of upper surface bright green; underparts yellowish green,
  marked with orange on the belly and abdomen; primaries blue,
  showing green on the outer webs, and having the shafts and tips
  brown; upper surface of tail green; under surface of tail yellow; bill
  dark.
  The sexes are similar.
  Length (skin). 9.50; wing. 6; tail, 5; tarsus, .35; bill, .90.
HABITAT. St. Thomas and St. Croix.

Conurus nanus (VIG.).

Psittacara nana VIG. Zool. Journ. V, p. 273 (1830).—LEAR, Parr. pl. 12.
  —WAGL. Mon. Psitt. p. 640 (1832).
Conurus nanus GRAY, Gen. Bds. II, No. 32 (1844-49).—SOUANCÉ, Icon.
  Perr. pl. 12. fig. 1.—ALBRECHT, J. f. O. 1862. p. 203.—SCL. Cat. Am.
  Bds. p. 349 (1862).—MARCH, Pr. Acad. Nat. Sci. Phila. 1863. p. 284.
  —FINSCH, Die Papag. Mon. Bearb. I, p. 520 (1867).—GRAY. Handl.
  Bds. II, p. 148 (1870).—SCL. & SALV. Nom. Avium Neotr. p. 112
  (1873).—A. & E. NEWTON. Handb. Jamaica, p. 110 (1881).—CORY,
  List Bds. W. I. p. 20 (1885).
Conurus flaviventer GOSSE, Bds. Jam. p. 263 (1847).

Sp. Char. *Male:*—Entire upper plumage bright green; underparts showing a brownish tinge on the throat and breast; dull olive on the
belly; upper surface of tail bright green shading into blue at the
tip; under surface of tail yellow; outer webs of primaries green,
inner webs blue, becoming brown at the edge; bill pale.

The sexes are apparently similar.

Length (skin), 10; wing, 5.25; tail, 5; bill, 1.

HABITAT. Jamaica.

Conurus chloropterus (Souancé).

Psittacara chloroptera SOUANCE. Rev. Mag. Zool. 1856, p. 59.

Conurus chloropterus SCL. P. Z. S. 1857, p. 234.—FINSCH, Die Papag.
Mon. Bearb. I, p. 46 (1867).—CORY. Bds. Haiti & San Domingo,
p. 113 (1855); *ib.* List Bds. W. I. p. 20 (1885).

Psittacus (Conurus) chloropterus BRYANT, Pr. Bost. Soc. Nat. Hist. XI,
p. 96 (1866).

Sp. Char. *Male:*—General plumage yellowish green, darkest on the tail
and wings; under surface of wings and tail greenish yellow; under
wing-coverts bright scarlet, showing upon the edge of the wing;
bill pale; feet dark brown; some specimens occasionally show
feathers tipped with red upon the back and wings, but generally so
slightly as to be hardly noticeable. Immature birds often show
yellow on the primaries.

The sexes are similar.

Length, 12; wing, 7; tail, 6; tarsus, .50.

HABITAT. San Domingo.

Conurus gundlachi CABAN.

? *Conurus evops* GUNDL. J. f. O. 1878, p. 184.

Conurus gundlachi CAB. Orn. Centralb. VI. p. 1 (1881); *ib.* J. f. O. 1882,
p. 119.—GUNDL. J. f. O. 1881, p. 401.

Sp. Char.—"This species is distinguished from *Conurus evops* in having
the wings nearly 3 ctm. longer, and by the extension of the red
coloring of the lower wing-coverts, also on the under row of large
wing-coverts, which in *evops* are yellowish olive as in most of the
species of *Conurus.*" (CABANIS, l. c., orig. descr., translated.)

HABITAT. Mona, near Porto Rico.

I have never seen this species, and the descriptions of it are
somewhat meagre. Dr. Gundlach says that he was told the
birds came from San Domingo to the Island of Mona, but adds
that this is only a supposition.

GENUS **Chrysotis** SWAINS.

Chrysotis "SWAINSON. Zool. Journ. 1837."

Chrysotis sallæi SCL.

"*Psittacus leucocephalus?* GMEL. Syst. Nat. I, p. 338 (1788)."

Chrysotis sallæi SCL. P. Z. S. 1857, p. 224; *ib.* Cat. Am. Bds. p. 353 (1862).
—SCL. & SALV. Nom. Avium Neotr. p. 113 (1873).—CORY, Bds.
Haiti & San Domingo, p. 115 (1885); *ib.* List Bds. W. I. p. 20
(1885).

Amazona sallæi SCHLEG. Mus. Pays-Bas, Psitt. p. 58 (1864).

Psittacus (Chrysotis) sallæi BRYANT, Pr. Bost. Soc. Nat. Hist. XI, p. 96
(1866).

Chrysotis ventralis GRAY, Handl. Bds. II, p. 164 (1870).

Chrysotis sallei CORY, Bull. Nutt. Orn. Club, VII, p. 154 (1881).

SP. CHAR. *Male:*—General plumage green; forehead white, extending in
front of the eye; top of head blue, narrowly tipped with black; a
patch of black on the cheeks; abdomen and basal half of tail-feathers
crimson, wanting upon the outer web of the outer tail-feather;
thighs pale blue in very adult birds, in most specimens green, the
blue being scarcely perceptible; primaries and secondaries dark
blue on the outer webs; inner webs dark brown; a small patch of
red on the throat, which does not appear to be constant.

The sexes are apparently similar.

Length, 10.50; wing, 8; tail, 4.50; tarsus, .60.

HABITAT. Haiti and San Domingo.

Chrysotis vittata (BODD.).

Psittacus vittatus BODD. Tabl. Pl. d' Aub. p. 49 (1783).—GRAY. Gen. Bds.
II, No. 24 (1844-49).—SUNDEV. Oefv. K. Vet. Akad. För. 1869, p.
599.

Psittacus dominicensis GMEL. Syst. Nat. I, p. 343 (1788).—VIEILL. Enc.
Méth. p. 1375.—WAGL. Mon. Psitt. p. 597 (1832).

Psittacus leucocephalus KUHL, Consp. p. 80 (1821).

Chrysotis dominicensis BP. Rev. Mag. Zool. 1854, p. 151.

Chrysotis vittata SCL. P. Z. S. 1857, p. 224.—GRAY, List Psitt. p. 83
(1859).—FINSCH, Die Papag. Mon. Bearb. II, p. 515 (1867).—GRAY.
Handl. Bds. II, p. 165 (1870).—SCL. & SALV. Nom. Avium Neotr.
p. 113 (1873).—CORY, List Bds. W. I. p. 20 (1885).

Amazona vittata SCHLEG. Mus. Pays-Bas, Psitt. p. 58 (1864).

Chrysotis vittatus TAYLOR, Ibis, 1864, p. 171.—GUNDL. J. f. O. 1874, p.
312; *ib.* Anal. Soc. Esp. Hist. Nat. VII, p. 228 (1878).

Chrysotis ———— BRYANT, Pr. Bost. Soc. Nat. Hist. X, p. 256 (1866).

Sp. Char. *Male:*—A narrow band of red on the forehead; upper plumage green; the feathers of the head and neck edged with dark brown, heaviest on the neck; underparts green, lighter than the back; yellowish green on the belly; the feathers on the breast edged with dull brown; primaries having the outer webs bright blue, lacking on the first.

The sexes are similar.

Length (skin), 10.50; wing, 7; tail, 4; tarsus, .60; bill, 1.35.

Habitat.   Porto Rico.

Chrysotis collaria (Linn.).

Psittacus collarius Linn. Syst. Nat. I, p. 149 (1766).—Gmel. Syst. Nat. I, p. 347 (1788).

Psittacus gutturalis Müll. Syst. Nat. Suppl. p. 78 (1776).

Psittacus leucocephalus, var. β. Lath. Ind. Orn. p. 118 (1790).—Wagl. Mon. Psitt. p. 600 (1832).

Psittacus leucocephalus Gray, Gen. Bds. II, No. 25 (1844-49).—Gosse, Bds. Jam. p. 269 (1847).

Pionus vinaceicollis Lafr. Rev. Zool. 1846, p. 241.

Psittacus vinaceicollis Gray, Gen. Bds. III, App. p. 20 (1849).

Chrysotis leucocephala Br. Naum. 1856, ———.

Chrysotis vinaceicollis Scl. P. Z. S. 1857. p. 225.

Chrysotis collaria Scl. P. Z. S. 1861, p. 79; *ib.* Cat. Am. Bds. p. 353 (1862).—Albrecht, J. f. O. 1862, p. 203.—March, Pr. Acad. Nat. Sci. Phila. 1863, p. 284.—Finsch, Die Papag. Mon. Bearb. II. p. 517 (1868).—Gray. Handl. Bds. II, p. 164 (1870).—Scl. & Salv. Nom. Avium Neotr. p. 113 (1873).—A. & E. Newton. Handb. Jamaica, p. 110 (1881).—Cory, List Bds. W. I. p. 20 (1885).

"*Chrysotis amazonicus*. juv. Bl. Cat. Bds. p. 11."

Sp. Char. *Male:*—Top of head blue, the feathers indistinctly showing black edges, fading into green on the nape; a narrow line of white on the forehead; the rest of upper surface bright green; throat dull red, some of the feathers narrowly edged with green; cheeks greenish blue; rest of underparts green; tail green, most of the feathers having the basal half of the inner webs dull green, faintly showing on the outer webs, the red entirely wanting on the central feathers.

Length (skin), 11.50; wing 7; tail, 4.50.

Habitat.   Jamaica.

Chrysotis leucocephala (Linn.).

Psittacus leucocephalus Linn. Syst. Nat. I, p. 100 (1758). — Aldrov. Orn. I, p. 670.—Gmel. Syst. Nat. I, p. 338 (1788).—Wagl. Mon. Psitt. p. 599 (1832).—Gundl. Repert. Fisico-Nat. Cuba, I, p. 297 (1866).

Psittacus martinicanus Briss. Orn. IV, p. 242 (1760).

Chrysotis leucocephalus SWAINS. Class. Bds. II. p. 301 (1837).—Br. Rev.
Mag. Zool. 1854.—CAB. J. f. O. 1856. p. 105.—BREWER, Pr. Bost.
Soc. Nat. Hist. VII. p. 307 (1860).—FINSCH. Die Papag. Mon.
Bearb. II, p. 520 (1868).—GRAY, Handl. Bds. II, p. 164 (1870). —
GUNDL. J. f. O. 1874. p. 161.

Amazona leucocephala SCHLEG. Mus. Pays-Bas, Psitt. p. 59 (1864).

Psittacus (Chrysotis) collaria (var. *bahamensis*) BRYANT, Pr. Bost. Soc.
Nat. Hist. VI. p.—— (1866).

Chrysotis leucocephala SCL. & SALV. Nom. Avium Neotr. p. 113 (1873).—
CORY, List Bds. W. I. p. 20 (1885).

Chrysotis collaria CORY, Bds. Bahama I. p. 123 (1880).

SP. CHAR. *Male:*—General plumage green, many feathers narrowly tipped
with dark brown; forehead and the top of the head to the eyes
white; throat bright red, sometimes touched with green; some of
the feathers on the belly brownish red; outer webs of primaries blue,
inner webs brown; tail green, showing blue on outer webs of
outer feathers, and red on the basal half of inner webs of all except
the central ones, which sometimes show a trace near the shaft; bill
yellowish white; iris brown.

The sexes are similar.

Length, 13; wing, 8; tail, 4.50; tarsus, .75; bill, 1.25.

HABITAT. Cuba and Bahamas.

Birds from the Island of Inagua, Bahamas, vary somewhat from Cuban examples. The Cuban bird shows deep
crimson on the belly; the tail is light green, having the basal
half of the inner web of the outer feathers deep red, rest of web
pale yellowish green, the red marking becoming less and less on
succeeding feathers until obsolete or nearly so on the two central
ones; red coloring slight or absent on under wing-coverts. The
Bahama bird differs from the above by showing very little red on
the belly, sometimes nearly absent; more red on the crissum and
under wing-coverts; tail bluish green, the red marking shown
only on the two outer feathers. It is possible that the differences
in question are not constant, but judging from the material examined I am of the opinion that the two forms are at least separable sub-specifically. If this should prove to be the case Dr.
Bryant's name *bahamensis* would be used for the Bahama bird.

Chrysotis agilis (LINN.).

Psittacus cayenensis BRISS. Orn. IV, p. 237 (1760).

Psittacus agilis LINN. Syst. Nat. I, p. 143 (1766).—GOSSE. Bds. Jam. p.
266 (1847).

Psittacus minor VIEILL. Nouv. Dict. XXV. p. 314.

Psittacus signatus SHAW, Gen. Zool. VIII. p. 510.—KUHL, Consp. p. 71 (1821).

Psittacus æstivus KUHL, Consp. p. 75 (1821).

Psittacus virescens BECHST. Kurze Ueb. p. 99.

Chrysotis signatus SWAINS. Class. Bds. II. p. 301 (1837).

Chrysotis agilis GRAY, List Psitt. p. 82 (1859).—SCL. Cat. Am. Bds. p. 354 (1862).—ALBRECHT, J. f. O. 1862, p. 203.—MARCH, Pr. Acad. Nat. Sci. Phila. 1863, p. 284.—FINSCH, Die Papag. Mon. Bearb. II, p. 531 (1868).—GRAY, Handl. Bds. II. p. 164 (1870).—SCL. & SALV. Nom. Avium Neotr. p. 113 (1873).—A. & E. NEWTON, Handb. Jamaica, p. 110 (1881).—CORY, List Bds. W. I. p. 20 (1885).

SP. CHAR. *Male:*—Top of head dark green, becoming somewhat lighter on the back; underparts light green; most of the feathers of the head, throat, and back delicately edged with black; a patch of black on the ear-coverts; primaries blue on the outer webs, edged with green on the basal portion, the blue color but slightly showing on the first four primaries; terminal portion of quills dark brown; tail green. One specimen before me has one of the wing-coverts bright red on one side, lacking in the other specimens

The sexes are similar.

Length (skin), 10; wing, 6.50; tail, 4; tarsus, .40; bill, 1.20.

HABITAT. Jamaica.

Chrysotis augusta (VIG.).

Psittacus havanensis KUHL, Consp. p. 79 (1821).—WAGL. Mon. Psitt. p. 740 (1832).

Psittacus augustus VIG. P. Z. S. 1836, p. 80.—GRAY & MITCH. Gen. Bds. pl. 104 (1844-49).

Chrysotis augustus GRAY & MITCH. Gen. Bds. No. 16 (1844-49).

Œnochrus augustus BP. Rev. Mag. Zool. 1854, p. 151.

Amazona augustus SCHLEG. Mus. Pays-Bas, p. 50 (1864).

Chrysotis augusta SCL. P. Z. S. 1865, p. 437.—FINSCH, Die Papag. Mon. Bearb, II, p. 557 (1868).—SCL. & SALV. Nom. Avium Neotr. p. 113 (1873).—LAWR. Pr. U. S. Nat. Mus. I, pp. 62, 487 (1878).—SCL. P. Z. S. 1881, p. 627.—CORY, List Bds. W. I. p. 20 (1885).

SP. CHAR.—Top of the head tinged with purplish, shading into green on the nape; cheeks and throat reddish purple, the feathers edged with bluish; the feathers of the sides of the neck and nape tipped with dull purple, forming a sort of collar; back and wings green: feathers on the rump slightly tipped with bluish; breast and underparts dull purple, the feathers pale at the tips, and showing delicate bluish edging when held in the light; flanks green; a patch of red on the outer webs of two of the secondaries, faintly showing on the

last primary, dull towards the terminal portion, becoming bright
scarlet on the basal half; a patch of red on the edge of the carpus;
quills brown; tail purplish-brown; under surface of tail green.

Length, 17; wing, 11; tail, 7; tarsus, .75; bill, 1.30.

HABITAT. Dominica.

Chrysotis guildingi (VIG.).

Psittacus guildingi VIG. P. Z. S. 1836, p. 80.—GRAY, Gen. Bds. No. 28
(1844-49).

Chrysotis guildingi BP. Rev. Mag. Zool. 1854, p. 151.—FINSCH, Die Papag.
Mon. Bearb. II. p. 559 (1868).—GRAY, Handl. Bds. II. p. 164 (1870).—
SCL. & SALV. Nom. Avium Neotr. p. 113 (1873).—LAWR. Pr. U. S.
Nat. Mus. I, pp. 193, 487 (1878).—LISTER, Ibis, 1880, p. 42.—SCL.
P. Z. S. 1881, p. 627.—CORY, List Bds. W. I. p. 20 (1885).

Amazona guildingi SCHLEG. Mus. Pays-Bas, p. 49 (1864).

SP. CHAR.—Top of the head and cheeks yellowish white, shading into
yellowish orange on the throat, and into bluish feathers with black
edgings on the sides of the neck and nape; back pale yellowish
brown, feathers edged with dull black; underparts pale reddish
brown, feathers shading into green at the ends, and tipped with
black on the belly; under surface of tail dull orange, pale yellow at
the tip, feathers banded across the middle by a broad stripe of green;
upper surface of tail-feathers yellow at base and tip, showing a band
of blue across the centre; primaries dark blue, the basal half of the
first five yellow, the rest shading into orange on the outer webs of
the basal half and showing less and less yellow on the inner webs;
under surface of wing yellow; greenish on the under wing-coverts;
bill pale.

The sexes are apparently similar.

Length, 19; wing, 10.75; tail, 7; tarsus, .75.

HABITAT. St. Vincent.

Chrysotis bouqueti (BECHST.).

Psittacus arausiacus MÜLL. Syst. Nat. Suppl. p. 79 (1766).

Psittacus autumnalis var. β. GMEL. Syst. Nat. I, p. 345 (1788).—LATH.
Ind. Orn. p. 124 (1790).

Psittacus bouqueti BECHST. Lath. Ueb. p. 99 (1793).—KUHL. Consp. p. 76
(1821).

Psittacus cyanocapillus VIEILL. Enc. Méth. p. 1373.—BURM. Syst. Ueb.
II, p. 186.

Psittacus cæralifrons SHAW, Gen. Zool. VIII, p. 515.—VOIGHT. Cuv.
Ueb. 1831, p. 741.

Chrysotis cyanocephalus SWAINS. Class. Bds. II, p. 301 (1837).

Chrysotis bouqueti GRAY, Gen. Bds. II, No. 11 (1844-49).—BP. Rev. Mag.

Zool. 1854.——.—SCL. P. Z. S. 1881, p. 627.—CORY, List Bds. W. I.
p. 20 (1885).
Chrysotis nichollsi LAWR. Pr. U. S. Nat. Mus. III. p. 254 (1880).

SP. CHAR.—Front of head and throat dull purplish blue, palest on the
throat; the feathers of the cheeks green, tipped with bluish; upper
parts dark green, and the feathers tipped with dark brown; quills
dark brown, tinged with green on the outer webs, and showing a
patch of red on the outer web of the tenth and eleventh feathers; a
patch of dull red mixed with yellow on the upper part of the breast
joining the throat; underparts green. the feathers narrowly edged
with bluish; tail-feathers showing red at the base of the inner webs,
succeeded by dark green at the middle. and tipped with light green.
    Length, 15; wing, 9.25; tail. 6.50; tarsus, .62; bill, 1.50.
HABITAT.   Dominica.

Chrysotis versicolor (MÜLL.).

Psittacus versicolor MÜLL. Syst. Nat. Suppl. p. 78 (1766).
Chrysotis cyanopsis FINSCH, Die Papag. Mon. Bearb. II, p. 528 (1868).—
    SCL. & SALV. Nom. Avium Neotr. p. 113 (1873).
Œnochrus versicolor GRAY, Handl. Bds. II, p. 165 (1870).
Chrysotis bouqueti SCL. P. Z. S. 1874, p. 323; *ib.* 1875, p. 61.—ALLEN,
    Bull. Nutt. Orn. Club. V, p. 169 (1880).
Chrysotis versicolor SCL. P. Z. S. 1881, p. 627.—CORY, List Bds. W. I. p.
    20 (1885).

SP. CHAR.—Forehead and in front of the eye dark blue; top of the head,
cheeks, and throat pale blue. the feathers tipped with black; a few
yellowish feathers at the base of the skull; breast red, the feathers
tipped with dark brown; underparts red. the feathers tipped with
green; back and wing-coverts green, tipped with black; primaries
dark blue on the basal portion of the outer webs; rest of the primaries dark brown; a broad patch of red on the outer webs on the
tenth and eleventh feathers; outer tail-feathers showing bright red
at the base, a band of blue extending half the length of the feather,
and the terminal portion being yellow; under surface of tail-feathers
green, having the terminal portion pale greenish yellow, and showing the red on the base of the outer feathers.
    Length, 18; wing, 11; tail, 8; tarsus, .75; bill, 1.40.
HABITAT.   Santa Lucia.

FAMILY STRIGIDÆ.

GENUS Strix LINN.

Strix LINNÆUS, Syst. Nat. I. p. 131 (1766).

Strix flammea furcata (TEMM.).

Strix furcata TEMM. Pl. Col. p. 432 (1832).—D'ORB. in La Sagra's Hist.
Nat. Cuba, Ois. p. 34 (1840).—GUNDL. J. f. O. 1855. p. 467.—
BREWER, Pr. Bost. Soc. Nat. Hist. VII. p. 306 (1860).—GUNDL.
Repert. Fisico-Nat. Cuba, I, p. 227 (1865); *ib.* J. f. O. 1871. p. 79.
Strix pratincola GOSSE, Bds. Jam. p. 23 (1847).—SCL. P. Z. S. 1861. p.
79.—ALBRECHT, J. f. O. 1862. p. 204.
Glyphidura furcata GRAY, Handl. Bds. I, p. 52 (1869).
Strix flammea var. *furcata* BD. BWR. & RIDGW. Hist. N. Am. Bds. III,
p. 12 (1874).
Strix flammea fasciata CORY, List Bds. W. I. p. 21 (1885).

This form varies from the usual coloration in having the secondaries and tail nearly white, the latter usually without bars.
Wing, 12.50; tail, 5.70; tarsus, 2.75.

HABITAT. Cuba and Jamaica.

Bubo virginianus is recorded from Jamaica in Sclater &
Salvin's 'Nomenclator Avium Neotropicalium.' I have been un-
able to find other West Indian records.

Strix flammea nigrescens LAWR.

Strix flammea var. *nigrescens* LAWR. Pr. U. S. Nat. Mus. I, p. 64 (1878).
Strix flammea LISTER, Ibis, 1880, p. 44.
Strix flammea nigrescens CORY, List Bds. W. I. p. 21 (1885).

SP. CHAR.—"Upper plumage of a fine blackish brown, rather sparsely
marked with small white spots; the tail is crossed with alternate
bands of brown and light dull ochraceous freckled with brown;
the wings are the color of the back, somewhat intermixed with
rufous; the under plumage is light reddish-ochraceous, marked with
small, round black spots (the color is lighter than the under plumage of the Costa Rica specimen); the ends of the ruff-feathers are
dark reddish-brown; feathers around the eye, black; the face is of
a light reddish fawn color. Bill white; iris deep chocolate, half an
inch in diameter. Length (fresh), 13 in.; wing, 10; tail, 4½; tarsus 2."
(LAWR., orig. descr., l. c.)

HABITAT. St. Vincent and Dominica.

Strix flammea pratincola.

Strix pratincola BONAP. List, 1838, p. 7.
Strix flammea var.? BRYANT, Pr. Bost. Soc. Nat. Hist. XI, p. 65 (1867)
(Bahamas).

Strix flammea var. *pratincola* Bd. Bwr. & Ridgw. Hist. N. Am. Bds.
    p. 13, III (1874).—Cory. Bds. Bahama I. p. 125 (1880).
Strix flammea pratincola Cory, List Bds. W. I. p. 21 (1885).

    This form occurs in the Bahama Islands.

Strix glaucops Kaup.

Athene dominicensis Bp. Consp. I, p. 38 (1850).—Sallé. P. Z. S. 1857. p.
    231.
Strix glaucops "Kaup, Contr. Orn. p. 118 (1852)"; *ib.* Tr. Z. S. IV, p.
    246.—Pelz. J. f. O. 1872, p. 23.—Cory, Bds. Haiti & San Domingo,
    p. 117 (1885); *ib.* List Bds. W. I. p. 21 (1885).
Strix (Athene) dominicensis Bryant, Pr. Bost. Soc. Nat. Hist. XI. p. 90
    (1867).
Strix flammea Sharpe. Cat. Bds. Brit. Mus. II, p. 292 (1875).
Strix dominicensis Cory, Bull. Nutt. Orn. Club, VII, p. 95 (1883).

Sp. Char. *Male:* — General plumage above dark brown, shading into
    orange-rufous on the side of the neck; quills showing inner webs
    brownish; outer webs dull orange-rufous, banded with brown; entire
    underparts pale orange-rufous, mottled with zigzag marking of light
    brown, whitening somewhat on the throat and abdomen; face deep
    gray; an ante-orbital spot of black; circle of feathers around the
    face dark chestnut, bordered with black on the throat; tarsus not
    feathered to the feet.
    The sexes are apparently similar.
    Length, 13.50; wing, 10; tail, 4.60; tarsus, 3.45.

Habitat.   Haiti and San Domingo.

Genus Pseudoscops Kaup.

Pseudoscops Kaup, Isis, 1848, p. 769.

Pseudoscops grammicus (Gosse).

Ephialtes grammicus Gosse, Bds. Jam. p. 19 (1847).
Pseudoscops grammicus Kaup, Isis, 1848, p. 769.—Scl. P. Z. S. 1861, p.
    80.—Albrecht, J. f. O. 1862, p. 204.—Gray, Handl. Bds. I, p. 51
    (1869).—Scl. & Salv. Nom. Avium Neotr. p. 116 (1873).—A. & E.
    Newton, Handb. Jamaica, p. 110 (1881).—Cory, List Bds. W. I. p.
    21 (1885).
Scops grammicus Bp. Consp. I, p. 46 (1850).—Strickl. Orn. Syn. p. 205.
Asio grammicus Sharpe, Cat. Bds. Brit. Mus. II, p. 242 (1875).

      "*Adult female.* General colour above sandy-buff, transversely
    vermiculated with dark brown, more distinctly on the back and
    scapular feathers, some of which are mesially streaked with dark

brown, some of the outermost rather more broadly barred with
sandy colour, but not exhibiting any trace of white or buff spots;
head and neck rather lighter sandy colour, the transverse black lines
very regular, especially on the ear tufts, which are coloured like the
rest of the head, but are nearly uniform rufous on their inner webs;
entire facial aspect foxy red, the hindermost of the ear-coverts
whitish, tipped with black, merging in the ruff, which is composed
of sandy-buff feathers, black at tip and at base; the stiff gular feathers sandy rufous, streaked and indistinctly barred with black; chin-
feathers buffy white; rest of the under surface deep sandy rufous,
with faint indications of dusky vermiculations of brown, the abdom-
inal plumes more or less verging on white and showing very little
of the mesial black streaks which are so distinct on the breast
feathers; leg feathers uniform tawny buff, as also the under tail-
coverts; under wing-coverts buff, slightly washed with sandy-rufous,
the lower series black, with fulvous bases, forming a bar across the
wing, and resembling the inner lining of the quills, which are ful-
vescent towards the base of the inner web, on which they are broadly barred with black; upper wing-coverts resembling the back, and
very coarsely vermiculated with dark brown, the sandy-buff bars
being pretty apparent here and there, and especially distinct on the
outer web of the spurious quills; primary-coverts dark brown, irregularly barred across with sandy-buff; quills banded alternately for
their entire length with blackish brown and sandy-buff, these bars
less distinct on the secondaries, the light interspaces obscured with
dark brown vermiculations, especially the innermost, which conse-
quently resemble the back; tail sandy-buff barred across with dark
brown, about eleven bars being traceable on the centre feathers, the
interspaces more or less mottled with vermiculations of brown, the
exterior rectrix paler and more fulvous, crossed with about thirteen
brown bars; cere blackish grey; bill pale blue-grey; feet dull lead-
colour; claws horny grey; iris hazel. Total length 12.2 inches,
wing 8.4, tail 5.1, tarsus 1.55." (SHARPE, l. c.)

HABITAT. Jamaica.

GENUS Asio BRISS.

Asio BRISSON, Orn. I, p. 28 (1760).

Asio stygius (WAGL.).

Nyctalops stygius WAGL. Isis, 1832, p. 1221.—GRAY. List Gen. Bds. p. 6.
—SCL. & SALV. Nom. Avium Neotr. p. 116 (1873).
Otus sygmapa D'ORB. in La Sagra's Hist. Nat. Cuba, Ois. p. 31
(1840).—GRAY, Gen. Bds. I, p. 40 (1844-49).—Br. Consp. I. p. 50
(1850).—CAB. J. f. O. 1855, p. 465.—GUNDL. Repert. Fisico-Nat.
Cuba, I, p. 226 (1865); *ib.* J. f. O. 1871. p. 374.

Otus stygius PUCHER. Rev. Mag. Zool. 1849, p 29.—GRAY, Gen. Bds. I, p.
40 (1844-49).—KAUP, Contr. Orn. 1852, p. 113.—BREWER, Pr. Bost.
Soc. Nat. Hist. VII, p. 306 (1860).—BD. BWR. & RIDGW. Hist. N.
Am. Bds. III, p. 17 (1874).
Otus melanopsis LICHT. Nom. Av. p. 6.
Asio stygius STRICKL. Orn. Syn. p. 207.—SHARPE, Cat. Bds. Brit. Mus.
II, p. 241 (1875).—CORY, List Bds. W. I. p. 21 (1885).
Asio signata STRICKL. Orn. Syn. p. 212.

SP. CHAR. "*Adult.* Above of a nearly uniform chocolate-brown, the
hinder neck and wig with a few longitudinal spots of light ochre,
rather more oval in shape on the latter, the rest of the upper surface having concealed spots and vermiculations of whitish buff,
rather larger and more distinct on the outer margin of the scapulars;
wing-coverts uniform with the back, with the same more or less
concealed vermiculations, these being absent on the primary-coverts,
which are uniform brown; quills chocolate-brown like the back,
with obsolete bars of lighter brown, more distinct on the secondaries, where they are often replaced by ochraceous spots or vermiculations, the innermost uniform with the back, and slightly freckled
with vermiculations of whitish buff or light ochre, the primaries with
a few spots of deep ochre on the outer web, only distinct or of any
size near the base; upper tail-coverts brown, with distinct bars of
bright ochre; tail deep chocolate-brown, with a whitish tip, and
crossed with five or six additional bars of bright ochre, these bars
more numerous (seven or eight) on the inner web when the tail is
spread; forehead and feathers above the eye brown, streaked with
silvery grey; ear-tufts 2 inches long, chocolate-brown, with outer
margins of light ochre; sides of face dingy brown, the cheeks
streaked with fulvous, and the ear-coverts fulvous at their bases;
ruff brown, mottled with light ochre, the hinder feathers almost entirely ochraceous, with brown margins and shaft-stripes; chin dingy
brown, mottled with pale ochraceous, the ruff across the throat composed of white feathers with dark brown centres; rest of under surface ochraceous, mottled with brown, this color more prevalent on
the upper breast where it occupies the centre of the feathers; the
lower breast and abdomen streaked with brown down the middle of
the feathers, with dark brown lateral bars to each, the interspaces
being oval spots of white; leg-feathers deep ochre, spotted with triangular brown markings; under tail-coverts deep ochre, the longest
ones streaked with brown; under wing-coverts deep ochre, the outermost spotted and margined with brown, the greater series light
ochraceous at base, dark brown at tip, thus resembling the inner
lining of the wing, which is almost entirely dark brown, excepting
a few irregular bars of ochraceous, these being almost entirely
absent near the primaries. Total length 20 inches, wing 13.7, tail
7.8, tarsus 1.4." (SHARPE, l. c.)

HABITAT. Cuba.

Asio accipitrinus (PALL.).

Strix acciptrina PALL. Reise Russ. Reich. I, p. 455 (1771).
Otus brachyotus LEMB. Aves Cuba, p. 21 (1850).
Brachyotus palustris CAB. J. f. O. 1855, p. 465 (Cuba).
Brachyotus cassinii BREWER, Pr. Bost. Soc. Nat. Hist. VII, p. 306 (1860)
   (Cuba).—GUNDL. Repert. Fisico-Nat. Cuba, I, p. 226 (1865); *ib.*
   J. f. O. 1871, p. 375 (Cuba).
Asio accipitrinus SHARPE. Cat. Bds. Brit. Mus. II, p. 234 (1875).—CORY,
   List Bds. W. I. p. 21 (1885).

Accidental in Cuba.

Asio portoricensis RIDGW.

Strix brachyotus SUNDEV. Oefv. K. Vet. Akad. För. 1869, p. 601 (Porto
   Rico) (?)
Brachyotus cassinii GUNDL. Anal. Soc. Esp. Hist. Nat. VII, p. 165, 1878;
   ib. J. f. O. 1878, p. 158 (Porto Rico).
Asio portoricensis RIDGW. Pr. U. S. Nat. Mus. IV, p. 366 (1881) (Porto
   Rico).—CORY, List Bds. W. I. p. 21 (1885).

SP. CHAR. "Above dusky brown, nearly or quite uniform on the dorsal
   region; the scapulars, however, narrowly bordered with pale ochraceous or dull buff; feathers of the head narrowly, and those of the
   nape broadly, edged with buff; rump and upper tail-coverts paler
   brown or fawn-color, the feathers marked near their tips by a cres-
   centic bar of dark brown. Tail deep ochraceous, crossed by about
   five distinct bands of dark brown, these very narrow on the lateral
   rectrices, but growing gradually broader toward the intermediæ,
   which are dark brown, with five or six pairs of ochraceous spots
   (corresponding in position to the ochraceous interspaces on the
   outer tail-feathers), these spots sometimes having a central small
   brown blotch. Wings with dark brown prevailing, but this much
   broken by a general and conspicuous spotting of ochraceous; primaries crossed with bands of dark brown and deep ochraceous, the
   latter broadest on the outer quills, the pictura of which is much
   as in *A. accipitrinus*, but with the lighter color usually less extended. Face with dull, rather pale, ochraceous prevailing; this
   becoming nearly white exteriorly, where bordered, around the side
   of the head, by a uniform dark brown post-auricular bar; eyes entirely surrounded by uniform dusky, this broadest beneath and
   behind the eye. Lower parts pale ochraceous or buff, the crissum,
   anal region, tarsi, and tibiæ entirely immaculate; jugulum and
   breast marked with broad stripes of dull brown, the abdomen, sides
   and flanks with narrow stripes or streaks of the same. Bill dusky;
   iris yellow. Wing, 11.25-12.00; tail, 5.25-5.50; culmen, .70; tarsus,
   1.85-2.00; middle toe, 1.20-1.30" (RIDGW., orig. descr.)

HABITAT. Porto Rico.

GENUS Gymnasio BONAP.

Gymnasio BONAPARTE, Rev. Mag. Zool. 1854. p. 543.

Gymnasio nudipes (DAUD.).

Strix nudipes DAUD. Traité d'Orn. II, p. 199 (1800).—VIEILL. Ois. Am.
    Sept. I, p. 45 (1807).
Noctua nudipes STEPH. Gen. Zool. XIII, p. 70.—LESS. Traité d'Orn. p.
    104.
Athene nudipes GRAY, Gen. Bds. I, p. 35 (1844).—STRICKL. Orn. Syn. p.
    173.
Surnia nudipes BP. Oss. Rég. An. Cuv. p. 59.
Surnium nudipes KAUP, Contr. Orn. p. 120 (1852).
Gymnasio nudipes BP. Rev. Mag. Zool. 1854. p. 543.—SHARPE, Cat. Bds.
    Brit. Mus. II, p. 149 (1875).—CORY, List Bds. W. I. p. 21 (1885).
Gymnoglaux nudipes A. & E. NEWTON, Ibis, 1859. p. 54; *ib.* 1860. p. 307.
    —CASSIN, Pr. Acad. Nat. Sci. Phila. 1860. p. 374.—SCL. & SALV.
    Nom. Avium Neotr. p. 116 (1873).—GUNDL. Anal. Soc. Esp. Hist.
    Nat. VII, p. 166 (1878); *ib.* J. f. O. 1878. p. 158.
Gymnoglaux newtoni LAWR. Ann. Lyc. N. Y. VIII, p. 258 (1867).
Gymnoglaux krugii CAB. J. f. O. 1875. p. 223.

SP. CHAR.—Entire upper surface reddish brown; feathers of the breast
    and belly pale, variously dotted and banded with light brown; face
    pale brown, showing whitish between the eyes; the feathers slight-
    ly marked with whitish on the cheeks and throat; under tail-coverts
    white, narrowly shafted with brown; tail dull brown; primaries
    brown, dotted with white, mixed with brownish white on the outer
    webs; lining of wing dull white, mottled with brown on the carpus.
        Length, 9; wing, 6.75; tail, 3; tarsus, 1.45; bill, 70.

HABITAT.   Porto Rico, St. John, St. Croix, and St. Thomas.

Gymnasio lawrenceii (SCL. & SALV.).

Noctua nudipes LEMB. Aves Cuba, p. 23, pl. 4. fig. 2 (1850).
Gymnoglaux nudipes CAB. J. f. O. 1855. p. 465.—LAWR. Ann. Lyc. N. Y.
    VII. p. 257 (1861).—GUNDL. Repert. Fisico-Nat. Cuba, I, p. 226
    (1865); *ib.* J. f. O. 1871. p. 376.
Ephialtes nudipes BREWER, Pr. Bost. Soc. Nat. Hist. VII. p. 306 (1860).
Gymnoglaux lawrenceii SCL. & SALV. P. Z. S. 1858. p. 328. pl. 29; *ib.*
    Nom. Avium Neotr. p. 117 (1873).
Gymnasia lawrenceii GRAY. Handl. Bds. I. p. 47 (1869).
Gymnasio lawrenceii SHARPE, Cat. Bds. Brit. Mus. II, p. 150 (1875).—
    CORY, List Bds. W. I. p. 21 (1885).

SP. CHAR.—Upper surface dark brown, mottled with white on the back
    and wing-coverts; face dull brownish white, palest on the throat;

breast brown mixed with white; rest of underparts dull white, the
feathers lined with brown; primaries dark brown marked with
white on the outer webs; tail dark brown, showing an indistinct
band of white on the under surface; bill horn-color.

Length, 8; wing, 5.50; tail, 3; tarsus, 1.25; bill, .50.

HABITAT. Cuba.

GENUS Glaucidium BOIE.

Glaucidium BOIE, Isis, 1826. p. 976.

Glaucidium siju (D'ORB.).

Noctua siju D'ORB. in La Sagra's Hist. Nat. Cuba. Ois. p. 33 (1840).—
GUNDL. Journ. Bost. Soc. Nat. Hist. VI, p. 318 (1857).
Athene siju GRAY, Gen. Bds. I. p. 25 (1844).—CASS. Cat. Strig. Phila.
Mus. p. 13.
Nyctale siju BP. Consp. I, p. 54 (1850).—STRICKL. Orn. Syn. p. 177.
Strix havanensis LICHT. Mus. Berol. unde.
Glaucidium havanense KAUP, Contr. Orn. p. 103 (1852).
Glaucidium siju CAB. J. f. O. 1855. p. 59.—BREWER, Pr. Bost. Soc. Nat.
Hist. VII. p. 306 (1860).—GUNDL. Repert. Fisico-Nat. Cuba, I. p. 226
(1865).—GRAY, Handl. Bds. I, p. 42 (1869).—GUNDL. J. f. O. 1871,
p. 375.—RIDGW. Pr. Bost. Soc. Nat. Hist. 1873. p. 65.—SCL. & SALV.
Nom. Avium Neotr. p. 117 (1873).—SHARPE, Ibis, 1875. p. 59; *ib.*
Cat. Bds. Brit. Mus. II, p. 193 (1875).—CORY, List Bds. W. I. p. 21
(1885).

SP. CHAR.—Very small. Top of head pale brown, the feathers delicately
dotted with dull white; face dull white mixed with brownish; throat
dull white, shading into mixed light brown and white on the breast;
underparts white, the feathers streaked with dark brown; thighs
rufous brown; under tail-coverts white; primaries dark brown,
dotted and blotched with white, heaviest on the basal portions; back
and wing-coverts dull brown, mottled with pale brown; upper surface of tail-feathers dark brown, narrowly banded with brownish
white; bill pale.

A female in my collection has the entire upper surface reddish
brown, with the feathers on the head unspotted.

Length, 7; wing, 4; tail, 2.40; tarsus, .75; bill, .45.

HABITAT. Cuba.

GENUS Speotyto GLOGER.

Speotyto GLOGER, Handb. Naturg. p. 226, 1842.

Speotyto dominicensis Cory.

Speotyto cunicularia Sharpe, Cat. Bds. Brit. Mus. II, p. 142 (1875).—
   Cory, Bds. Haiti & San Domingo, p. 118 (1885).
Athene cunicularia Brace, Pr Bost. Soc. Nat. Hist. XIX, p. 240 (1877)
   (?)
Speotyto cunicularia dominicensis Cory, Bull. Nutt. Orn. Club, VI, p.
   154 (1881); *ib.* List Bds. W. I. p. 22 (1885).

Sp. Char. *Male:*—General plumage brown; the head marked with streaks
   of dull white; feathers of the nape showing a sub-terminal bar of
   dull white; back mottled and barred with dusky white; quills brown
   tipped with dull white and barred with pale brown; secondaries
   marked on the outer web; tail brown, tipped with buff white and
   banded; ear-coverts brown; cheeks dull white; throat and upper
   neck dull white, separated from each other by a mark of sandy buff,
   barred with brownish; underparts dull white, barred with brown,
   the bars becoming narrower on the lower part of the body; thighs
   buff; under wing-coverts yellowish buff, sometimes spotted with
   brown near the outer edge, and becoming dull white on the edge of
   the wing; tarsus feathered in front to the foot; iris yellow.

   The sexes are similar.

   Length, 8; wing, 6; tail, 2.50; tarsus, 1.50.

Habitat. Haiti. Bahamas?

It is possible that the species mentioned by Brace was the
Florida form. I have never seen a specimen from the Bahama
Islands.

Speotyto guadeloupensis (Ridgw.).

Speotyto cunicularia var. *guadeloupensis* Ridgw. in Bd. Bwr. & Ridgw.
   Hist. N. Am. Bds. III, p. 90 (1874).—Coues, Bds. N. W. p. 322
   (1874).
Speotyto guadeloupensis Sharpe, Cat. Bds. Brit. Mus. II, p. 147 (1875).—
   Cory, List Bds. W. I. p. 21 (1885).

Sp. Char.—"Primaries without broad or regular bars of whitish on either
   web; primary-coverts plain brown. Brown markings on the lower
   parts regularly transverse, and equal in extent to the white. White
   spots on the upper parts very small, reduced to mere specks on the
   dorsal region.

   "Wing, 6.40; tail, 3.40; culmen, .60; tarsus, 1.82; middle toe, .85.
   Outer tail-feathers and inner webs of the primaries with the light
   (ochraceous) bars only about one fourth as wide as the brown (dis-
   appearing on the inner quills)." (Ridgw., orig. descr., l. c.)

Habitat. Guadeloupe and St. Nevis.

Speotyto amaura LAWR.

Speotyto amaura LAWR. Pr. U. S. Nat. Mus. 1, p. 234 (1878).—CORY. List. Bds. W. I. p. 21 (1885).

SP. CHAR. *Male:*—"Upper plumage of a fine deep, brown color, marked with roundish spots of light fulvous; the spots are smallest on the crown, hind neck, and smallest wing-coverts; they are conspicuously large on the other wing-coverts, the dorsal region, scapulars, and tertials; the quills are blackish brown, with indented marks of pale reddish fulvous on the outer webs of the primaries, and large round-ish paler spots on the inner webs; under wing-coverts reddish ful-vous sparsely mottled with black; tail dark brown, of the same color as the back, crossed with four bars (including the terminal one), of light reddish fulvous, which do not quite reach the shaft on each web; bristles at the base of the bill black, with the basal portion of their shafts whitish; front white, superciliary streak pale fulvous; cheeks dark brown, the feathers tipped with fulvous; upper part of throat pale whitish buff, the lower part grayish white, with a buffy tinge, separated by a broad band of dark brown across the middle of the throat, the feathers of which are bordered with light fulvous; the sides of the neck and the upper part and sides of the breast are dark brown, like the back, the feathers ending with fulvous, the spots being larger on the breast; the feathers of the abdomen are pale fulvous, conspicuously barred across their centres with dark brown; on some of the feathers the terminal edgings are of the same color; the flanks are of a clear light fulvous, with bars of a lighter brown; under tail-coverts fulvous, with indistinct bars of brown; thighs clear fulvous, with nearly obsolete narrow dusky bars; the feathers of the tarsi are colored like the thighs and extend to the toes; bill clear light yellow with the sides of the upper mandible blackish, toes dull yellowish-brown.

"Length (fresh), 8½ in.; wing, 6¾; tail, 3⅓; tarsus, 1½.

"The female differs but little from the male in plumage; the bars on the abdomen appear to be a little more strongly defined, and at the base of the culmen is a small red spot. There are two females in the collection, the other also having the red spot; in one the tarsi are feathered to the toes, in the other only for two-thirds their length.

"Length of one (fresh), 8 in.; wing, 6½; tail, 2¾; tarsus, 1¼.

"Length of the other, 8½; wing, 6½; tail, 3; tarsus, 1½.

"Compared with *guadeloupensis*, the prevailing color is dark brown, instead of a rather light earthy-brown, and the spots on the interscapular region are much larger; it is more strikingly barred below, the other having the breast more spotted; the bars on the tail are four instead of six. In the Antigua bird each feather of the breast is crossed with but one bar, while those of the other are crossed with two." (LAWR., orig. descr., l. c.)

HABITAT. Antigua.

Family FALCONIDÆ.

Genus **Pandion** Sav.

Pandion Savigny, "Descr. de l'Egypt, Ois. p. 95, 1809."

Pandion haliaëtus carolinensis (Gmel.).

Falco carolinensis "Gmel. Syst. Nat. I, p. 263 (1788)."
Falco cayennensis "Gmel. Syst. Nat. I, p. 263 (1788)."
Pandion carolinensis Gosse. Bds. Jam. p. 19 (1847).—Bryant, Pr. Bost.
    Soc. Nat. Hist. VII, p. 105 (1859) (Bahamas).—Brewer, *ib.* p. 306
    (1860) (Cuba).—Albrecht, J. f. O. 1862, p. 204 (Jamaica).—
    March, Pr. Acad. Nat. Sci. Phila. 1863, p. 152 (Jamaica).—Gundl.
    Repert. Fisico-Nat. Cuba, I, p. 222 (1865); *ib.* J. f. O. 1871, p. 364
    (Cuba); *ib.* Anal. Soc. Esp. Hist. Nat. VII, p. 158 (1878) (Porto
    Rico); *ib.* J. f. O. 1878, p. 158 (Porto Rico).
Pandion haliaëtus Lemb. Aves Cuba, p. 12 (1850).—Lawr. Pr. U. S.
    Nat. Mus. I, p. 65 (1878) (Dominica); *ib.* p. 194 (St. Vincent); *ib.*
    p. 236 (Antigua); *ib.* p. 273 (Grenada).—Cory, Bds. Bahama I. p.
    131 (1881).—A. & E. Newton, Handb. Jamaica, p. 110 (1881).—
    Cory, Bds. Haiti & San Domingo, p. 125 (1885).—Wells, List
    Bds. Grenada, p. 6 (1886).
Pandion haliaëtus carolinensis Cory, List Bds. W. I. p. 22 (1885).
Pandion ridgwayi Maynard, Am. Ex. & Mart. Jan. 15, 1887.
    Common throughout the Bahamas and Antilles.

Genus **Circus** Lacép.

Circus Lacépède, Mém. de l' Inst. III, p. 506, 1801.

Circus hudsonius (Linn.).

Falco hudsonius Linn. Syst. Nat. I, p. 128 (1766).
Circus cyaneus D'Orb. in La Sagra's Hist. Nat. Cuba, Ois. p. 19 (1840)
Circus hudsonicus Brewer, Pr. Bost. Soc. Nat. Hist. VII, p. 306 (1860)
    (Cuba).—Cory, List Bds. W. I. p. 22 (1885).
Circus hudsonius Gundl. Repert. Fisico-Nat. Cuba, I, p. 224 (1865).—Bry-
    ant, Pr. Bost. Soc. Nat. Hist. XI, p. 65 (1867) (Bahamas).—Gundl.
    J. f. O. 1871, p. 369 (Cuba).
Circus cyaneus var. *hudsonius* Cory, Bds. Bahama I. p. 128 (1880).
    Cuba and Bahamas.

Genus **Rupornis** Kaup.

Rupornis Kaup, Classif. Säug. u. Vög. 1844.

Rupornis ridgwayi Cory.

Rupornis ridgwayi Cory, Journ. Bost. Zool. Soc. II. p. 46 (1883); *ib.*
    Auk, I. p. 4 (1884); *ib.* Bds. Haiti & San Domingo. p. 121 (1885);
    ib. List Bds. W. I. p. 22 (1885).

Sp. Char. *Male:*—Above slaty brown; shafts of the feathers of the head
    and upper back dark brown; underparts slaty, faintly touched with

rufous on the belly and abdomen; chin dull white; shoulders and thighs rufous, the latter much the brighter, and faintly pencilled with indistinct pale lines; wings and tail dark brown, imperfectly banded with dull white, and showing various shadings of a rufous tinge; all the outer primaries imperfectly banded with white, gradually becoming fainter on the outer webs, until just perceptible on the sixth, the rest of the primaries and secondaries with the outer web dark brown, and the inner webs thickly banded with white, showing traces of rufous.

Length, 13.75; wing, 9.15; tail, 6.; tarsus, 2.75; bill, 1.20.

Female:—Top of the head and neck brownish ash, becoming darker on the back; the feathers of the back and tertiaries edged with rufous; underparts dark rufous, the feathers narrowly banded with white; thighs showing the rufous much brighter, the feathers banded with very fine pale lines; crissum white, with rufous bands near the tips; under part of breast slaty, shading into dull white on the throat; the shafts of the feathers on the throat and breast dark brown, showing in hair-like lines; the rest as in the male.

Length, 14.50; wing 10.; tail, 6.45; tarsus, 2.65; bill, 1.25.

Immature Male:—In general appearance much like *Buteo pennsylvanicus*. Underparts dull white, the feathers slightly tinged with rufous, the centre of the surface feathers showing a stripe of brown, giving the body a striped appearance; thighs rufous, but paler than in the adult; above much resembling the adult; the white wing- and tail-bands replaced by rufous bands on the terminal half of the feathers.

HABITAT. San Domingo.

Mr. Gurney mentions *Rupornis magnirostris* from the Island of Martinique (Ibis, 1876, p. 482), but says that it might have possibly belonged to one or the other of the two Central American forms, which at that time had not been separated from it.

GENUS **Buteo** CUVIER.

Buteo "Cuv. Leç. d'Anat. Comp. I, tabl. ii, Ois. 1799-1800."

Buteo borealis (GMEL.).

Falco borealis GMEL. Syst. Nat. II, p. 226 (1788).

Buteo borealis GOSSE Bds. Jam. p. 11 (1847).—LEMB. Aves Cuba, p. 18 (1850).—BREWER, Pr. Bost. Soc. Nat. Hist. VII, p. 306 (1860) (Cuba).—ALBRECHT, J. f. O. 1862. p. 203 (Jamaica).—MARCH, Pr. Acad. Nat. Sci. Phila. 1863, p. 151 (Jamaica).—GUNDL. Repert. Fisico-Nat. Cuba I, p. 223 (1865).—BRYANT, Pr. Bost. Soc. Nat. Hist. XI, p. 64 (1867) (Bahamas).—GUNDL. J. f. O. 1871, p. 365 (Cuba); *ib* 1878, p. 158 (Porto Rico); *ib.* Anal. Soc. Esp. Hist. Nat. VII. p.

159 (1878) (Porto Rico).—CORY, Bds. Bahama I. p. 131 (1880).—
A. & E. NEWTON, Handb. Jamaica, p. 110 (1881).—CORY, List Bds.
W. I. p. 22 (1885).

Recorded from Cuba, Jamaica, Porto Rico. and Bahamas.
Mr. J. H. Gurney writes me he has an example of this species
from Haiti.

Buteo latissimus (WILS.).

Falco latissimus WILS. Am. Orn. I, p. 92 (1812).
Buteo latissimus LEMB. Aves Cuba, p. 19 (1850).- CORY. Ibis, 1886, p.
  473 (St. Vincent)
Buteo pennsylvanicus BREWER, Pr. Bost. Soc. Nat. Hist. VII, p. 306 (1860)
  (Cuba).—GUNDL. Repert. Fisico-Nat Cuba, I, p. 223 (1865); *ib.*
  J. f. O. 1871, p. 366 (Cuba).—LAWR. Pr. U. S. Nat. Mus. I, p. 194
  (1878) (St Vincent); *ib.* p. 236 (Antigua); *ib.* p. 273 (Grenada).—
  GUNDL. Anal. Soc. Esp. Hist. Nat. VII, p 160 (1878) (Porto Rico);
  ib. J. f. O. 1878, p. 158 (Porto Rico).—ALLEN, Bull Nutt. Orn.
  Club, V, p. 169 (1880) (Santa Lucia).—LISTER, Ibis, 1880, p. 43
  (St. Vincent).—CORY, List Bds. W. I. p. 22 (1885).—WELLS, List
  Bds. Grenada, p. 6 (1886).

Common winter visitant, and possible resident in the Lesser
Antilles.

Recorded from Cuba, Porto Rico, and Lesser Antilles.

GENUS Accipiter BRISS.

Accipiter BRISSON, Orn. I. p. 310. 1760

Accipiter gundlachi LAWR.

Astur cooperi LEMB. Aves Cuba, p. 17 (1850).—CAB J. f. O. 1854.
Nisus pileatus LEMB. Aves Cuba, Suppl. p. 125 (1850).
Accipiter cooperi BREWER, Pr. Bost. Soc Nat Hist VII, p. 306 (1860).
Accipiter pileatus BREWER, Pr. Bost Soc Nat. Hist. VII. p. 306 (1860)(?)
Accipiter mexicanus BREWER, Pr. Bost. Soc. Nat Hist VII, p 306 (1860).
Accipiter gundlachi LAWR. Ann. Lyc. N. Y. 1862, p. 252.—GUNDL.
  Repert. Fisico-Nat. Cuba, I, p. 224 (1865).—SCL. & SALV. Nom.
  Avium Neotr. p. 120 (1873).—SHARPE. Cat. Bds. Brit. Mus. I, p.
  137 (1874).—CORY. List Bds W. I. p. 22 (1885).
Cooperastur gundlachi GRAY, Handl. Bds. I, p. 33 (1869).
Nisus cooperi var. *gundlachi* BD BWR. & RIDGW. Hist. N. Am. Bds. III,
  p. 22 (1874).
Nisus gundlachi RIDGW. Studies Am. Falc. p. 104 (1876)

"*Adult male:*—Front, crown and occiput sooty-black; upper
plumage dull bluish ash, the feathers of the back with brownish

margins ; tail of the same color as the back, partly tinged with dull rufous and crossed with four brown bars, three of which are imperfect,
being but little developed on the outer webs, the outer bar, however,
crosses both webs, and is narrowly tipped with white ; quill feathers brown, having the shafts, as are also those of the tail-feathers,
reddish brown ; cheeks dusky ash ; space forward of the eye pale
dull rufous ; a line of whitish feathers runs along the edge of the
crown and extends over the eye ; throat ashy white tinged with
rufous ; sides of the neck, upper part of the breast and a band running to the hind neck, grayish ash ; lower portion of the breast and
upper part of the abdomen rufous, the feathers very narrowly edged
with dull white, lower part of abdomen of a paler rufous, with transverse bars of dull white ; long feathers of the sides grayish ash
tinged with rufous and destitute of bars or spots ; sides just above
the junction of the tail plain rufous ; thighs of a bright but rather
pale rufous, the feathers having darker sub-marginal ends, terminating with very narrow edgings of dull white ; under wing-coverts
and axillars bright rufous barred with white ; the feathers of the
throat, breast and sides have their shafts dark brown ; upper tail-
coverts grayish ash, lower white ; bill horn color, with a whitish
mark on the tooth and also on the edge of the lower mandible near
its base ; legs greenish yellow.

"Length about 18 inches ; wing from flexure 9¾ ; tail 7¾ ; tarsus 2¾."
(LAWR., orig. descr., l. c.)

HABITAT. Cuba.

Accipiter fringilloides VIG.

Accipiter fringilloides VIG. Zool. Journ. III, p. 434 (1828).—DENNY,
   P. Z. S. 1847, p. 38.—BREWER, Pr. Bost. Soc. Nat. Hist. VII, p.
   306 (1860).—GRAY, Handl. Bds. I, p. 32 (1869).—GUNDL. J. f. O.
   1871, p. 368.—SCL. & SALV. Nom. Avium Neotr. p. 120 (1873).—
   SHARPE, Cat. Bds. Brit. Mus. I, p. 135 (1874).—CORY, Bds. Haiti
   & San Domingo, p. 120 (1885) ; *ib.* List Bds. W. I. p. 22 (1885).
Nisus fringilloides D'ORB. in La Sagra's Hist. Nat. Cuba, Ois. p. 18
   (1840).—LEMB. Aves Cuba, p. 128 (1850).—RIDGW. Studies Am.
   Falc. p. 117 (1876).
Nisus fuscus LEMB. Aves Cuba, p. 128 (1850).—GUNDL. J. f. O. 1854.—
   CORY, Bull. Nutt. Orn. Club, VI, p. 154 (1881).
Accipiter fuscus BREWER, Pr. Bost. Soc. Nat. Hist. VII, p. 306 (1860).
Nisus fuscus var *fringilloides* BD. BWR. & RIDGW. Hist. N. Am. Bds. III,
   p. 223 (1874).

SP. CHAR. *Female:*—Resembles *Accipiter fuscus,* but plumage much paler ;
   above brown, the concealed portions of the feathers showing much
   white ; concealed feathers of the back regularly marked with broad
   spots of white ; tail pale brown, showing five somewhat indistinct

bands of darker brown; under surface of tail dull white, regularly
banded with brown; breast and belly white, the shafts of the feathers dark brown, showing hair-like lines over the whole surface: these
lines are in many cases bordered with pale brown, giving the appearance of arrow-shaped markings; under tail-coverts white; quills
brown, barred with white on the inner webs; under surface of wings
white, barred with brown.

Length, 11.50; wing 7.; tail, 5.50; tarsus, 1.75.

HABITAT.   Cuba, Haiti, and San Domingo.

Dr Gundlach has a fine adult male of this species in his collection.   It is smaller than the female, as would be expected, and
has the cheeks and sides of the throat tinged a beautiful orange
brown, the color also showing in the breast marking.

The female described was killed a few miles from Port au
Prince, Haiti, during March, 1881.   It was the only one seen.

Accipiter velox (WILS.).

Falco velox "WILS. Am. Orn. V, p. 116 (1812)."

Accipiter fuscus BRYANT, Pr. Bost. Soc. Nat. Hist. VII, p. 105 (1859)
(Bahamas).—CORY, Bds. Bahama I. p. 128 (1880); *ib.* List Bds.
W. I. p. 22 (1885).

Accidental in the Bahamas.

GENUS Urubitinga LESS.

Urubitinga LESSON, Rev. Zool. 1839, p. 132.

Urubitinga anthracina (LICHT.).

Falco anthracinus LICHT. in Mus. Berol. undè Nitzsch. Pteryl. p. 83
(1840).

Morphnus urubitinga LEMB. Aves Cuba, p. 14 (1850).—ALBRECHT, J. f. O.
1862, p. 204 (Jamaica).

Hypomorphus gundlachi BREWER, Pr. Bost. Soc. Nat. Hist. VII, p. 306
(1860) (Cuba).

Hypomorphus gundlachi GUNDL. Repert. Fisico-Nat. Cuba, I, p. 223
(1865); *ib.* J. f. O. 1871, p. 365 (Cuba).

Urubitinga anthracina SHARPE, Cat. Bds. Brit. Mus. I, p. 215 (1874)
(Cuba).—LAWR. Pr. U. S. Nat. Mus. I, p. 194 (1878) (St. Vincent).
—LISTER, Ibis, 1880, p 43 (St. Vincent).—CORY, List Bds. W. I. p.
22 (1885).—WELLS, List Bds. Grenada, p. 6 (1886) (?) —CORY,
Ibis, 1886, p. 473 (St. Vincent).

Records from Cuba, Jamaica, St. Vincent, and Grenada (?)

GENUS Falco LINN.

Falco LINN.EUS. Syst. Nat. I, p. 124, 1766.

Falco peregrinus anatum (BONAP.).

Falco anatum BP. Geog. & Comp. List. p. 4 (1834).—GOSSE, Bds. Jam.
p. 16 (1847).—BRYANT, Pr. Bost. Soc. Nat. Hist. VII. p. 105 (1859)
(Bahamas); *ib.* BREWER, p. 306 (:860)(Cuba).—ALBRECHT, J. f. O.
1862. p. 204 (Jamaica).—MARCH, Pr. Acad. Nat. Sci. Phila. 1863.
pp. 152, 304 (Jamaica).—GUNDL. Repert. Fisico-Nat. Cuba, I. p.
225 (1865); *ib.* J. f. O. 1871. p. 371 (Cuba); *ib.* 1878. p. 158 (Porto
Rico); *ib.* Anal. Soc. Esp. Hist. Nat. VII, p. 161 (1878) (Porto Rico).
Falco peregrinus LEMB. Aves Cuba, p. 11 (1850).—BRYANT, Pr. Bost.
Soc. Nat. Hist. XI, p. 64 (1867) (Bahamas).—A. & E. NEWTON,
Handb. Jamaica, p. 110 (1881).—CORY, List Bds. W. I. p. 22 (1885).
Falco communis SUNDEV. Oefv. Af. K. Vet. Akad. För. 1869, p. 586 (St.
Bartholomew).—CORY. Bds. Bahama I. p. 129 (1880).
Falco communis var. *anatum* LAWR. Pr. U. S. Nat. Mus. I, p. 487 (1878)
(Antigua); *ib.* p. 240 (Barbuda).

Many records from the Antilles ; specimens have been taken in
the Bahamas. Cuba. Jamaica, Antigua, Barbuda, Porto Rico, and
St. Bartholomew.

Falco columbarius LINN.

Falco columbarius LINN. Syst. Nat. I, 10th ed. p. 90 (1758); *ib.* 12th ed.
p. 128 (1766).—D'ORB. in La Sagra's Hist. Nat. Cuba, Ois. p 23
(1840).—GOSSE, Bds. Jam. p. 17 (1847).—SUNDEV. Oefv. Af. K.
Vet. Akad. För. 1869. p. 601 (Porto Rico).—A. & E. NEWTON,
Handb. Jamaica, p. 110 (1881).—CORY, Bds. Haiti & San Domingo,
p. 123 (1885); *ib.* List Bds. W. I. p. 22 (1885)
Hypotriorchis columbarius BREWER. Pr. Bost. Soc. Nat. Hist. VII, p. 306
(1860) (Cuba).—SCL. P. Z. S. 1861. p. 79 (Jamaica).—ALBRECHT,
J. f. O. 1862. p. 203 (Jamaica).—MARCH, Pr. Acad. Nat. Sci. Phila.
1863. p. 152 (Jamaica).—GUNDL. Repert. Fisico-Nat. Cuba. I. p.
225 (1865); *ib.* J. f. O. 1871. p. 372 (Cuba); *ib.* 1878. p. 158 (Porto
Rico); *ib.* Anal. Soc. Esp. Hist. Nat. VII. p. 162 (1878)(Porto Rico).
Æsalon columbarius WELLS. List Bds. Grenada, p. 6 (1886).

Recorded from San Domingo, Porto Rico, Cuba, Jamaica,
Grenada, and St. Thomas.

Falco sparverius LINN.

Falco sparverius LINN. Syst. Nat. I, 10th ed. p. 90 (1758); *ib.* 12th ed. p.
128 (1766).—GMEL. Syst. Nat. I, p. 284 (1788).—LATH. Ind. Orn.

p. 42 (1790).—VIEILL. Enc. Méth. III, p. 1234 (1820).—WAGL. Isis,
1831, p. 517.—AUD. Bds. Am. I, p. 94 (1839).—CASSIN, in Baird's
Bds. N. Am. p. 13 (1860).—SALV. P. Z. S. 1867, p. 158.—SUNDEV.
Oefv. Af. K. Vet. Akad. För. 1869, p. 586.—SCHLEG. Rev. Accipitr.
p. 45 (1873).—COUES, Key N. Am. Bds. p. 537 (1884).—CORY, Bds.
Bahama I. p. 103 (1880); *ib.* List Bds. W. I. p. 22 (1885).
Falco noveboracensis GMEL. Syst. Nat. I, p. 284 (1788).
Tinnunculus sparverius VIEILL. Ois. Am. Sept. pls. XII, XIII (1807).—
Br. Consp. I, p. 27 (1850).—CASSIN, Pr. Acad. Nat. Sci. Phila. 1855,
p. 278.—STRICKL. Orn. Syn. I. p. 99 (1855).—BRYANT, Pr. Bost.
Soc. Nat. Hist. VII, p. 105 (1859); *ib.* BREWER, p. 306 (1860).—SCL.
& SALV. Nom. Avium Neotr. p. 121(1873).—GUNDL. J. f. O. 1878, p.
158 (?).—GURNEY, List Bds. Prey, p. 98 (1884).—WELLS, List Bds.
Grenada, p. 6 (1886).
Cerchneis sparverius BP. List Eur. & N. Am. Bds. p. 5 (1838).
Falco isabellinus SWAINS. An. Menag. p. 281 (1838).
Tinnunculus phalæna LESS. Mam. et Ois. p. 178 (1847).
Pœcilornis sparverius KAUP, Mon. Falc. Cont. Orn. p. 53 (1850).—GRAY.
Handl. Bds. I. p. 23 (1869).
Tinnunculus sparverius var. *isabellinus* RIDGW. Pr. Acad. Nat. Sci. Phila.
1870, p. 149.—BD. BWR. & RIDGW. Hist. N. Am. Bds. III, p. 171
(1874).
Cerchneis sparveria SHARPE. Cat. Bds. Brit. Mus. I, p. 437 (1874).
Cerchneis isabellina SHARPE. Cat. Bds. Brit. Mus. I. p. 441 (1874)
Falco (Tinnunculus) sparverius BD. BWR. & RIDGW. Hist. N. Am. Bds.
III, p. 169 (1874).
Tinnunculus isabellinus GURNEY. Ibis, 1881, p. 561; *ib.* List Bds. Prey.
p. 99 (1884).
Falco sparverius isabellinus COUES, Key N. Am. Bds. p. 538 (1884).

Several forms of this species occur in the West Indies, but vary
much in different localities I have a specimen in my cabinet
from San Domingo which is apparently true *F. sparverius.*

Falco dominicensis GMEL.

Falco dominicensis GMEL. Syst. Nat. I, p. 288 (1788).—BRYANT, Pr. Bost.
Soc. Nat. Hist. XI. p. 90 (1866).
Falco sparverius D'ORB. in La Sagra's Hist. Nat. Cuba, Ois. p. 25 (1840).
—SALLÉ, P. Z. S. 1857, p. 231—SUNDEV. Oefv. Af. K. Vet. Akad.
För. 1869. p. 586.
Tinnunculus dominicensis STRICKL. Orn. Syn. p. 100 (1855).—BREWER,
Pr. Bost. Soc. Nat. Hist. VII. p. 306 (1860).—GUNDL. Repert. Fisico-
Nat. Cuba, I, p. 225 (1865).—GRAY, Handl. Bds. I, p. 24 (1869).—
GUNDL. J. f. O. 1871. p. 373; *ib.* Anal. Soc. Esp. Hist. Nat. VII, p.
163 (1878).—GURNEY, List Bds. Prey. p. 99 (1884).
Tinnunculus sparverius CASSIN, Pr. Acad. Nat. Sci. Phila. 1860, p. 374.

Falco leucophrys RIDGW. Pr. Acad. Nat. Sci. Phila. 1870, p. 147.

Tinnu..culus leucophrys RIDGW. Pr. Acad. Nat Sci. Phila. 1870, p. 149.—
SCL. & SALV. Nom. Avium Neotr. p. 121 (1873).—BD. BWR. &
RIDGW. Hist. N. Am. Bds. III. p. 161 (1874).

Tinnunculus sparverius var. *dominicensis* RIDGW. Pr. Acad. Nat. Sci.
Phila. 1870, p. 149.

Cerchneis leucophrys SHARPE, Cat. Bds. Brit. Mus I, p. 442 (1874).

Falco sparverius var. *dominicensis* BD. BWR. & RIDGW. Hist. N. Am. Bds.
III, p. 167 (1874).

Tinnunculus sparverius (?) CORY, Bull. Nutt. Orn. Club, VI, p. 154 (1881).

Falco sparverius isabellinus CORY. Bds. Haiti & San Domingo, p. 124
(1875).

Falco sparverius dominicensis CORY, List Bds. W. I. p. 22 (1884).

SP. CHAR. *Male:*—Top of head slate color; forehead whitish; throat
white; a maxillary and auricular black stripe; breast rufous; back
dark rufous brown; tail rufous brown, tipped with white, and having a sub-terminal band of black; outer web of outer tail-feather
white; wing-coverts slate color; abdomen and belly white; a patch
of black on the side of the neck.

Female:—Top of head slate color, showing a patch of rufous; entire upper parts rufous brown, banded with dull black; underparts
very pale rufous, delicately streaked and spotted with brown; throat
white.

Length, 10.; wing, 7.; tail, 5. tarsus, 1.20.

HABITAT. Cuba? Haiti, San Domingo, and Porto Rico.

Falco sparverioides VIG.

Falco sparverioides VIG. Zool. Journ. III, p. 436 (1828).—D'ORB. in La
Sagra's Hist. Nat. Cuba, Ois. p. 30 (1840).—RIDGW. Pr. Acad. Nat.
Sci. Phila. 1870, p. 149.—COUES, Key N. Am. Bds. p. 538 (1884).

Tinnunculus sparverioides GRAY, Gen. Bds. I, p. 21 (1844).—Br. Consp.
I, p. 27 (1850).—STRICKL. Orn. Syn. p. 100 (1855).—LAWR. Ann.
Lyc. N. Y. 1860, p. 247.—BREWER, Pr. Bost. Soc. Nat. Hist. VIII,
p. 306 (1860).—SCL. & SALV. Nom. Avium Neotr. p. 121 (1873).—
GURNEY, Ibis, 1881, p. 565; *ib.* List Bds. Prey, p. 100 (1884).

Pœciloruis sparverioides KAUP, Contr. Orn. p. 53 (1850).—Br. Rev. Mag.
Zool. 1854, p. 537.—GRAY, Handl. Bds. I, p. 24 (1869.)

Cerchneis sparverioides SHARPE, Cat. Bds. Brit. Mus. I, p. 443 (1874).

Falco (Tinnunculus) sparverioides BD. BWR. & RIDGW. Hist. N. Am.
Bds. III, p. 162 (1874).

Falco sparverius sparverioides CORY, List Bds. W. I. p. 22 (1885).

SP. CHAR. *Male:*—Above entirely slate blue in the adult bird; most specimens seen have the back chestnut brown mixed with slaty; rump,
upper tail-coverts, and tail chestnut brown; tail with a sub-terminal

band of black; inner secondaries gray; sides of the face and throat white; a streak of black on sides of throat; slight mark on the nape and a patch near the ear-coverts black: breast pale chestnut. and becoming whitish, tinged with chestnut on the belly and vent; flanks showing a grayish tinge, and a few faint black spots.

Length (skin), about 10; wing, 6.50; tail, 4.70; tarsus, 1.50; bill, .60.

HABITAT. Cuba.

Falco caribbæarum GMEL.

Falco caribbæarum GMEL. Syst. Nat. I. p. 284 (1788).

Falco æsalon. var. β. LATH. Ind. Orn. I, p. 49 (1790).

Cerchneis carribæarum (?) SHARPE, Cat. Bds. Brit. Mus. I, p. 442 (1874).

Tinnunculus sparverius var. *antillarum* LAWR. Pr. U. S. Nat. Mus. I, p. 487 (1878).— ALLEN, Bull. Nutt. Orn. Club, V, p. 169 (1880).

Tinnunculus antillarum GURNEY, Ibis, 1881, p. 547.

Tinnunculus caribbæarum GRISDALE, Ibis, 1882. p. 491.— GURNEY, List Bds. Prey, p. 99 (1884).— RIDGW. Pr. U. S. Nat. Mus. VII, p. 172 (1884).

Falco sparverius caribbæarum CORY, List Bds. W. I. p. 22 (1885); *ib.* Ibis, 1886, p. 474.

SP. CHAR. *Male:*—General plumage above chestnut brown, heavily banded with black; forehead grayish; top of head chestnut brown, showing faint lines of black; underparts dull white, tinged with rufous on the breast, and spotted and streaked with black, heaviest on the sides of the body; primaries heavily blotched with white on the inner webs: under surface of tail brown, showing numerous bands of black, a wide subterminal band of black, and narrowly tipped with grayish white.

Length (skin) 9.50: wing, 6; tail, 4.50; tarsus, 1; bill .55.

HABITAT. Lesser Antilles.

GENUS Elanoides VIEILL.

Elanoides "VIEILLOT, Nouv. Dict. XXIV, p. 101, 1818. Type *Falco furcatus* = *F. forficatus* LINN."

Elanoides forficatus (LINN.).

Falco forficatus LINN. Syst. Nat. I, p. 89 (1758).

Nauclerus furcatus GOSSE, Bds. Jam. p. 19 (1847).— BREWER, Pr. Bost. Soc. Nat. Hist. VII, p. 306 (1860) (Cuba).— ALBRECHT, J. f. O. 1862, p. 204 (Jamaica).— MARCH, Pr. Acad. Nat. Sci. Phila. 1863, p. 153 (Jamaica).— GUNDL. Repert. Fisico-Nat. Cuba, I, p. 225 (1865); *ib.* J. f. O. 1871, p. 370 (Cuba).—A. & E. NEWTON, Handb. Jamaica, p. 110 (1881).

Elanoides forficatus CORY, List Bds. W. I. p. 22 (1885).

Recorded from Cuba and Jamaica.

Genus **Rostrhamus** Less.

Rostrhamus Lesson, Traité d'Orn. p. 55, 1831.

Rostrhamus sociabilis (Vieill.).

Herpetotherus sociabilis Vieill. Nouv. Dict. XVIII, p. 318 (1818).
Rostrhamus sociabilis D'Orb in La Sagra's Hist. Nat. Cuba, Ois. p. 15
  (1840).— Brewer. Pr. Bost. Soc. Nat. Hist. VII, p. 306 (1860)
  (Cuba).— Gundl. Repert. Fisico-Nat. Cuba, I, p. 222 (1865); *ib.*
  J. f. O. 1871, p. 362 (Cuba).
Rostrhamus hamatus Brewer, Pr. Bost. Soc. Nat. Hist. VII, p. 306 (1860)
  (Cuba).
Rostrhamus sociabilis Cory, List Bds. W. I. p. 22 (1885).

Cuba.

Genus **Regerhinus** Kaup.

Regerhinus Kaup, Mus. Senck. III. p. 262, 1845.

Regerhinus wilsonii (Cass.).

Cymindis wilsonii Cassin, Journ. Acad. Nat. Sci. Phila. new ser. I. p. 21.
  pl. vii (1847).— Bp. Consp. I, p. 21 (1850).— Lawr. Ann. Lyc. N.
  Y. VII. p. 257 (1860).— Scl. & Salv. Nom. Avium Neotr. p. 122
  (1873).
Regerhinus wilsonii Kaup, Arch. f. Naturg. 1850, p. 40.— Brewer. Pr.
  Bost. Soc. Nat. Hist. VII, p. 306 (1860).— Gundl. J. f. O. 1871. p.
  360.— Ridgw. Studies Am. Falc. p. 159 (1876).— Cory. List Bds.
  W. I. p. 23 (1885).
Cymindis uncinatus Lemb. Aves Cuba. Suppl. (1850).—Brewer. Pr. Bost.
  Soc. Nat. Hist. VII, p. 306 (1860).
Regerhinus uncinatus Cab. J. f. O. 1854. p. 80.
Regerhinus wilsoni Gray, Handl. Bds. I, p. 28 (1869).
Leptodon wilsoni Sharpe, Cat. Bds. Brit. Mus. I, p. 333 (1874).

  "*Male:*— Body above entirely dark brown, paler on the head;
beneath white, every feather from the chin to the under tail-coverts
crossed by several bars of bright rufous, and these colours extending upwards into a collar around the neck; 4th, 5th, and 6th primaries longest and nearly equal, external webs nearly black, internal
webs of outer primaries white at base, and for nearly half their
length, remaining part reddish inclining to chestnut, every primary
(on its inner web) having two irregularly shaped black marks, and
tipped with black. Tail of the same colour as the back, but paler,
white at base, and crossed by about four broad bars, which are
nearly black, the second bar from the tip accompanied by a narrow
rather indistinct bar of rufous; tip of tail narrowly edged with
white. Bill very large, larger than that of any other species of

this genus, yellowish white, inclining to bluish horn-colour at
base. Total length 17 inches.

"*Female:*— Body above entirely light bluish ash-colour, paler on
the head, beneath barred with the same, the bars having a ferruginous tinge" (CASSIN, l. c.).

HABITAT. Cuba.

Regerhinus uncinatus (TEMM.).

Falco uncinatus TEMM. Pl. Col. 103, 104, 105 (1824).
Cymindis uncinatus LESS. Man. d'Orn. I, p. 91 (1828).— GRAY, Gen. Bds.
   I, p. 25 (1844).— BP. Consp. I. p. 21 (1850).— BURM. Th. Bras. II,
   p. 108 (1856).— LÉOT. Ois. Trinid. p. 36 (1866).— GRAY, Handl.
   Bds. I, p. 136 (1869).—PELZ. Orn. Bras. pp. 5, 398 (1871).—SCHLEG.
   Rev. Accipitr. p. 136 (1873).— SHARPE, P. Z. S. 1873, p. 419.— SCL.
   & SALV. Nom. Avium Neotr. p. 122 (1873).— WELLS, List Bds.
   Grenada, p. 6 (1886).
Falco vitticaudus MAX. Beitr. III, p. 178 (1830).
Cymindis cuculoides SWAINS. Classif. Bds. II, p. 209 (1837).
Regerhinus uncinatus KAUP. Mus. Senckenb. III. p. 262 (1845).— CAB. in
   Schomb. Reis. Guian. III, p. 736 (1848).—GUNDL. J. f. O. 1871, p.
   284.— CORY, List Bds. W. I. p. 23 (1885).
Rostrhamus uncinatus STRICKL. Orn. Syn. p. 136 (1855).
Cymindis pucherani LÉOT. Ois. Trinid. p. 40 (1866).— GRAY, Handl. Bds.
   I, p. 25 (1869).— FINSCH, P. Z. S. 1870, p. 557.
Cymindis boliviensis BURM. P. Z. S. 1868, p. 635.— GRAY, Handl. Bds. I,
   p. 28 (1869).
Cymindis vitticaudus PELZ. Orn. Bras. pp. 6, 398 (1871).
Leptodon uncinatus SHARPE. Cat. Bds. Brit. Mus. I, p. 330 (1874).

"*Young:*— Above brown, the dorsal feathers and wing-coverts
margined with pale rufous, the upper tail-coverts broadly barred
and tipped with buff; quills dark brown, with rufous-buff tips, the
primaries barred with dark brown above, the secondaries more or
less distinctly barred with rufous or rufous buff; the under surface
of the wing ashy brown, barred with darker brown, the bases of the
feathers creamy buff, washed with rufous near the tips; tail ashy
brown, tipped with whitish, barred across with dark brown bars,
the interspaces on the inner web creamy buff, more or less mottled
with brown above, at the base barred above and below with creamy
buff, like the upper tail-coverts; crown of the head dark brown,
with no pale margins; sides of the face and a collar around the
neck white, slightly spotted with pale brown, the ear-coverts inclining to bluish grey; under surface of body white, the throat indistinctly spotted, and the breast narrowly barred with pale brown,
the bars almost linear on the under tail-coverts, those on the

thigh-feathers broader and more rufous; under wing-coverts and axillaries white, barred with pale rufous. Total length 17 inches, culmen 1·65, wing 10·4, tail 8·6, tarsus 1·45.

"Another specimen still quite young, agrees with the foregoing in the coloration of the wings and tail, but has the edgings to the feathers of the upper surface very much broader, and a broad white tip to the tail; the sides of the face and collar round the neck are creamy white, without any brown spots; the under surface of the body is also more free from spots, with here and there a feather appearing broadly barred with tawny rufous, indicative of the next change in the plumage.

"*Mature:*—Altogether different from the preceding stage. Above leaden brown, the head more slaty, the sides of the face and chin clear slaty blue; around the neck a rufous collar; quills brown, with narrow apical margins of pale rufous or buffy white, the outer secondaries rufous for nearly their whole extent, the under surface of the wing greyish, creamy white near the base, all the quills barred above and below with blackish brown; tail ashy grey, crossed by two very broad bars of black, tipped with creamy white, before which an indistinct subterminal line of ashy grey is visible, some of the outer upper tail-coverts and base of tail slightly mottled with whitish; under surface of body tawny rufous, crossed with broad bars of ochraceous buff, the under wing-coverts similarly marked, the lower ones ochraceous buff, with greyish black cross-bars.

"The next change seems to be in the undersurface, where the ochre-coloured become quite white, and whitish bars appear on the grey throat. From this stage (to judge by our specimens) it changes by a partial moult, and by a gradual change of feather at the same time; for the bars on the breast lose by degrees their rufous tint and become grey, while the back also becomes slaty grey instead of brown; the nuchal collar gradually disappears. This gradual development seems to be satisfactorily traced, with the exception of the tail, which, instead of agreeing with that of the rufous or "mature" stage, has four rather narrow black bars, like the young specimen first described. This can only be accounted for by the fact that Hawks have really no fixed laws of change in plumage, and that it is impossible for anyone to define exactly the regular sequence of the variations. No two birds are exactly alike; for one has the head more advanced another the tail, vice versa. Thus the bird last noticed as donning his grey dress is very far advanced as regards his body-plumage, but has not moulted his tail, whereas those in the rufous dress are not so forward in their body-plumage, but have already the tail of the adult (one being in the act of moulting).

"*Adult female:*—Slaty blue above and below; no trace of a nuchal collar; under surface narrowly but irregularly barred with white.

the under tail-coverts clear buff; under wing-coverts grey, thickly
barred with buffy white · quills blackish, shaded with slaty grey
above, the secondaries entirely of this colour, the under surface
greyish white, with black bars and tips, less conspicuous on the
upper surface; tail alternately crossed with two bands of black
above, with a broad intermediate band of ashy grey between, narrowly tipped with ashy grey, barred with ochraceous buff and black
below, the bars very broad. Total length 17 inches, culmen 1·6,
wing 11·7, tail 7·5, tarsus 1·4.

"*Adult Male:*—A little smaller than the female. Total length 16
inches, culmen 1·55, wing 11, tail 7·5, tarsus 1·4."   (SHARPE, l. c.)

I have quoted Mr. Sharpe's admirable description of this
species in full; as the series of specimens at my command is
totally inadequate to enable me to properly describe the various
stages of plumage.

The bird is recorded from Grenada, and is probably accidental
in the Antilles.

GENUS Polyborus VIEILL.

Polyborus VIEILLOT, Analyse, p. 22, 1816.

Polyborus cheriway (JACQ.).

Falco cheriway JACQ. Beitr. p. 17, tab. 4 (1784).
Polyborus vulgaris D'ORB. in La Sagra's Hist. Nat. Cuba, Ois. p. 9 (1840).
    —BREWER, Pr. Bost. Soc. Nat. Hist. VII, p. 306 (1860) (Cuba).
Polyborus tharus BREWER, Pr. Bost. Soc. Nat. Hist. VIII, p. 306 (1860)
    (Cuba).—GUNDL. Repert. Fisico-Nat. Cuba, I, p. 221 (1865).
Polyborus cheriway BREWER, Pr. Bost. Soc. Nat. Hist. VIII, p. 306 (1860)
    (Cuba).—GUNDL. J. f. O. 1871, p. 284 (Cuba).—CORY, List Bds.
    W. I. p. 23 (1885).
Polyborus brasiliensis BREWER, Pr. Bost. Soc. Nat. Hist. VII, p. 306 (1860)
    (Cuba)
Polyborus auduboni GUNDL. J. f. O. 1871, p. 357 (Cuba) (?).
    Accidental in Cuba.

FAMILY CATHARTIDÆ.

GENUS Cathartes ILLIGER.

Cathartes ILLIGER, Prodr. p. 236, 1811.

Cathartes aura (LINN.).

Vultur aura LINN. Syst. Nat. I. p. 86 (1758).
Cathartes aura D'ORB. in La Sagra's Hist. Nat. Cuba, Ois. p. 4 (1840).—

Gosse. Bds. Jam. p. 1 (1847).—Bryant, Pr. Bost. Soc. Nat. Hist.
VII, p. 104 (1859) (Bahamas); ib. Brewer, p. 306 (1860) (Cuba).
— Albrecht, J. f. O. 1862, p. 203 (Jamaica).— March, Pr. Acad.
Nat. Sci. Phila. 1863, p. 150 (Jamaica).—Gundl. Repert. Fisico-Nat.
Cuba, I, p. 221 (1865); ib. J. f. O. 1871, p. 253 (Cuba).— Cory,
Bds. Bahama I. p. 134 (1880).— A. & E. Newton. Handb. Jamaica,
p. 111 (1881).— Cory, List Bds. W. I. p. 23 (1885).

Recorded from the Bahamas, Cuba, and Jamaica.

Genus Catharista Vieill.

Catharista Vieillot, Analyse, p. 21, 1816.

Catharista atrata (Bartr.).

Vultur atratus Bartr. Trav. Car. p. 285 (1792).
Cathartes atratus March. Pr. Acad. Nat. Sci. Phila. 1863. p. 151 (Ja-
maica).—A. & E. Newton. Handb. Jamaica. p. 111 (1881).
Catharista atrata Cory, List Bds. W. I. p. 23 (1885).

This species is claimed to have occurred in Jamaica. No
other West Indian record.

Family COLUMBIDAE.

Genus Columba Linn.

Columba Linnæus, Syst. Nat. 1735, and Syst. Nat. ed. 10, p. 162 (1758).

Columba leucocephala Linn.

Columba leucocephala Linn. Syst. Nat. I, p. 164 (1758).—Nutt. Man.
Orn. I, p. 625 (1832).—Gosse, Bds. Jam. p. 299 (1847).—Sallé,
P. Z. S. 1857, p. 235.—March, Pr. Acad. Nat, Sci. Phila. 1863, p.
301.—Bryant. Pr. Bost. Soc. Nat. Hist. XI, p. 96 (1866).—Sundev.
Oefv. K. Vet. Akad. För. 1869, pp. 585, 600.—Sci. & Salv. Nom.
Avium Neotr. p. 132 (1873).—Bd. Bwr. & Ridgw. Hist. N. Am.
Bds. III, p. 363 (1874).—Lawr. Pr. U. S. Nat. Mus. I, p. 487 (1878)

Cory. Bds. Bahama I. p. 137 (1880).—A. & E. NEWTON, Handb.
Jamaica, p. 114 (1881).—CORY, Bull. Nutt. Orn. Club, VI. p. 154
(1881); *ib.* Bds. Haiti & San Domingo, p. 134 (1885); *ib.* List Bds.
W. I. p. 23 (1885).—COUES, Key N. Am. Bds. p. 565 (1884).

Patagiœnas leucocephalas REICH. Syst. Av. (1851).—BP. Consp. II. p. 54
(1854).—A. & E. NEWTON, Ibis, 1859, p. 253.

Patagiœnas leucocephala REICH. Syst. Nat. p. 25 (1851).—BP. Consp. II.
p. 54 (1854).—CAB. J. f. O. 1856. p. 107.—BREWER, Pr. Bost. Soc.
Nat. Hist. VII, p. 307 (1860).—SCL. P. Z. S. 1861. p. 80.—ALBRECHT,
J. f. O. 1862, p. 204.—*ib.* GUNDL. 1874, p. 288; *ib.* Anal. Soc. Esp.
Hist. Nat. VII, p. 345 (1878).

SP. CHAR. *Male:*—Above grayish blue, showing slight reflections; crown
pure white, bordered at the nape by a band of dark purple, and below it a cape extending upon each side of the neck of metallic green,
showing blue in some lights, the feathers bordered with black;
quills dark brown, becoming lighter upon the secondaries; underparts grayish blue; crissum plumbeous; tail very dark brown.

The female resembles the male, but is somewhat paler.

Length, 12.50; wing. 7.25; tail, 2 25; tarsus, .80.

HABITAT.   Bahamas and Antilles.

Columba corensis GMEL.

Columba corensis GMEL. Syst. Nat. I, p. 783 (1788).—SALLÉ, P. Z. S.
1857, p. 235.—A. & E. NEWTON, Ibis, 1859, p. 252.—CASSIN, Pr.
Acad. Nat. Sci. Phila. 1860. p. 377.—BRYANT, Pr. Bost. Soc. Nat.
Hist. XI, p. 96 (1866).—SUNDEV. Oefv. K. Vet. Akad. För. 1869, p.
601.—SCL. & SALV. Nom. Avium Neotr. p. 132 (1873).—BD. BWR.
& RIDGW. Hist. N. Am. Bds. III, p. 360 (1874).—LAWR. Pr. U. S.
Nat. Mus. I, p. 487 (1878).—LISTER, Ibis, 1880, p. 42.—ALLEN. Bull.
Nutt. Orn. Club, V, p. 169 (1880).—CORY, Bds. Haiti & San Do-
mingo, p. 136 (1885); *ib.* List Bds. W. I. p. 23 (1885).—WELLS,
List Bds. Grenada, p. 6 (1886).

Columba portoricensis TEMM. Hist. Gen. Pigeons, I, pl. 15 (1813).—
D'ORB. in La Sagra's Hist. Nat. Cuba, Ois. p. 172 (1840).

Columba monticolor VIEILL. Nouv. Dict. XXVI. p. 355 (1818).

Columba imbricata WAGL. Syst. Nat. No. 48 (1827).

Patagiœnus corensis BP. Consp. II, p. 54 (1854).—CAB. J. f. O. 1856, p.
108.—BREWER, Pr. Bost. Soc. Nat. Hist. VII, p. 307 (1860).—
GUNDL. Repert Fisico-Nat. Cuba, I. p. 299 (1866); *ib.* J. f. O. 1874,
p. 289; *ib.* Anal. Soc. Esp. Hist. Nat. VII, p. 344 (1878).

SP. CHAR. *Male:*—General plumage slaty; top of head, throat and breast
pale purple; a broad cape extending from the sides of the neck, over
the upper back, of beautifully rounded feathers, showing bright,
metallic purple when held in the light, each feather narrowly edged
with dark brown at the base of the skull.

The sexes are similar, the female being slightly paler.

Length. 13.50; wing. 7.50; tail, 5.50; tarsus, 1.

HABITAT. Antilles.

Columba caribæa LINN.

Columba caribæa LINN. Syst. Nat. I (1766).—GMEL. Syst. Nat. I, p. 773
(1788).—LATH. Ind. Orn. p. 603 (1790).—TEMM. Hist Gen. Pigeons, 450 (1813-15).—SHAW, Gen. Zool. XI, p. 37 (1819).—SUNDEV.
Oefv. K. Vet. Akad. För. 1869, p. 601.—SCL. & SALV. Nom. Avium
Neotr. p. 132 (1873).—BD. BWR. & RIDGW. Hist. N. Am. Bds. III, p.
359 (1874).—A. & E. NEWTON, Handb. Jamaica, p. 114 (1881).—
CORY, List Bds. W. I. p. 23 (1885).

Columba caribbæa DENNY, P. Z. S. 1847. p. 39.

Columba caribbea GOSSE, Bds. Jam. p. 291 (1847).

Patagiœnas caribæa BP. Consp. II, p. 54 (1854).—SCL. P. Z. S. 1861, p. 80.
—ALBRECHT. J. f. O. 1862, p. 204.—REICH. Handb. p. 65, tab. 230.—
GRAY, Handl. Bds. II. p. 234 (1870).

Columba carribea MARCH, Pr. Acad. Nat. Sci. Phila. 1863. p. 301.

SP. CHAR. *Male:*—Forehead and cheeks showing a faint olive, the rest of
the head a dull purplish tinge; chin dull white; feathers of the nape
and upper back showing golden green reflections when held in the
light; rest of upper parts dull olive; breast showing a dull purplish
tinge; rest of underparts pale reddish brown; upper surface of tail
dark slaty brown, almost black, to within two inches of the tip,
which is very pale brown; upper tail-coverts nearly covering the
dark brown of the basal portion; under surface of tail dull white;
primaries dark brown, narrowly edged with white on the outer
webs, showing brightest on the second, third, and fourth feathers.

The sexes are similar.

Length (skin), 14; wing. 8.50; tail, 6; tarsus, .90.

HABITAT. Jamaica and Porto Rico.

Columba inornata VIG.

Columba inornata VIG. Zool. Journ. 1827. p. 446.—D'ORB. in La Sagra's
Hist. Nat. Cuba, Ois. p. 173 (1840).—DENNY, P. Z. S. 1847, p. 39 —
SCL. P. Z. S. 1861, p. 80.—MARCH, Pr. Acad. Nat. Sci. Phila. 1863.
p. 301.—SCL. & SALV. Nom. Avium Neotr. p. 132 (1873).—BD.
BWR. & RIDGW. Hist. N. Am. Bds. III. p. 360 (1874).—A. & E.
NEWTON, Handb. Jamaica, p. 114 (1881).—CORY, Bds. Haiti & San
Domingo. p 136 (1885); *ib.* List Bds. W. I. p. 23 (1885).

Columba rufina GOSSE, Bds. Jam. p. 296 (1847).

Chlorœnas inornata BP. Consp. II, p. 55 (1854).—CAB. J. f. O. 1856, p.
106.—BREWER, Pr. Bost. Soc. Nat. Hist. VII. p. 307 (1860).—
SCLATER, P. Z. S. 1861, p. 80.—ALBRECHT, J. f. O. 1862, p. 204.—

GUNDL. Repert. Fisico-Nat. Cuba, I. p. 298 (1866); ib. J. f. O. 1874,
pp. 286. 312 · ib. Anal. Soc. Esp. Hist. Nat. VII. p. 343 (1878).

SP. CHAR. Male:—Head, neck, underparts, and some of the wing-coverts
dull purple; rest of plumage slaty; edges of outer webs of some of
the wing-coverts white, distinctly marking the wings; chin dull
white.

The sexes are similar.

Length, 14.50; wing, 8.50; tail, 5.50; tarsus, 1.10.

HABITAT. Greater Antilles.

GENUS Engyptila SUNDEV.

Engyptila SUNDEVALL, Stockholm Acad. Handl. 1835.

Engyptila jamaicensis (LINN.).

Columba jamaicensis LINN. Syst. Nat. I, p. 283 (1766).—GMEL. Syst. Nat.
I, p. 782 (1788).—TEMM. Hist. Gen. Pigeons, p. 495 (1813-15).
Columba frontalis TEMM. Hist. Gen. Pigeons (1813-15).
Goura jamaicensis SHAW, Gen. Zool. XI, p. 126 (1819).
Peristera jamaicensis GOSSE, Bds. Jam. p. 313 (1847).—ALBRECHT. J. f. O.
1862, p. 204.
Leptoptila jamaicensis SLOANE, Jam. pl. 262.—BP. Icon. Pig. t. 119; ib.
Consp. II, p. 73 (1854).—MARCH, Pr. Acad. Nat. Sci. Phila. 1863. p.
302.—GRAY, Handl. Bds. II. p. 242 (1870).—SCL. & SALV. Nom.
Avium Neotr. p. 133 (1873).
Engyptila jamaicensis A. & E. NEWTON, Handb. Jamaica, p. 114 (1881).—
CORY. List Bds. W. I. p. 23 (1885).

SP. CHAR. Male:—Forehead dull white, shading into slaty gray on the
top of the head; a cape of metallic purple, blue and gray, when
held in the light; rest of upper surface olive; throat dull white,
becoming slaty on the underparts; flanks and belly dull white;
under surface of wings bright rufous; tail feathers slaty, tipped with
white, except the two central ones, which are pale brown; primaries
pale brown.

Length (skin), 10; wing, 6; tail, 4.25; tarsus, 1; bill .75.

HABITAT. Jamaica.

Engyptila wellsi LAWR.

Engyptila wellsi LAWR. Auk, I. No 2, p. 180 (1884).—CORY, List Bds.
W. I. p. 23 (1885); ib. Revised List (1886).—WELLS, List Bds.
Grenada, p. 7 (1886).

SP. CHAR. Female:—" The front is whitish, with a slight tinge of fawn
color on the anterior portion, and is of a bluish cast on the posterior ·

the crown and occiput are dark brown; the hind neck is of a rather lighter brown; the back, wings, and upper tail-coverts are of a dull olivaceous green; the first outer tail-feather is brownish-black, narrowly tipped with white; the second is dark brown for two-thirds its length, terminating in blackish; all the other tail-feathers are dark umber brown above, and black underneath; the chin is white; the neck in front and the upper part of the breast are of a reddish fawn color; the middle and lower parts of the breast and the abdomen are creamy white; the sides are of a light fulvous color; the under tail-coverts are white, tinged with fulvous; the quills have their outer webs of a clear warm brown; the inner webs and under wing-coverts are of a rather light cinnamon color; the bill is black; the tarsi and toes are bright carmine red.

"Mr. Wells says the sexes are alike.

"Length, 12.25 inches; wing, 6.00; tail, 4.00; bill, .63; tarsus, 1.25." (LAWR., l. c., orig. discr.)

HABITAT. Grenada.

Engyptila collaris CORY.

Engyptila collaris CORY, Auk, III, pp. 498, 502 (1886).

SP. CHAR.—Forehead dull white; top of the head dark gray, showing a metallic tinge of purple on the nape; a cape of metallic purple showing greenish red reflections where it joins the back; back dark brownish olive; throat dull white; breast dull vinaceous, shading into dull white on the belly; sides dull red brown; under wing-coverts and under surface of wing rufous brown; primaries brown, having the inner webs heavily marked with rufous brown; tail slaty brown, two or three outer feathers tipped with white; feet red; bill black; iris dull white.

Length, 9.50; wing, 5.75; tail 3.50; tarsus, 1.25; bill, .75.

HABITAT. Grand Cayman.

GENUS Zenaidura BONAP.

Zenaidura "BONAPARTE, Consp. II, 1854. p. 84."

Zenaidura macroura (LINN.).

Columba macroura LINN. S. N. ed. 10, p. 164 (1758), part.

Columba carolinensis LINN. Syst. Nat. I, p. 286 (1766).—D'ORB. in La Sagra's Hist. Nat. Cuba, Ois. p. 176 (1840).—SUNDEV. Oefv. K. Vet. Akad. För. 1869, p. 601 (Porto Rico).

Zenaidura carolinensis SALLÉ, P. Z. S. 1857. p. 235 (San Domingo).— CORY, Bull. Nutt. Orn. Club, VI. p. 154 (1881) (Haiti); *ib.* Bds. Haiti & San Domingo, p. 129 (1885); *ib.* List Bds. W. I. p. 23 (1885).

Perissura carolinensis BREWER, Pr. Bost. Soc. Nat. Hist. VII, p. 307 (1860)

(Cuba).—GUNDL. Repert. Fisico-Nat. Cuba, I, p. 301 (1865); *ib.*
J. f. O. 1874, p. 298 (Cuba).
Columba (*Zenaidura*) *caroliuensis* BRYANT, Pr. Bost. Soc. Nat. Hist. XI,
p. 96 (1867) (San Domingo).

Recorded from Haiti, San Domingo, Cuba, and Porto Rico.

GENUS Ectopistes SWAINS.

Ectopistes SWAINSON. Zool. Jour. III, p. 362, 1827.

Ectopistes migratorius (LINN.).

Columba migratoria LINN. Syst. Nat. I. p. 285 (1766).
Ectopistes migratoria SWAINS. Zool. Journ. III. p. 362 (1827).—GUNDL.
J. f. O. 1856, p. 112 (Cuba); *ib.* Repert. Fisico-Nat. Cuba, I. p. 302
(1865); *ib.* J. f. O. 1874, p. 300 (Cuba).—CORY, List Bds. W. I. p.
24 (1885).

Accidental in Cuba.

GENUS Zenaida BONAP.

Zenaida BONAPARTE, Geog. and Comp. List. p. 41, 1838.

Zenaida zenaida (BONAP.).

Columba zenaida BP. Journ. Acad. Nat. Sci. Phila. V, p. 30 (1825).—
WAGL. Isis, 1829, p. 744.—D'ORB. in La Sagra's Hist. Nat. Cuba,
Ois. p. 177 (1840).—AUD. Bds. Am. V, p. 1 (1842).—BRYANT, Pr.
Bost. Soc. Nat. Hist. X, p. 257 (1866).
Zenaida amabilis BP. List, 1838; *ib.* Consp. II, p. 82 (1854).—GOSSE,
Bds. Jam. p. 307 (1847).—BRYANT, Pr. Bost. Soc. Nat. Hist. VII. p.
120 (1859).—A. & E. NEWTON, Ibis, 1859. p. 253.—BREWER, Pr.
Bost. Soc. Nat. Hist. VII, p. 120 (1859).—A. & E. NEWTON. Ibis,
1859, p. 253.—BREWER. Pr. Bost. Soc. Nat. Hist. VII, p. 307 (1860).—
CASSIN, Pr. Acad. Nat. Sci. Phila. 1860. p. 378.—SCL. P. Z. S.
1861, p. 80.—ALBRECHT, J. f. O. 1862, p. 204.—MARCH. Pr. Acad.
Nat. Sci. Phila. 1863. p. 302.—GUNDL. Repert. Fisico-Nat. Cuba, I,
p. 301 (1865); *ib.* J. f. O. 1874. pp. 298, 312; *ib.* Anal. Soc. Esp.
Hist. Nat. VII. p. 346 (1878).—SCL. & SALV. Nom. Avium Neotr.
p. 132 (1873).—A. & E. NEWTON, Handb. Jamaica. p. 114 (1881).—
CORY, Bds. Bahama I. p. 138 (1880); *ib.* Bds. Haiti & San Domingo,
p. 128 (1885); *ib.* List Bds. W. I. p. 24 (1885).—COUES, Key N. Am.
Bds. p. 569 (1884).

SP. CHAR. *Male:*—Above olive brown; top of the head and underparts
pale purplish brown; sides of the body and under wing-coverts
bluish; tail-feathers, with the exception of the central ones bluish,

with a black band about an inch from the tip; a slight streak of
metallic blue below the ear; quills dark brown; secondaries tipped
with white; feet red.

The sexes are similar.

Length, 10; wing, 6; tail, 4.30; tarsus 80; bill, .55.

HABITAT.　Antilles.

Zenaida spadicea CORY.

Zenaida spadicea CORY, Auk, III, pp. 498, 502 (1886).

SP. CHAR.—General upper plumage dark olive brown, rufous brown on
the forehead and showing a tinge of very dull purple on the crown,
apparently wanting in some specimens; a sub-auricular spot of
dark metallic blue; sides of the neck and nape rich metallic purple;
chin pale buff, shading into rich rufous chestnut on the throat and
breast; belly brown, showing a slight vinaceous tinge; upper surface of tail brown, the feathers showing a sub-terminal band of
black, and all the feathers except the central ones tipped with gray;
primaries dark brown, almost black, faintly tipped with dull white;
the secondaries broadly tipped with white; under wing-coverts gray;
bill black; feet red.

Length, 9.60; wing, 6; tail, 3.75; tarsus, .75; bill, .50.

HABITAT.　Grand Cayman.

A specimen of *Zenaida* taken in Little Cayman differs somewhat from *Z. spadicea*, being lighter colored and having the
metallic feathers of the neck somewhat differently colored—paler
and less in extent. I have separated the Little Cayman bird provisionally and with much hesitation, and have proposed the name
Zenaida richardsoni for it (see Auk, IV, p. 7, 1887), should
further investigation prove them specifically separable.

Zenaida martinicana BONAP.

Zenaida martinicana, BP. Consp. II, p. 82 (1854).—TAYLOR, Ibis, 1864,
p. 171.—GRAY, Handl. Bds. II, p. 241 (1870).—SCL. & SALV. Nom.
Avium Neotr. p. 132 (1873).—LAWR. Pr. U. S. Nat. Mus. I, p. 487
(1878).—ALLEN, Bull. Nutt. Orn. Club, V. p. 169 (1880).—LISTER,
Ibis, 1880, p. 43.—GRISDALE, Ibis, 1882, p. 492.—CORY, List Bds.
W. I. p. 24 (1885).—WELLS, List Bds. Grenada, p. 7 (1886).

Columba (Zenaida) martinicana SUNDEV. Oefv. K. Vet. Akad. För. 1869,
p. 585.

SP. CHAR.—Top of head, cheeks, and upper throat pale rufous brown;
narrow line of dark blue on the cheek; chin dull white; feathers of

the sides of the neck tipped with metallic purple; throat tinged with pale lavender; underparts dull bluish white; back olive brown, shading into chestnut brown on the rump; central tail-feathers brown; rest of tail-feathers slate color at the base, succeeded by a band of black, and tipped with white; quills dark brown, showing an indistinct white edging on the outer primaries, tipped with white; bill black.

Length, 10; wing, 6; tail, 4; tarsus, .75; bill, .50.

HABITAT.   Lesser Antilles.

Zenaida rubripes LAWR.

Zenaida rubripes LAWR. Auk. II, p. 357 (1885).—CORY. Revised List Bds. W. I. p. 24 (1886).—WELLS, List Bds. Grenada, p. 7 (1886).

SP. CHAR.   *Female :*—"The front is of a light brown tinged with vinaceous; the upper plumage is olivaceous-brown, with a dull reddish tinge, which is most observable on the back; the hind part and sides of the neck are grayish, the latter glossed with golden changing to light violet: the two central tail-feathers are olive brown; the outer web of the first lateral feather is pale rufous: the bases of the four outer ones are brownish-cinereous, with their ends largely pale rufous, the two colors separated by a black bar; the other tail-feathers are dark cinereous with a subterminal black bar, on the under side the color of the basal portion of the tail-feathers is blackish cinereous; the primaries are dark umber-brown, the secondaries brownish-black, both narrowly edged with white; the tertials are the color of the back, and are marked with four conspicuous oval spots of black; the under wing-coverts are light bluish-ash, the flanks dark ashy-blue; behind the eye is a small spot of black, and another below the ears; sides of the head and the chin pale vinaceous, the latter lighter in color; the under plumage is of a reddish cinnamon color, rather dull on the throat and breast, but somewhat brighter on the abdomen and under tail-coverts; bill black; tarsi and toes carmine red.

"The color of the feet in the dried specimen is quite bright; in the living bird it is doubtless much more so. The tail has fourteen rectrices.

"Length, fresh. 9.50 inches; wing, 5.25; tail, 3.38; bill, .62; tarsus, .75." (LAWR.. l. c., orig. descr.)

HABITAT.   Grenada.

GENUS Melopelia BONAP.

Melopelia BONAPARTE, Consp. II, p. 81, 1854.

Melopelia leucoptera (LINN.).

Columba leucoptera LINN. Syst. Nat. I, p. 164 (1758).
Turtur leucoptera GOSSE, Bds. Jam. p. 304 (1847).

Zenaida leucoptera SCL. P. Z. S. 1861 p. 80, (Jamaica).—ALBRECHT.
    J. f. O. 1862, p. 204 (Jamaica).
Melopelia leucoptera MARCH, Pr. Acad. Nat. Sci. Phila. 1863. p. 302
    (Jamaica).—GUNDL. Repert. Fisico-Nat. Cuba, I, p. 301 (1866); *ib.*,
    J. f. O. 1874. p. 297 (Cuba).—A. & E. NEWTON, Handb. Jamaica
    p. 114 (1881).—CORY, Bds. Haiti & San Domingo, p. 131 (1885); *ib.*
    List Bds. W. I. p. 24 (1885).

This species has been recorded from Cuba, Jamaica, and San
Domingo.

GENUS Columbigallina BOIE.

Columbigallina BOIE, Isis, 1826, p. 977.

Columbigallina passerina (LINN.).

Columba passerina LINN. Syst. Nat. I, p. 285 (1766).—GMEL. Syst. Nat. I,
    p. 787 (1788).—D'ORB. in La Sagra's Hist. Nat. Cuba. Ois. p. 179
    (1840).—SUNDEV. Oefv. K. Vet. Akad. För. 1869, pp. 586, 601.
Columbigallina passerina ZELEDON, Pr. U. S. Nat. Mus. VIII, p. 112
    (1885).—CORY, Ibis, 1886, pp. 472, 474; *ib.* Auk. III, p. 502 (1886).
"*Columba (Goura) passerina* BP. Obs. Wils. 1825, No. 181.—NUTT. Man.
    I, p. 635 (1832)."
"*Charmepelia passerina* SWAINS. Zool. Journ. III. p. 358 (1827)."
Chamæpelia passerina BP. List, 1838, p. 41.—GOSSE, Bds. Jam. p. 311
    (1847).—SALLÉ, P. Z. S. 1857, p. 236.—BRYANT, Pr. Bost. Soc. Nat.
    Hist. VII, p. 120 (1859); *ib.* XI, p. 96 (1866).—WELLS, List Bds. Gre-
    nada. p. 7 (1886).—BREWER, *ib.* VII, p. 307 (1860).—SCL. P. Z. S.
    1861. p. 80; *ib.* 1874. p. 175.—ALBRECHT, J. f. O. 1862, p. 204.—
    MARCH. Pr. Acad. Nat. Sci. Phila. 1863. p. 302.—SCL. & SALV. Nom.
    Avium Neotr. p. 133 (1873).—BD. BWR. & RIDGW. Hist. N. Am.
    Bds. III, p. 389 (1874).—GUNDL. Anal. Soc. Esp. Hist. Nat. VII, p.
    349 (1878).—LAWR. Pr. U. S. Nat. Mus. I, p. 487 (1878).—LISTER,
    Ibis, 1880, p. 43.—ALLEN, Bull. Nutt. Orn. Club, V, p. 169 (1880).
    —A. & E. NEWTON, Handb. Jaimaca. p. 114 (1881).—CORY, Bds.
    Bahama I. p. 139 (1880); *ib.* Bull. Nutt. Orn. Club, VI. p. 154
    (1881); *ib.* Bds. Haiti & San Domingo, p. 127 (1885); *ib.* List Bds.
    W. I. p. 24 (1885).—COUES, Key N. Am. Bds. p. 569 (1884).—
    RIDGW. Pr. U. S. Nat. Mus. VII, p. 172 (1884).
Chamæpelia trochila A. & E. NEWTON, Ibis, 1859, p. 253.—CASSIN, Pr.
    Acad. Nat. Sci. Phila. 1860, p. 378.—SCL. P. Z. S. 1872, p. 633.
Columba (Chamæpelia) passerina BRYANT, Pr. Bost. Soc. Nat. Hist. X,
    p. 257 (1866).
Chamæpelia bahamensis MAYNARD, Am. Exch. & Mart, Jan. 15. 1887.

SP. CHAR. *Male:*—Above grayish olive, showing a bluish tinge upon the
    nape and crown; underparts reddish purple, becoming ashy on the

sides; under wing-coverts and quills showing reddish brown, the latter margined and tipped with dark brown; middle tail-feathers like the back, the others dark brown, two outer feathers tipped with white; upper surface of wing showing large spots of bluish purple; bill and feet yellowish, the former becoming dark at the tip.

The sexes are similar.

Length, 6.30; wing, 3.30; tail, 2.60; tarsus, .50; bill, .50.

HABITAT.   Bahamas and Antilles.

GENUS Geotrygon GOSSE.

Geotrygon GOSSE, Bds. Jam. p. 316, 1847.

Geotrygon cristata (TEMM.).

Columba cristata TEMM. Hist. Gen. Pigeons, p. 449 (1813-15).—SHAW, Gen. Zool. XI. p. 40 (1819).
Geotrygon sylvatica GOSSE, Bds. Jam. p. 316 (1847).—ALBRECHT, J. f. O. 1862, p. 204.
Geotrygon cristata BP. Consp. II, p. 70 (1854).—MARCH, Pr Acad. Nat. Sci. Phila. 1863, p. 300.—GRAY, Handl. Bds. II, p. 243 (1870).— SCL. & SALV. Nom. Avium Neotr. p. 134 (1873).—A. & E. NEW-TON, Handb. Jamaica, p. 114 (1881).—CORY, List Bds. W. I. p. 24 (1885).

SP. CHAR. *Male:*—Forehead black, shading into grayish olive on the top of the head; a malar stripe of pale rufous; breast, sides of the neck, and upper back forming a broad collar of beautiful metallic purple; held in the light it shows bright golden green; back and wing-coverts dark purple, tinged with blue, showing chestnut in some lights; rump dark green; under surface of wings rufous brown; belly slate color; sides and flanks rufous brown; the first six primaries bright rufous, shading into green on the tips and inner webs; secondaries green; upper surface of tail green.

Length (skin), 11; wing, 6.75; tail, 4; tarsus, 1.05; bill, .90.

HABITAT.   Jamaica.

Geotrygon mystacea (TEMM.).

Columba mystacea TEMM. Hist. Gen. Pigeons, p. 473 (1814-15).—SHAW, Gen. Zool. XI, p. 56 (1819).—"REICH. Syst. Av. t. 257, b. f. 3382."
Geotrygon mystacea BP. Consp. II, p. 71 (1854).—GRAY, Handl. Bds. II. p. 243 (1870).—SCL. & SALV. Nom. Avium Neotr. p. 134 (1873).— LAWR. Pr. U. S. Nat. Mus. I, p. 487 (1878).—SCL. P. Z. S. 1879, p. 765.—ALLEN, Bull. Nutt. Orn. Club, V, p. 169 (1880).—CORY, List Bds. W. I. p. 24 (1885); *ib.* Ibis, 1886, p. 475.

SP. CHAR.—Forehead brownish, shading into green on the top of the head; sides of the neck and upper back bright metallic green, becoming bright purple with bluish reflections on reaching the back; stripe of white on the cheeks, passing from the lower mandible; rest of upper parts dark olive green; upper portion of throat dull white; becoming brown with greenish reflections on the breast; underparts dull purplish white, becoming dull white on the belly; under tail-coverts chestnut brown, tipped with white; primaries deep rufous chestnut, olive at tips; tail-feathers, except the two central ones, chestnut, shading to dull olive at the tip.

Length, 11; wing, 6.50; tail, 5; tarsus, 1.25; bill, .70.

HABITAT. Guadeloupe, Santa Lucia, and Grand Terre.

Geotrygon caniceps GUNDL.

Columba caniceps GUNDL. Journ. Bost. Soc. Nat. Hist. VI, p. 315 (1852).
Geotrygon caniceps GUNDL. J. f. O. 1856, p. 110.—BREWER. Pr. Bost. Soc. Nat. Hist. VII, p. 307 (1860).—GUNDL. Repert. Fisico-Nat. Cuba, I, p. 300 (1866); *ib.* J. f. O. 1874, p. 295.—SCL. & SALV. Nom. Avium Neotr. p. 134 (1873).—CORY, List Bds. W. I. p. 24 (1885).

SP. CHAR. *Male:*—Forehead whitish, shading into slate color on the top of the head; the feathers of the nape showing greenish and purple reflections when held in the light; back purple; rump steel blue, showing greenish reflections in the light; throat pale, becoming slaty on the breast, with slight reflections in the light; underparts pale slate color, showing rufous on the abdomen, and deep rufous brown on the crissum; under surface of wing reddish brown; primaries olive brown, showing rufous brown on the inner webs.

The sexes are similar.

Length (skin), 10.50; wing, 6; tail, 3.50; tarsus, 1.20.

HABITAT. Cuba.

Geotrygon montana (LINN.).

Columba montana LINN. Syst. Nat. I, p. 281 (1766).—GMEL. Syst. Nat. I, p. 772 (1788).—SUNDEV. Oefv. K. Vet. Akad. För. 1869, p. 601.
Peristera montana GRAY, Gen. Bds. II, p. 475 (1844-49).
Geotrygon montana GOSSE, Bds. Jam. p. 320 (1847).—BP. Consp. II, p. 72 (1854).—CAB. J. f. O. 1856, p. 109.—SALLÉ, P. Z. S. 1857, p. 235.—BREWER, Pr. Bost. Soc. Nat. Hist. VII, p. 307 (1860).—SCL. P. Z. S. 1861, p. 80.—ALBRECHT, J. f. O. 1862, p. 204.—MARCH, Pr. Acad. Nat. Sci. Phila. 1863, p. 300.—TAYLOR, Ibis, 1864, p. 171.—GUNDL. J. f. O. 1874, p. 294; *ib.* Anal. Soc. Esp. Hist. Nat. VII, p. 348 (1878).—LAWR. Pr. U. S. Nat. Mus. I, p. 487 (1878).—LISTER, Ibis, 1880, p. 43.—ALLEN, Bull. Nutt. Orn. Club, V, p. 169 (1880).—A. & E. NEWTON, Handb. Jamaica, p. 114 (1881).—CORY, Bds. Haiti & San

Domingo, p. 132 (1885); *ib.* List Bds. W. I. p. 24 (1885); *ib.* Ibis,
1886. p. 473.—WELLS, List Bds. Grenada, p. 7 (1886).
Columba (Geotrygon) montana BRYANT Pr. Bost. Soc. Nat. Hist. XI p.
96 (1866); *ib.* X, p. 257 (1866).

SP. CHAR. *Male:*—Above purplish brown, becoming light brown on the
wings; throat dull white, becoming pale purple on the breast; belly
pale brown, becoming brownish white on the under tail-coverts.

Female:—Upper parts greenish brown · forehead light brown, the
color extending upon the cheeks and sides of the head; breast chestnut brown.

Length, 9.25; wing 6, tail, 3.25; tarsus. 1.

HABITAT. Antilles.

Geotrygon martinica (GMEL.).

Columba martinica GMEL. Syst. Nat. I, p 781 (1788).
Columba montana AUD. Orn. Biog. II, p. 382 (1834).—NUTT. Man. I, 2nd
ed. p. 756 (1840) (not of LINN.).
Zenaida montana BP. Geog. & Comp. List, 1838.
Orcopeleia martinicana REICH. Syst Nat. Av. p. 25 (1851).
Geotrygon martinica BP. Consp. II. p. 74 (1854).—CAB. J. f. O. 1856, p.
108.—BREWER, Pr. Bost Soc. Nat. Hist. VII. p. 307 (1860).—SCL.
& SALV. Nom. Avium Neotr. p. 134 (1873).—GUNDL. J. f. O. 1874,
p. 293; *ib.* Anal. Soc. Esp. Hist. Nat. VII, p. 347 (1878).—CORY,
Bds. Bahama I. p. 141 (1880); *ib.* Bds. Haiti & San Domingo, p.
133 (1885); *ib.* List Bds. W. I. p. 24 (1885).—COUES, Key N. Am.
Bds. p. 571 (1884).
Orcopeleia martinica BAIRD, Bds. N. Am. p. 607 (1858).—GUNDL. Repert.
Fisico-Nat. Cuba, I, p. 299 (1866).—GRAY, Handl. Bds. II, p. 242
(1870).—BD. BWR. & RIDGW. Hist. N. Am. Bds. III. p. 393 (1874).
Columba (Geotrygon) martinica BRYANT, Pr. Bost. Soc. Nat. Hist. XI, p.
96 (1866).

SP. CHAR. *Male:*—Above chestnut rufous; crown and neck with metallic
reflections of green and purple: back showing brilliant purple reflections, becoming less distinct on the rump; a band of white from
the base of the lower mandible. under the eye, to the side of the
neck, bordered below by a streak of dull purple; underparts showing
the breast pale purple, becoming dull white on the throat and abdomen; primaries bright rufous, becoming darker at the tips; tail
rufous; legs light red; bill red, tip horn color; iris light brown.

The sexes appear to be similar.

Length, 10.75; wing, 6; tail 4.25; tarsus, 1.05; bill, .90.

HABITAT. Bahamas and Antilles.

GENUS Starnœnas BONAP.

Starnœnas BONAPARTE, Geog. & Comp. List, 41, 1838.

Starnœnas cyanocephala (Linn.).

Columba cyanocephala Linn. Syst. Nat. I. p. 282 (1766).—D'Orb. in La Sagra's Hist. Nat. Cuba. Ois. p. 174 (1840).
Starnœnas cyanocephala Gosse. Bds. Jam. p. 324 (1847).—Can. J. f. O. 1856. p. 108 (Cuba).—Brewer. Pr. Bost. Soc. Nat. Hist. VII. p. 307 (1860) (Cuba).—Albrecht, J. f. O. 1862. p. 204 (Jamaica).—Gundl. Repert. Fisico-Nat. Cuba, I. p. 299 (1865); *ib.* J. f. O. 1874, p. 291 (Cuba).—Cory, List Bds. W. I. p. 24 (1885).

Sp. Char.—Top of the head bright blue; a narrow line of black extending through the eye, meeting at the nape, immediately joining a band of white which passes under the eye from the lower mandible and chin; throat glossy black, narrowly banded with white on the last black feathers of the lower throat, forming a white edging to the black throat; the feathers on the sides of the neck narrowly tipped with blue; upper parts purplish brown on the back, shading into olive brown on the lower back and rump; wings and tail brown; breast tinged with purple, shading into rufous brown on the belly; under surface of tail-feathers dark brown, almost black; basal portion of bill and feet deep red.

Length, 11; wing, 6; tail, 4; tarsus, 1.25; bill, .50.

Habitat. Cuba.

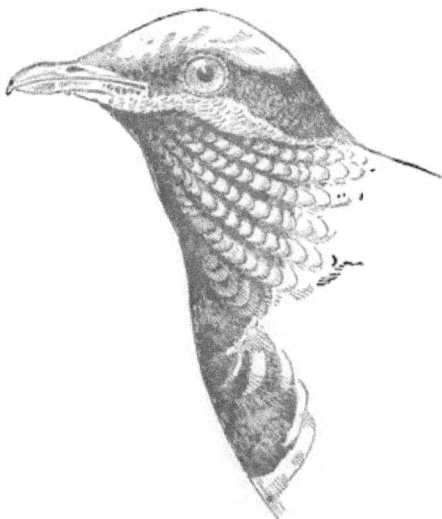

This species is common in parts of the interior. On several occasions I have seen the living birds offered for sale in the markets of Havana. Albrecht records it from Jamaica.

Turtur risoria of authors is claimed to have been introduced into the West Indies many years ago ; I have a specimen in my cabinet labelled " San Domingo." It has also been recorded from St. Bartholomew, Cuba, and Jamaica.*

BY CHARLES B. CORY.

[*Continued from p.* 120.]

FAMILY PHASIANIDÆ.

GENUS **Numida** LINN.

Numida LINNÆUS. Syst. Nat. I, 1766.

Numida meleagris LINN.

Numida meleagris LINN. Syst. Nat. I. p. 273 (1766).—GOSSE, Bds. Jam.
p. 325 (1847).—DENNY, P. Z. S. 1847, p. 39 (Jamaica).—SALLÉ,
P. Z. S. 1857, p. 236 (San Domingo).—SCL. P. Z. S. 1861, p. 80
(Jamaica).—ALBRECHT, J. f. O. 1862, p. 204 (Jamaica).—MARCH.
Pr. Acad. Nat. Sci. Phila. 1863. p. 303 (Jamaica).—BRYANT, Pr.
Bost. Soc. Nat. Hist. XI, p. 97 (1866) (San Domingo).—GUNDL
Repert. Fisico-Nat. Cuba. I. p. 397 (1866).—SUNDEV. Oefv. K. Vet.
Akad. For. 1869. p. 601 (Porto Rico).—LAWR. Pr. U. S. Nat. Mus.
I. p. 487 (1878) (Barbuda).—A. & E. NEWTON, Handb. Jamaica, p.
117 (1881).—CORY, Bds. Haiti & San Domingo, p. 16 (1885) ; *ib.*
List Bds. W. I. p. 24 (1885).

Common in Cuba, San Domingo, Jamaica, Porto Rico, and Barbuda.

Ortalida ruficauda is mentioned as occurring in the Grenadines, and is supposed to have been introduced (*Lawr.* Pr. U. S. Nat. Mus. I, p. 278 (1878).

FAMILY TETRAONIDÆ.

GENUS **Colinus** LESS.

Colinus LESSON, Man. d'Orn. II, p. 190. 1828.

Colinus cubanensis (GOULD).

Ortyx virginianus D'ORB. in La Sagra's Hist. Nat. Cuba, Ois. p. 182
(1840).—SUNDEV. Oefv. K. Vet. Akad. For. 1869. p. 601.

---

* *Turtur risoria* SUNDEV. Oefv. K. Vet. Acad. För. 1869, p. 586 (St. Bartholomew).
Turtur risorius MARCH, Pr. Acad. Nat. Sci. Phila. 1863, p. 302 (Jamaica).
Turtur risorus A. & E. NEWTON, Handb. Jamaica, p. 117 (1881).

Ortyx cubanensis GOULD. Mon. Odont. (1850).—CAB. J. f. O. 1856. p. 337.
—BREWER, Pr. Bost. Soc. Nat. Hist. VII, p. 307 (1860).—ALBRECHT.
J. f. O. 1861, p. 213.—GUNDL. Repert. Fisico-Nat. Cuba, I, p. 303
(1866).—GRAY, Handl. Bds. II, p. 273 (1870).—GUNDL. J. f. O. 1874,
p. 300; *ib.* 1875, p. 293; *ib.* 1878, p. 161; *ib.* Anal. Soc. Esp. Hist.
Nat. VII, p. 350 (1878).
Ortyx cubensis SCL. & SALV. Nom. Avium Neotr. p. 137 (1873).
Ortyx virginianus var. *cubanensis* BD. BWR. & RIDGW. Hist. N. Am. Bds.
III, p. 468 (1874).
Colinus cubanensis CORY, List Bds. W. I. p. 24 (1885).

SP. CHAR.—Upper portions of throat and superciliary stripe white; band
of neck passing from the mandible, under the eye, down the sides of
the neck; breast and lower portion of throat black; back chestnut,
variegated with dull brown; the feathers on the nape heavily spotted
with white; under parts variegated, dull brown, rufous, white, and
dark brown; sides of the body dull rufous, heavily spotted with
white and black; primaries dull brown.

The female differs from the male in having the white stripe and
throat tawny buff, and in lacking the chestnut on the breast to a
great extent.

Length, 8; wing, 4; tail, 2.50; tarsus, 1; bill, .45.

HABITAT. Cuba and Porto Rico.

Colinus virginianus (LINN.).

Tetrao virginianus LINN. Syst. Nat. I, p. 277 (1766).
Ortyx virginiana GOSSE, Bds. Jam. p. 328 (1847).—CORY, Bull. Nutt.
Orn. Club, VI, p. 154 (1881) (Haiti).
Ortyx virginianus A. & E. NEWTON, Ibis, 1859, p. 254 (St. Croix).—BRY-
ANT, Pr. Bost. Soc. Nat. Hist. VII, p. 120 (1859) (Bahamas).—SCL.
P. Z. S. 1861, p. 80 (Jamaica).—ALBRECHT, J. f. O. 1862, p. 205
(Jamaica).—MARCH, Pr. Acad. Nat. Sci. Phila. 1863, p. 303
(Jamaica).—LAWR. Pr. U. S. Nat. Mus. I, pp. 237, 487 (1878) (An-
tigua).—CORY, Bds. Bahama I. p. 142 (1880).—A. & E. NEWTON,
Handb. Jamaica, p. 117 (1881).—CORY, Bds. Haiti & San Domingo,
p. 138 (1885).
Colinus virginianus CORY, List Bds. W. I. p. 24 (1885).

SP. CHAR. *Male:*—Above rich brownish red, mottled with black; crown
black, shading into brown at the base of the skull, and mottled with
black and white on the nape; a white superciliary line passing from
nostril to nape; throat white, bordered broadly with black; upper
breast and sides reddish brown, shading into white on the belly, the
feathers thickly banded with black; crissum reddish brown; tertials
and some of the wing-coverts edged with yellowish white; bill entirely black.

Female:—Resembles the male; the white of the head and throat
replaced by tawny, without black edging.

Length, 8.50; wing, 4.50; tail, 2.50; tarsus, 1; bill, .52.

HABITAT. Bahamas, Haiti, San Domingo, Jamaica, St. Croix,
and Antigua.

The forms represented in the different islands vary considera-
bly, and it is possible that they represent good geographical races.
The Bahama bird differs from that found in Florida in having
heavy chestnut stripings on the side much broader than in the
Florida birds. The black on the throat is more restricted; the
lower throat showing considerable chestnut, separating the black
from the upper breast; the red on the back is paler. The feathers on the underparts are very heavily banded with black, about
equalling some specimens of *floridanus* in this respect, but the
underparts are never mottled gray as in some specimens of *cubanensis*. The Bahama bird differs even more from that found in
San Domingo, which has the underparts covered with narrow
black arrow-shaped markings, somewhat obsolete in the female;
the male having a patch of black on the throat succeeded by pale
chestnut; the general chestnut coloring is paler than in the Bahama bird.

GENUS Eupsychortyx GOULD.

Eupsychortyx GOULD, Mon. Odontophorinæ. p. 15-16, 1850. Type, *Tetrao
cristatus* LINN.

Eupsychortyx sonninii (TEMM.).

Perdix sonninii TEMM. Pig. et Gall. III, p. 451 (1815); *ib.* Pl. Col. 75
(1820-29).
Eupsychortyx sonnini NEWTON, Ibis, 1860, p. 308 (St. Thomas). — CAS-
SIN, Pr. Acad. Nat. Sci. Phila. 1860, p. 378 (St. Thomas). — CORY,
Revised List Bds. W. I. p. 24 (1885).
Ortyx sonninii NEWTON (REINHARDT), Ibis, 1861, p. 114 (St. Thomas).

CHAR. *Male:*—Face dull white; head crested; feathers of the crest dull
buff brown; throat and superciliary stripe passing down the sides of
the neck dull brownish red; sides of the neck mottled with black
and white; upper back mottled with reddish brown, buff, and black;
rest of upper surface marked with chestnut, black, and gray, margined with buff; tail slaty dotted and marked with buff and dark
brown; primaries brown; chest grayish, mottled with brown; rest

of under surface, sides, and under tail-coverts chestnut brown, the
feathers dotted with white; bill black.

Female:—Top of the head and crest brown; throat and superciliary stripe dark gray, tinged with pale brown; flank marking
paler than in the male; general plumage somewhat paler than in
the male, and the black patches somewhat heavier.

Length, 7.50; wing. 4.30; tail. 2.55; tarsus, 1.30; bill, .54.

HABITAT. St. Thomas.

In 1860 Professor Newton mentions this species as occurring in
St. Thomas. The bird undoubtedly still exists in the Island of
St. Thomas. I have lately seen a specimen in the collection of
Mr. Geo. N. Lawrence, from that Island. It was probably in-
troduced from South America.

Cassin writes (l.c.), "Mr. Swift has had the kindness to inform
me that this species was introduced into the Island of St. Thomas
some years since, from Venezuela, and that it has now become of
frequent occurrence, quite naturalized, and rearing young freely
throughout the Island. The present specimens are exactly the
species figured by Mr. Gould under this name, and identical
with specimens in Acad. Mus. labelled 'Venezuela' and 'Cu-
mana.'"

FAMILY ŒDICNEMIDÆ.

GENUS Œdicnemus TEMM.

Œdicnemus TEMMINCK, Man. d'Orn. 1815.

Œdicnemus dominicensis CORY.

Œdicnemus dominicensis CORY. Journ. Bost. Zool. Soc. II, p. 46 (1883);
ib. Auk, I, p. 4 (1884); *ib.* Bds. Haiti and San Domingo, p. 140
(1885); *ib.* List Bds. W. I. p. 25 (1885).—THOMPSON, Auk, II, p.
110 (1885).

SP. CHAR. *Male:*—Top of the head, back, wing-coverts. and tail brown;
feathers with very pale edgings, giving a mottled appearance to the
back; the tail-feathers showing a band of dull white, succeeded by
a broad black tip; breast slaty becoming dull white on the throat;
abdomen white tinged with very pale rufous; a line of black passing
from the top of the eye, along the sides of the head to the neck;
under surface of wings white, becoming dark brown at the tips;
the shafts of the feathers on the breast and throat dark brown, form

ing numerous hair-like lines on the surface of the plumage; legs
and feet greenish yellow; upper mandible black; under mandible
green at the base, shading into black at the tip.

The sexes appear to be similar.

Length, 14.50; wing, 8.50; tail, 3.75; tarsus, 3.75; bill, 1.50.

Habitat. San Domingo.

ŒDICNEMUS DOMINICENSIS CORY.

This peculiar species is apparently restricted to the eastern portion of the island of Santo Domingo. Mr. M. A. Frazar, who discovered the species, says, "Although I found it feeding during the day, it seems somewhat nocturnal in its habits, as I saw them in comparative abundance in the large savanna through which I passed at midnight on my return from La Vega. . . . Their note is a repetition of the syllable 'tu' repeated very rapidly." He also says the natives tame the birds and keep them in the houses for the purpose of killing the spiders and bugs which abound in large numbers, and which it greedily kills and eats. It is known to the natives by the name of *Boukera.*

FAMILY CHARADRIIDÆ.

GENUS Charadrius LINN.

Charadrius LINNÆUS, Syst. Nat. I, 10th ed. p. 150, 1758; *ib.* 12th ed. p. 253, 1766.

Charadrius dominicus MÜLL.

Charadrius dominicus MÜLL. Syst. Nat. Suppl. p. 116 (1776).—CASSIN, Pr. Acad. Nat. Sci. Phila. 1864, p. 241 (Jamaica).—A. & E. NEWTON, Handb. Jamaica, p. 115 (1881).—CORY, List Bds. W. I. p. 25 (1885). —WELLS, List Bds. Grenada. p. 7 (1886).

Charadrius virginianus GOSSE, Bds. Jam. p. 333 (1847).—GUNDL. J. f. O. 1856, p. 423 (Cuba).—A. & E. NEWTON, Ibis, 1859, p. 255 (St. Croix).—MARCH, Pr. Acad. Nat. Sci. Phila. 1864, p. 66 (Jamaica).— GUNDL. Repert. Fisico-Nat. Cuba, I, p. 358 (1866).—LAWR. Ann. Lyc. N. Y. VIII, p. 98 (1867) (Sombrero).—GUNDL. J. f. O. 1875, p. 332 (Cuba).—LAWR. Pr. U. S. Nat. Mus. I, p. 67 (1878) (Dominica); *ib.* p. 197 (St. Vincent); *ib.* p. 238 (Antigua); *ib.* p. 241 (Barbuda); *ib.* p. 276 (Grenada); *ib.* p. 461 (Gaudeloupe).—GUNDL. Anal. Soc. Esp. Hist. Nat. VII, p. 381 (1878) (Porto Rico).

Charadrius marmoratus LEMB. Aves Cuba, p. 105 (1850).—BREWER, Pr. Bost. Soc. Nat. Hist. VII, p. 308 (1860) (Cuba).

Charadrius virginiacus ALBRECHT, J. f. O. 1862, p. 205 (Jamaica).

Charadrius pluvialis SUNDEV. Oefv. K. Vet. Akad. For. 1869, p. 588 (St. Bartholomew).

Charadrius pluvialis americanus SUNDEV. Oefv. K. Vet. Akad. For. 1869, p. 602 (Porto Rico).

Charadrius fulvus var. *virginiacus* CORY, Bds. Bahama I. p. 145 (1886).

Antilles in Winter.

Charadrius squatarola (LINN.).

Tringa squatarola LINN. Syst. Nat. I, 10th ed. p. 149 (1758); *ib.* 12th ed. p. 252 (1766).

Tringa helvetica LINN. Syst. Nat. I. p. 250 (1766).

Vanellus squatarolus D'ORB. in La Sagra's Hist. Nat. Cuba, Ois. p. 242 (1840).

Squatarola helvetica GOSSE, Bds. Jam. p. 333 (1847).—BRYANT, Pr. Bost.
    Soc. Nat. Hist. VII, p. 121 (1859) (Bahamas).—ALBRECHT, J. f. O.
    1862, p. 205 (Jamaica).—MARCH, Pr. Acad. Nat. Sci. Phila. 1864,
    p. 66 (Jamaica).—GUNDL. Repert. Fisico-Nat Cuba, I, p. 358
    (1866); *ib.* J. f. O. 1875, p. 232 (Cuba).—LAWR. Pr. U. S. Nat. Mus.
    I, p. 197 (1878) (St. Vincent).—GUNDL. Anal. Soc. Esp. Hist. Nat.
    VII, p. 380 (1878) (Porto Rico).—CORY, Bds. Bahamas I. p. 144
    (1880).—A. & E. NEWTON, Handb. Jamaica, p. 115 (1881).—CORY,
    List Bds. W. I. p. 25 (1885).—WELLS, List Bds. Grenada, p. 7
    (1886).

Charadrius helveticus BREWER, Pr. Bost. Soc. Nat. Hist. VII, p. 308 (1860)
    (Cuba).

Found in winter in the Bahamas, Cuba, Jamaica, Porto Rico,
St. Vincent, and Grenada.

GENUS ÆGIALITIS BOIE.

Ægialitis BOIE, Isis, 1822, p. 558.

Ægialitis vocifera (LINN.).

Charadrius vociferus LINN. Syst. Nat. I, 10th ed. p. 150 (1758).—D'ORB.
    in La Sagra's Hist. Nat. Cuba, Ois. p. 246 (1840).—BREWER, Pr.
    Bost. Soc. Nat. Hist. VII, p. 308 (1860) (Cuba).—SUNDEV. Oefv.
    K. Vet. Akad. For. 1869. p. 602 (Porto Rico).

Ægialitis vociferus GOSSE, Bds. Jam. p. 330 (1847).—ALBRECHT, J. f. O.
    1862, p. 205 (Jamaica).—MARCH, Pr. Acad. Nat. Sci. 1864,
    p. 66 (Jamaica).—CORY, Bds. Bahama I. p. 145 (1880); *ib.* Bds.
    Haiti & San Domingo, p. 141 (1885); *ib.* List Bds. W. I. p. 25 (1885).

Oxyechus vociferus GUNDL. J. f. O. 1856, p. 124 (Cuba); *ib.* Repert. Fisico-
    Nat. Cuba, I, p. 359 (1866); *ib.* J. f. O. 1875. p. 333 (Cuba); *ib.*
    Anal. Soc. Esp. Hist. Nat. VII, p. 382 (1878) (Porto Rico).

Ægialites vociferus SALLÉ, P. Z. S. 1857, p. 236 (San Domingo).—BRY-
    ANT, Pr. Bost. Soc. Nat. Hist. VII, p. 121 (1859) (Bahamas).—NEW-
    TON, Ibis. 1860, p. 307 (St. Thomas).

Charadrius (Ægialitis) vociferus BRYANT, Pr. Bost. Soc. Nat. Hist. XI,
    p. 97 (1867) San Domingo).

Ægialitis vocifera A. & E. NEWTON, Handb. Jamaica, p. 115 (1881).

Oxyechus vociferus WELLS, List Bds. Grenada, p. 7 (1886).

Recorded from the Bahamas and Greater Antilles.

Ægialitis wilsonia (ORD).

Charadrius wilsonius ORD, ed. Wils. IX, p. 77 (1825). — LEMB. Aves
    Cuba, p. 106 (1850).—BREWER, Pr. Bost. Soc. Nat. Hist. VII, p.
    308 (1860) (Cuba).

Ægialitis wilsonius BRYANT, Pr. Bost. Soc. Nat. Hist. VII, p. 121 (1859)
(Cuba).
Ægialitis wilsonius CASSIN, Pr. Acad. Nat. Sci. Phila. 1860, p. 378 (St.
Thomas).—MARCH, Pr. Acad. Nat. Sci. Phila. 1864, p. 66 (Jamaica).
—CORY, Bds. Bahama I. p. 147 (1880); *ib.* Bds. Haiti & San Do-
mingo, p. 143 (1885); *ib.* List Bds. W. I. p. 25 (1885).
Ochthodromus wilsonius GUNDL. Repert. Fisico-Nat. Cuba, I, p. 359
(1866); *ib.* J. f. O. 1874, p. 313 (Porto Rico); *ib.* 1875, p. 333 (Cuba);
ib. Anal. Soc. Esp. Hist. Nat. VII, p. 381 (1878) (Porto Rico).
Ægialitis wilsonia A. & E. NEWTON, Handb. Jamaica, p. 115 (1881).

Common in the Bahamas and Greater Antilles.

Ægialitis semipalmata (BONAP.).

Charadrius semipalmatus "BP. Obs. Wils. No. 219 (1825)."—L'EMB. Aves
Cuba, p. 107 (1850).—BREWER, Pr. Bost. Soc. Nat. Hist. VII, p.
308 (1860) (Cuba).—SUNDEV. Oefv. K. Vet. Akad. For. 1869, p.
588 (St. Bartholomew); *ib.* p. 602 (Porto Rico).
Ægialitis semipalmata GOSSE, Bds. Jam. p. 333 (1847).—MARCH, Pr.
Acad. Nat. Sci. Phila. 1864, p. 66 (Jamaica).—SCL. P. Z. S. 1876, p.
14 (Santa Lucia).—LAWR. Pr. U. S. Nat. Mus. I, p. 197 (1878) (St.
Vincent); *ib.* p. 241 (Barbuda); *ib.* p. 261 (Guadeloupe).—ALLEN,
Bull. Nutt. Orn. Club, V, p. 169 (1880) (Santa Lucia).—A. & E.
NEWTON, Handb. Jamaica, p. 115 (1881).—CORY, Auk, III, p. 502
(1886) (Grand Cayman).
Ægialitis semipalmatus BRYANT, Pr. Bost. Soc. Nat. Hist. VII, p. 121
(1859) (Bahamas).
Ægialeus semipalmatus GUNDL. Repert. Fisico-Nat. Cuba, I, p. 359 (1866);
ib. J. f. O. 1875, p. 335 (Cuba); *ib* Anal. Soc. Esp. Hist. Nat. VII,
p. 384 (1878) (Porto Rico).
Ægialitis semipalmatus GUNDL. J. f. O. 1862, p. 88 (Cuba).—ALBRECHT,
J. f. O. 1862, p. 205 (Jamaica).—LAWR. Ann Lyc. N. Y. VIII, p. 100
(1867) (Sombrero).—CORY, Bds. Bahama I. p. 148 (1880); *ib.* Bds.
Haiti & San Domingo, p. 144 (1885); *ib.* List Bds. W. I. p. 25
(1885).—TRISTRAM, Ibis, 1884, p. 168 (San Domingo).
Ægialites semipalmata WELLS, List Bds. Grenada, p. 7 (1886).

Abundant in winter in many parts of the West Indies. Re-
corded from Bahamas, Cuba, Jamaica, Haiti, San Domingo,
Porto Rico, Santa Lucia, St. Vincent, Barbuda, Guadeloupe,
St. Bartholomew, Sombrero, Grenada, and Grand Cayman.

Ægialitis nivosa CASS.

Ægialitis nivosa CASSIN, in Baird's Bds. N. Am. p. 696 (1858).
Ægialitis tenuirostris LAWR. Ann. Lyc. N. Y. VII, p. 455 (1862) (Cuba).

Ægialeus tenuirostris GUNDL. Repert. Fisico-Nat. Cuba, I, p. 359 (1866);
    ib. J. f. O. 1875, p. 336 (Cuba).
Ægialitis nivosus CORY, List Bds. W. I. p. 25 (1885).

Accidental in Cuba.

Ægialitis meloda (ORD).

Charadrius melodus ORD, ed. Wils. VII, p. 71 (1824).—BREWER, Pr. Bost.
    Soc. Nat. Hist. VII, p. 308 (1860) (Cuba).
Ægialitis melodus GOSSE, Bds. Jam. p. 330 (1847).—SCL. P. Z. S. 1861, p.
    80 (Jamaica).—GUNDL. J. f. O. 1862, p. 88 (Cuba).—ALBRECHT,
    J. f. O. 1862, p. 205 (Jamaica).—CORY, Bds. Bahama I. p. 148 (1880);
    ib. List Bds. W. I. p. 25 (1885).
Ægialites melodus BRYANT, Pr. Bost. Soc. Nat. Hist. VII, p. 121 (1859)
    (Bahamas).
Ægialeus melodus GUNDL. Repert. Fisico-Nat. Cuba, I, p. 359 (1866);
    ib. J. f. O. 1075, p. 386 (Cuba); *ib.* Anal. Soc. Esp. Hist. Nat. VII,
    p. 385 (1878) (Porto Rico).

Winter visitant to the Bahamas and Greater Antilles.

FAMILY HÆMATOPODIDÆ.

GENUS **Hæmatopus** LINN.

Hæmatopus LINNÆUS, Syst. Nat. I, 10th ed. p. 152, 1758; *ib.* 12th ed. p.
    257, 1766.

Hæmatopus palliatus TEMM.

Hæmatopus palliatus TEMM. Man. d'Orn. II, p. 532 (1820).—LEMB. Aves
    Cuba, p. 104 (1850).—GUNDL. J. f. O. 1856, p. 423 (Cuba).—BRYANT,
    Pr. Bost. Soc. Nat. Hist. VII, p. 121 (1859) (Bahamas); *ib.* BREWER,
    p. 308 (1860) (Cuba).—GUNDL. Repert. Fisico-Nat. Cuba, I, p. 358
    (1866).—SUNDEV. Oefv. K. Vet. Akad. For. 1869, p. 588 (St. Bar-
    tholomew).—GUNDL. Anal. Soc. Esp. Hist. Nat. VII, p. 379 (1878)
    (Porto Rico).—CORY, Bds. Bahama I. p. 150 (1880); *ib.* Bds. Haiti
    & San Domingo. p. 145 (1885); *ib.* List Bds. W. I. p. 25 (1885).

Records of the occurrence of this species in the Bahamas,
Cuba, Haiti, San Domingo, Porto Rico, and St. Bartholomew.

FAMILY APHRIZIDÆ.

GENUS **Arenaria** BRISS.

Arenaria BRISSON, Orn. V, p. 132, 1760.

Arenaria interpres (LINN.).

Tringa interpres LINN. Syst. Nat. I. 10th ed. p. 148 (1758); *ib.* 12th ed.
p. 248 (1766).
Strepsilas interpres GOSSE, Bds. Jam. p. 333 (1847).—LEMB. Aves Cuba,
p. 100 (1850).—BRYANT, Pr. Bost. Soc. Nat. Hist. VII, p. 121 (1859)
(Bahamas); *ib.* BREWER, p. 308 (1860) (Cuba).—ALBRECHT, J. f. O.
1862, p. 205 (Jamaica).—MARCH, Pr. Acad. Nat. Sci. Phila. 1864, p.
66 (Jamaica).—GUNDL. Repert. Fisico-Nat. Cuba I, p. 357 (1866).—
LAWR. Ann. Lyc. N. Y. VIII, p. 100 (1867) (Sombrero).—SUNDEV.
Oefv. K. Vet. Akad. For. 1869. p. 588 (St. Bartholomew); *ib.* p. 602
(Porto Rico).—LAWR. Pr. U. S. Nat. Mus. I, p. 67 (1878) (Domin-
ica); *ib.* p. 197 (St. Vincent).—GUNDL. J. f. O. 1875. p. 331 (Cuba);
ib. Anal. Soc. Esp. Hist. Nat. VII, p. 379 (1878) (Porto Rico).—
CORY, Bds. Bahama I. p. 151 (1880).—A. & E. NEWTON, Handb.
Jamaica, p. 115 (1881).—TRISTRAM. Ibis. 1884, p. 168 (San Domin-
go).—CORY, List Bds. W. I. p. 25 (1885).—WELLS, List Bds.
Grenada, p. 7 (1886).
Arenaria interpres CORY, Auk, III, p. 302 (1886) (Grand Cayman).

Bahamas and Antilles in winter.

FAMILY RECURVIROSTRIDÆ.

GENUS Himantopus BRISS.

Himantopus BRISSON. Orn. V, p. 33. 1760.

Himantopus mexicanus (MÜLL.).

Charadrius mexicanus MÜLL. Syst. Nat. Suppl. p. 117 (1776).
Himantopus nigricollis GOSSE, Bds. Jam. p. 386 (1847).—LEMB. Aves
Cuba p. 102 (1850).—A. & E. NEWTON, Ibis, 1859. p. 258 (St. Croix).
—BRYANT, Pr. Bost. Soc. Nat. Hist. VII, p. 121 (1859) (Bahamas);
ib. BREWER, p. 308 (1860) (Cuba).—MARCH, Pr. Acad. Nat. Sci.
Phila. 1864, p. 67 (Jamaica).—SUNDEV. Oefv. K. Vet. Akad. For.
1869. p. 602 (Porto Rico).—LAWR. Pr. U. S. Nat. Mus. I, p. 197
(1878) (St. Vincent); *ib.* p. 238 (Antigua); *ib.* p. 242 (Barbuda).—
CORY, Bds. Bahama I. p. 153 (1880).—A. & E. NEWTON, Handb.
Jamaica, p. 115 (1881).—CORY, List Bds. W. I. p. 26 (1885).
Macrotarsus nigricollis GUNDL. J. f. O. 1856, p. 422 (Cuba); *ib.* Repert.
Fisico-Nat. Cuba. I. p. 357 (1866); *ib.* J. f. O. 1874. p. 113 (Porto
Rico); *ib.* Anal. Soc. Esp. Hist. Nat. VII, p. 377 (1878) (Porto
Rico).
Himantopus mexicanus SALLÉ, P. Z. S. 1857. p. 237 (San Domingo).—
BRYANT, Pr. Bost. Soc. Nat. Hist. XI. p. 97 (1867) (San Domingo).
—CORY, Bds. Haiti & San Domingo, p. 146 (1885).

Common in the Bahamas and Antilles.

Genus Recurvirostra Linn.

Recurvirostra Linnæus, Syst. Nat. I, 10th ed. p. 151, 1758.

Recurvirostra americana Gmel.

Recurvirostra americana Gmel. Syst. Nat. I, p. 693 (1788).—Gosse,
Bds. Jam. p. 387 (1847).—Brewer, Pr. Bost. Soc. Nat. Hist. VII, p.
308 (1860) (Cuba).—Gundl. J. f. O. 1862, p. 88 (Cuba).—Albrecht,
J. f. O. 1862. p. 206 (Jamaica).—March, Pr. Acad. Nat. Sci. Phila.
1864, p. 67 (Jamaica).—Gundl. Repert. Fisico-Nat. Cuba, I, p. 357
(1866); *ib.* J. f. O. 1875, p. 330 (Cuba)—A. & E. Newton, Handb.
Jamaica, p. 115 (1881).—Cory, List Bds. W. I. p. 26 (1885).

Recorded from Cuba and Jamaica.

Family SCOLOPACIDÆ.

Genus Gallinago Leach.

Gallinago Leach, Syst. Cat. Brit. Mam. & Bds. p. 31, 1816.

Gallinago delicata (Ord).

Scolopax delicata Ord, Wils. Orn. ix, 1825, p. ccxviii.
Scolopax wilsoni Sundev. Oefv. K. Vet. Akad. For. 1869, p. 587 (St.
Bartholomew); *ib.* p. 601 (Porto Rico).
Scolopax gallinago D'Orb. in La Sagra's Hist. Nat. Cuba, Ois. p. 231
(1840).
Gallinago wilsoni Gosse. Bds. Jam. p. 353 (1847).— Bryant, Pr. Bost.
Soc. Nat. Hist. VII, p. 121 (1859) (Bahamas); *ib.* Brewer, p. 3 8
(1860) (Cuba).— Gundl. J. f. O. 1862, p. 85 (Cuba).— March, Pr.
Acad. Nat. Sci. Phila. 1864, p. 67 (Jamaica).—Gundl. Repert. Fis-
ico-Nat. Cuba, I, p. 353 (1866); *ib.* J. f. O. 1875, p. 321 (Cuba). —
Lawr. Pr. U. S. Nat. Mus. I, p. 197 (1878) (St. Vincent); *ib.* p.
238 (Antigua); *ib.* p. 242 (Barbuda). — Gundl. Anal. Soc. Esp.
Hist. Nat. VII, p. 368 (1878) (Porto Rico).—Cory, Bds. Bahama I.
p. 156 (1880). — A. & E. Newton, Handb. Jamaica, p. 116 (1881).
—Cory, List Bds. W. I. p. 26 (1885).
Gallinago wilsonii A. & E. Newton, Ibis, 1859. p. 258 (St. Croix).—Scl.
P. Z. S. 1861, p. 80 (Jamaica).—Albrecht. J. f. O. 1862, p. 205
(Jamaica).·
Gallinago media wilsoni Wells, List Bds. Grenada, p. 8 (1886).

Bahamas and Antilles during migrations.

Philohela minor (Gmel.).

Scolopax minor A. & E. Newton, Handb. Jamaica, p. 116 (1881).
Rusticola minor Gosse, Bds. Jam. p. 354 (1847) (Hill). — March, Pr.
Acad. Nat. Sci. Phila. 1864, p. 68 (Jamaica).

Recorded by Gosse and others from Jamaica. The bird might occasionally wander to Cuba, and possibly Jamaica, as it is not uncommon in some parts of Florida in winter.

GENUS Macrorhamphus LEACH.

Macrorhamphus "LEACH, Cat. Brit. Birds, 1816."

Macrorhamphus griseus (GMEL.).

Scolopax grisea GMEL. Syst. Nat. I, p. 568 (1788).
Limnodramus griseus LEMB. Aves Cuba, p. 91 (1850).
Macrorhamphus griseus GUNDL. J. f. O. 1862, p. 85 (Cuba); *ib.* Repert.
  Fisico-Nat. Cuba, I, p. 353 (1866); *ib.* J. f. O. 1875, p. 322 (Cuba).
Macrorhamphus griseus CORY, Bds. Bahama I. p. 157 (1880). — A. & E.
  NEWTON, Handb. Jamaica, p. 116 (1881). — CORY, List Bds. W. I.
  p. 26 (1885).

Recorded from Bahamas, Cuba, and Jamaica.

Macrorhamphus scolopaceus (SAY).

Limosa scolopacea SAY, Long's Exp. 1823, p. 170.
Macrorhamphus scolopaceus ALBRECHT, J. f. O. 1861, p. 213 (Cuba); *ib.*
  GUNDL. 1862, p. 85 (Cuba); *ib.* Repert. Fisico-Nat. Cuba, I, p. 354
  (1866), *ib.* J. f. O. 1875, p. 322 (Cuba).
Macrorhamphus scolopaceus CORY, List Bds. W. I. p. 26 (1885).

Cuba and Antilles during migrations.

GENUS Micropalama BAIRD.

Micropalama BAIRD, Birds N. Am. p. 726, 1858.

Micropalama himantopus (BONAP.).

Tringa himantopus Br. Ann. Lyc. N. Y. II, p. 157 (1826). — BREWER, Pr.
  Bost. Soc. Nat. Hist. VII, p. 308 (1860) (Cuba).
Totanus himantopus LEMB. Aves Cuba, p. 95 (1850). — MARCH, Pr. Acad.
  Nat. Sci. Phila. 1864, p. 67 (Jamaica). — GUNDL. Repert. Fisico-
  Nat. Cuba, I, p. 336 (1866). — A. & E. NEWTON, Handb. Jamaica,
  p. 116 (1881).
Ereunetes himantopus SUNDEV. Oefv. K. Vet. Akad. For. 1869, p. 587 (St.
  Bartholomew); *ib.* p. 602 (Porto Rico).
Micropalma himantopus MOORE, Pr. Bost. Soc. Nat. Hist. XIX, p. 241
  (1877) (Bahamas). — CORY. List Bds. W. I. p. 26 (1885).
Micropalama himantopus GUNDL. J. f. O. 1875, p. 326 (Cuba); *ib.* Anal.
  Soc. Esp. Hist. Nat. VII, p. 373 (1878). — CORY, Revised List Bds.
  W. I. p. 26 (1885). — WELLS, List Bds. Grenada, p. 8 (1886).

Found throughout the Antilles.

Genus **Ereunetes** Illig.

Ereunetes Illiger, Prodromus, p. 262, 1811.

Ereunetes pusillus (Linn.).

Tringa pusilla Linn. Syst. Nat. I, p. 252 (1766).
Pelidna pusilla Gosse, Bds. Jam. p. 348 (1847)?
Hemipalama semipalmata Lemb. Aves Cuba, p. 96 (1850). — Brewer,
    Pr. Bost. Soc. Nat. Hist. VII. p. 308 (1860) (Cuba).
Hemipalama minor Lemb. Aves Cuba, p. 97 (1850).—Brewer, Pr. Bost.
    Soc. Nat. Hist. VII, p. 308 (1860) (Cuba).
Tringa semipalmata Bryant, Pr. Bost. Soc. Nat. Hist. VII, p. 121
    (1859) (Bahamas).
Ereunetes pusillus Cassin, Pr. Acad. Nat. Sci. Phila. 1860, p. 195 (Jamai-
    ca). — Gundl. Repert. Fisico-Nat. Cuba, I. p. 356 (1866); *ib.* J. f.
    O. 1875, p. 327 (Cuba); Anal. Soc. Esp. Hist. Nat. VII, p. 374
    (1878) (Porto Rico). — Sundev. Oefv. K. Vet. Akad. For. 1869, p.
    587 (St. Bartholomew): *ib.* p. 602 (Porto Rico). — Cory, Bds.
    Bahama I. p. 157 (1880). — A. & E. Newton, Handb. Jamaica, p.
    116 (1881). — Cory, List Bds. W. I. p. 26 (1885). — Wells, List
    Bds. Grenada, p. 8 (1886). — Cory, Ibis, 1886, p. 502 (Grand Cay-
    man).
Ereunetes petrificatus March, Pr. Acad. Nat. Sci. Phila. 1864, p. 68
    (Jamaica).— Lawr. Pr. U. S. Nat. Mus, I, p. 488 (1878) (Domin-
    ica); *ib.* p. 238 (Antigua); *ib.* p. 242 (Barbuda); *ib.* p. 488 (Guade-
    loupe).

Throughout the Antilles during migrations.

Ereunetes occidentalis Lawr., if it be considered different
from the preceding species, must be given a place in the West
India Avifauna.

Genus **Tringa** Linn.

Tringa Linnæus, Syst. Nat. I, 10th ed. p. 148, 1758; *ib.* 12th ed. p. 247,
    1766.

Tringa minutilla Vieill.

Tringa minutilla Vieill. Nouv. Dict. XXXIV, p. 452 (1819). — Lawr.
    Pr. U. S. Nat. Mus. I, p. 197 (1878) (St. Vincent). — Allen, Bull.
    Nutt. Orn. Club, V, p. 169 (1880) (Santa Lucia). — Cory, Bds.
    Bahama I. p. 158 (1880).—A. & E. Newton, Handb. Jamaica, p.
    116 (1881). — Cory, List Bds. W. I. p. 26 (1885); *ib.* Ibis, 1886, p.
    502 (Grand Cayman).

Tringa temminckii D'ORB. in La Sagra's Hist. Nat. Cuba, Ois. p. 240
(1840).
Pelidna pusilla GOSSE, Bds. Jam. p. 338 (1847)? — GUNDL. J. f. O. 1862,
p. 87 (Cuba).
Actodromas wilsonii A. & E. NEWTON, Ibis, 1859, p. 258 (St. Croix).
Tringa wilsoni BRYANT, Pr. Bost. Soc. Nat. Hist. VII, p. 121 (1859)
(Bahamas).—SCL. P. Z. S. 1861, p. 80 (Jamaica).
Tringa pusilla BREWER, Pr. Bost. Soc. Nat. Hist. VII, p. 308 (1860)
(Cuba).
Tringa wilsonii ALBRECHT, J. f. O. 1862, p. 205 (Jamaica).
Actodromas minutilla MARCH, Pr. Acad. Nat. Sci. Phila. 1864, p. 67
(Jamaica). — GUNDL. Repert. Fisico-Nat. Cuba, I, p. 357 (1866);
ib. Anal. Soc. Esp. Hist. Nat. VII, p. 376 (1878) (Porto Rico).

Winter visitant to the Bahamas and Antilles.

Tringa maculata VIEILL.

Tringa maculata VIEILL. Nouv. Dict. XXXIV, p. 456 (1819).—A. & E.
NEWTON, Ibis, 1859, p. 258 (St. Croix). — BRYANT, Pr. Bost. Soc.
Nat. Hist. XI, p. 69 (1867) (Bahamas); *ib.* MOORE, XIX, p. 241
(1877) (Bahamas). —SUNDEV. Oefv. K. Vet. Akad. For. 1869, p. 587
(St. Bartholomew). — LAWR. Pr. U. S. Nat. Mus. I, p. 461 (1879)
(Guadeloupe). — CORY, Bds. Bahama I. p. 159 (1880). — A. & E.
NEWTON, Handb. Jamaica, p. 116 (1881). — CORY, List Bds. W. I.
p. 26 (1885); *ib.* Ibis, 1886, p. 502 (Grand Cayman).
Tringa pectoralis LEMB. Aves Cuba, p. 98 (1850).—BREWER, Pr. Bost.
Soc. Nat. Hist. VII, p. 308 (1860) (Cuba). — SUNDEV. Oefv. K.
Vet. Akad. For. 1869, p. 602 (Porto Rico).
Pelidna pectoralis GUNDL. J. f. O. 1862, p. 87 (Cuba).
Actodromas maculata GUNDL. Repert. Fisico-Nat. Cuba, I, p. 356 (1866);
ib. Anal. Soc. Esp. Hist. Nat. VII, p. 375 (1878) (Porto Rico).—
WELLS, List Bds. Grenada, p. 8 (1886).
Actodromas maculatus GUNDL. J. f. O. 1875, p. 328 (Cuba).
Tringa maculosa MOORE, Pr. Bost. Soc. Nat. Hist. XIX, p. 241 (1877)
(Bahamas).

Antilles in winter.

Tringa fuscicollis VIEILL.

Tringa fuscicollis VIEILL. Nouv. Dict. XXXIV, p. 461 (1819). — SCL. P.
Z. S. 1876, p. 14 (Santa Lucia).—ALLEN, Bull. Nutt. Orn. Club, V,
p. 169 (1880) (Santa Lucia). — A. & E. NEWTON, Handb. Jamaica,
p. 116 (1881).—CORY, List Bds. W. I. p. 26 (1885).
Tringa schinzii LEMB. Aves Cuba, p. 98 (1850).—BREWER, Pr. Bost. Soc.
Nat. Hist. VII, p. 308 (1860) (Cuba).
Pelidna schinzii GUNDL. J. f. O. 1856, p. 421 (Cuba); *ib.* 1862, p. 87
(Cuba).

Tringa bonapartii SCL. P. Z. S. 1861, pp. 70, 80 (Jamaica).—ALBRECHT,
    J. f. O. 1862, p. 205 (Jamaica). — MOORE, Pr. Bost. Soc. Nat. Hist.
    XIX, p. 241 (1877) (Bahamas).
Actodromas bonapartii GUNDL. Repert. Fisico-Nat. Cuba, I, p. 356
    (1866); *ib.* J. f. O. 1875, p. 328 (Cuba).
Tringa bonapartei CORY, Bds. Bahama I. p. 159 (1880).

Antilles in winter.

Tringa canutus LINN.

Tringa canutus LINN. Syst. Nat. I, 10th ed. p. 149 (1758); *ib.* 12th ed. p.
    251 (1766). — GOSSE, Bds. Jam. p. 354(1847). — MARCH, Pr. Acad.
    Nat. Sci. Phila 1864. p. 67 (Jamaica).—A. & E. NEWTON, Handb.
    Jamaica, p. 116 (1881). — CORY, Revised List Bds. W. I. p. 26
    (1886).

Recorded from Jamaica.

Tringa ferruginea BRÜNN.

Tringa ferruginea BRÜNN. Orn. Bor. 1764, p. 53. —A. O. U. Check-list,
    N. Am. Bds. p. 152 (1886).
Anclyohcilus subarquata KAUP. Sk. Ent. Eur. Theirw. 1829, p. 50.
Ancylochilus subarquatus WELLS, List Bds. Grenada, p. 8 (1886).

Recorded from Grenada.

GENUS Calidris CUV.

Calidris CUVIER, Anat. Comp. V, 1805.

Calidris arenaria (LINN.).

Tringa arenaria LINN. Syst. Nat. I, p. 251 (1766).
Calidris arenaria GOSSE, Bds. Jam. p. 354 (1847).—GUNDL. J. f. O. 1856,
    p. 422 (Cuba). — BREWER, Pr. Bost. Soc. Nat. Hist. VII. p 308
    (1860) (Cuba). — MARCH, Pr. Acad. Nat. Sci. Phila. 1864, p. 67
    (Jamaica). — GUNDL. Repert. Fisico-Nat. Cuba, I, p. 357 (1866);
    ib. J. f. O. 1875, p. 329 (Cuba). — LAWR. Pr. U. S. Nat. Mus. I. p.
    197 (1878) (St. Vincent); *ib.* p. 461 (Guadeloupe).—GUNDL. Anal.
    Soc. Esp. Hist. Nat. VII. p. 376 (1878) (Porto Rico).—CORY, Bds.
    Bahama I. p. 160 (1880). — A. & E. NEWTON. Handb. Jamaica. p.
    116 (1881).—CORY, List. Bds. W. I. p. 26 (1885).
Arenaria calidris LEMB. Aves Cuba, p. 101 (1850).

Antilles in winter.

GENUS Limosa BRISS.

Limosa BRISSON, Orn. 1760.

Limosa fedoa (LINN.).

Scolopax fedoa LINN. Syst. Nat. I, p. 244 (1766).
Limosa fedoa LEMB. Aves Cuba, p. 90 (1850).—BREWER, Pr. Bost. Soc.
Nat. Hist. VII, p. 308 (1860) (Cuba). — GUNDL. J. f. O. 1862, p. 84
(Cuba); *ib.* Repert. Fisico-Nat. Cuba. I, p. 353 (1866): *ib.* J. f. O.
1875. p. 320 (Cuba): *ib.* Anal. Soc. Esp. Hist. Nat. VII. p. 368 (1878)
(Porto Rico).—WELLS, List Bds. Gredada, p. 8 (1886).

Recorded from the Greater Antilles.

Limosa hæmastica (LINN.).

Scolopax hæmastica LINN. Syst. Nat. I, 10th ed. p. 147 (1758).
Limosa hudsonica LEMB. Aves Cuba, Suppl. (1850 .—GUNDL. J. f. O.
1862, p. 84 (Cuba); *ib.* Repert. Fisico-Nat. Cuba, I. p. 353 (1866);
ib. J. f. O. 1875, p. 320 (Cuba).
Limosa hæmastica CORY, List Bds. W. I. p. 26 (1885).

Recorded from Cuba.

GENUS Symphemia RAF.

Symphemia RAFINESQUE, Jour. de Phys. 1819.

Symphemia semipalmata (GMEL.).

Scolopax semipalmata GMEL. Syst. Nat. I. p. 959 (1788).
Totanus semipalmatus LEMB. Aves Cuba, p. 92 (1850). — CORY, Bds.
Bahama I, p. 160 (1880).
Catoptrophorus semipalmatus GOSSE, Bds. Jam. p. 354 (1847).—BREWER,
Pr. Bost. Soc. Nat. Hist. VII, p. 308 (1860) (Cuba).
Symphemia semipalmata BRYANT, Pr. Bost. Soc. Nat. Hist. VII, p. 122
(1859) (Bahamas).—MARCH, Pr. Acad. Nat. Sci. Phila. 1864, p. 67
(Jamaica).—GUNDL. Repert. Fisico-Nat. Cuba, I, p. 354 (1866); *ib.*
J. f. O. 1875, p. 322 (Cuba).—LAWR. Pr. U. S. Nat. Mus. I, p. 238
(1878) (Antigua); *ib.* p. 242 (Barbuda).—GUNDL. Anal. Soc. Esp.
Hist. Nat. VII, p. 369 (1878) (Porto Rico).—A. & E. NEWTON,
Handb. Jamaica, p. 116 (1881).—CORY, Revised List Bds. W. I.
p. 27 (1886).—WELLS, List Bds. Grenada, p. 8 (1886).
Catophtrophorus speculiferus BREWER, Pr. Bost. Soc. Nat. Hist. VII, p.
308 (1860) (Cuba).
Symphemia semipalmatus CORY, List Bds. W. I. p. 27 (1885).

Common in the Bahamas and Antilles.

GENUS **Totanus** BECHST.

Totanus BECHSTEIN, Orn. Taschenb. Deutschl. p. 282, 1803.

Totanus melanoleucus (GMEL.).

Scolopax melanoleucus GMEL. Syst. Nat. I. p. 659 (1788).

Totanus melanoleucus GOSSE, Bds. Jam. p. 352 (1847).—BRYANT, Pr. Bost. Soc. Nat. Hist. XI, p. 69 (1867) (Bahamas).—SUNDEV. Oefv. K. Vet. Akad. For. 1869. p. 588 (St. Bartholomew) ; *ib.* p. 602 (Porto Rico).—ALLEN, Bull. Nutt. Orn. Club, V, p. 169 (1880).—CORY, Bds. Bahama I. p. 161 (1880).—A. & E. NEWTON, Handb. Jamaica, p. 116 (1881).—CORY, List Bds. W. I. p. 27 (1885).—WELLS, List Bds. Grenada, p. 8 (1886).

Totanus vociferus LEMB. Aves Cuba, p. 93 (1850).—BREWER, Pr. Bost. Soc. Nat. Hist. VII, p. 308 (1860) (Cuba).

Gambetta melanoleuca SCL. P. Z. S. 1861, p. 80 (Jamaica).—ALBRECHT, J. f. O. 1862, p. 205 (Jamaica).—MARCH, Pr. Acad. Nat. Sci. Phila. 1864, p. 68 (Jamaica).—GUNDL. Repert. Fisico-Nat. Cuba, I, p. 354 (1866) ; *ib.* J. f. O. 1875, p. 323 (Cuba).—LAWR. Pr. U. S. Nat. Mus. I, p. 238 (1878) (Antigua).—GUNDL. Anal. Soc. Esp. Hist. VII. p. 370 (1878) (Porto Rico).

Glottis melanoleuca GUNDL. J. f. O. 1862, p. 85 (Cuba).

Records from Bahamas, Cuba, Jamaica, Porto Rico, St. Bartholomew, Antigua, and Grenada.

Totanus flavipes (GMEL.).

Scolopax flavipes GMEL. Syst. Nat. I, p. 659 (1788).

Totanus flavipes D'ORB. in La Sagra's Hist. Nat. Cuba. Ois. p. 234 (1840). —GOSSE, Bds. Jam. p. 351 (1847).—BREWER, Pr. Bost. Soc. Nat. Hist. VII, p. 308 (1860) (Cuba).—ALBRECHT, J. f. O. 1862. p. 205 (Jamaica).—BRYANT, Pr. Bost. Soc. Nat. Hist. XI, p. 69 (1867) (Bahamas).—SUNDEV. Oefv. K. Vet. Akad. For. 1869, p. 588 (St. Bartholomew) ; *ib.* p. 602 (Porto Rico).—ALLEN, Bull. Nutt. Orn. Club, V, p. 169 (1880) (Santa Lucia).—CORY, Bds. Bahama I. p. 162 (1880).—A. & E. NEWTON, Handb. Jamaica, p. 116 (1881).— CORY, List Bds. W. I. p. 27 (1885).—WELLS, List Bds. Grenada, p. 8, 1886.—CORY, Ibis, 1886, p. 502 (Grand Cayman).

Gambetta flavipes MARCH. Pr. Acad. Nat. Sci. Phila. 1864, p. 68 (Jamaica). —GUNDL. Repert. Fisico-Nat. Cuba. I, p. 354 (1866).—LAWR. Ann. Lyc. N. Y. VIII. p. 100 (1867) (Sombrero) ; *ib.* Pr. U. S. Nat. Mus. I, p. 197 (1878) (St. Vincent) ; *ib.* p. 242 (Barbuda).—GUNDL. Anal. Soc. Esp. Hist. Nat. VII, p. 371 (1878) (Porto Rico).

Antilles in Winter.

Totanus solitarius (Wils.).

Tringa solitaria Wils. Am. Orn. VII, p. 53 (1813).

Totanus solitarius D'Orb. in La Sagra's Hist. Nat. Cuba, Ois. p. 238
(1840).—Sundev. Oefv. K. Vet. Akad. For. 1869, p. 587 (St. Bartholomew); *ib.* p. 602 (Porto Rico).—A. & E. Newton, Handb.
Jamaica, p. 116 (1881).

Totanus chloropygius Gosse, Bds. Jam. p. 350 (1847).—Brewer, Pr. Bost.
Soc. Nat. Hist. VII, p. 308 (1860) (Cuba).—Gundl. J. f. O. 1862,
p. 86 (Cuba).

Phyacophilus solitarius Scl. P. Z. S. 1861, p. 80 (Jamaica).—Albrecht,
J. f. O. 1862, p. 205 (Jamaica).—March, Pr. Acad. Nat. Sci. Phila.
1864, p. 67 (Jamaica).—Gundl. Repert. Fisico-Nat. Cuba, I, p. 355
(1886); *ib.* J. f. O. 1874, p. 313 (Porto Rico); *ib.* 1875, p. 324 (Cuba);
ib. Anal. Soc. Esp. Hist. Nat. VII, p. 372 (1878) (Porto Rico).—
Lawr. Pr. U. S. Nat. Mus. I, p. 238 (1878) (Antigua); *ib.* p. 242
(Barbuda); *ib.* p. 461 (Gaudeloupe).—Allen, Bull. Nutt. Orn.
Club, V, p. 169 (1880) (Santa Lucia).—Cory, List Bds. W. I. p.
27 (1885).

Records from Cuba, Jamaica, Porto Rico, Santa Lucia, Antigua, Barbuda, Gaudeloupe, and St. Bartholomew.

Genus Actitis Illiger.

Actitis Illiger, Prodr. 1811, p. 262.

Actitis macularia (Linn.).

Tringa macularia Linn. Syst. Nat. I, p. 249 (1766).—Bryant, Pr. Bost.
Soc. Nat. Hist. X, p. 257 (1866) (Porto Rico).

Actitis macularius Gosse, Bds. Jam. p. 349 (1847).

Totanus macularius Lemb. Aves Cuba, p. 94 (1850).

Tringoides macularia Brewer, Pr. Bost. Soc. Nat. Hist. VII, p. 308
(1860) (Cuba).

Tringoides macularius Scl. P. Z. S. 1861, p. 80 (Jamaica).—Albrecht,
J. f. O. 1862, p. 205 (Jamaica).—Gundl. Repert. Fisico-Nat. Cuba,
I, p. 355 (1866); *ib.* J. f. O. 1875, p. 325 (Cuba).—Scl. P. Z. S. 1872,
p. 653 (Santa Lucia).—Lawr. Pr. U. S. Nat. Mus. I, p. 67 (1878)
(Dominica); *ib.* p. 197 (St. Vincent); *ib.* p. 276 (Grenada); *ib.* p.
360 (Martinique).—Gundl. Anal. Soc. Esp. Hist. Nat. VII, p. 372
(1878) (Porto Rico).—Allen, Bull. Nutt. Orn. Club, V, p. 169
(1880) (Santa Lucia).—Lister, Ibis, 1880, p. 44 (St. Vincent).—
Cory, Bds. Bahama I. p. 162 (1880).—A. & E. Newton, Handb.
Jamaica, p. 115 (1881).—Cory, Bull. Nutt. Orn. Club, VI, p. 154
(1881) (Haiti); *ib.* Bds. Haiti & San Domingo, p. 148 (1885); *ib.*
List Bds. W. I. p. 27 (1885).—Wells, List Bds. Grenada. p. 8
(1886).

Actitis macularia SUNDEV. Oefv. K. Vet. Akad. For. 1869, p. 587 (St.
Bartholomew); *ib.* p. 602 (Porto Rico).

Antilles, common.

GENUS Bartramia LESS.

Bartramia LESSON, Traite d' Orn. p. 553. 1831.

Bartramia longicauda (BECHST.).

Tringa longicauda BECHST. Vog. Nachtr. übers. Lath. Ind. Orn. p. 453
(1812).
Totanus longicauda D'ORB. in La Sagra's Hist. Nat. Cuba, Ois. p. 237
(1840).
Tringoides bartramius BREWER, Pr. Bost. Soc. Nat. Hist. VII, p. 308
(1860) (Cuba).
Euligia bartramia GUNDL. J. f. O. 1862. p. 86 (Cuba).
Actiturus bartramius MARCH. Pr. Acad. Nat. Sci. Phila. 1864, p. 67 (Ja-
maica)?
Actiturus longicaudus GUNDL. Repert. Fisico-Nat. Cuba, I, p. 355 (1866);
ib. J. f. O. 1875, p. 326 (Cuba).
Actiturus longicaudatus GUNDL. J. f. O. 1881, p. 401 (Cuba).
Actiturus longicauda A. & E. NEWTON, Handb. Jamaica, p. 115 (1881).
Bartramia longicauda CORY, List Bds. W. I. p. 57 (1885).—WELLS, List
Bds. Grenada, p. 8 (1886).

Records from Cuba, Jamaica, and Grenada. I have seen a
specimen taken in the Bahama Islands.

GENUS Tryngites CAB.

Tryngites CABANIS, J. f. O. 1856, p. 418.

Tryngites subruficollis (VIEILL.).

Tringa subruficollis VIEILL. Nouv. Dict. XXXIV, p. 465 (1819).
Tringa rufescens VIEILL. Nouv. Dict. XXXIX, p. 470 (1819).—LEMB.
Aves Cuba, p. 99 (1850).—BREWER, Pr. Bost. Nat. Hist. VII, p.
308 (1860) (Cuba).
Tringites rufescens GUNDL. Repert. Fisico-Nat. Cuba, I, p. 355 (1866); *ib.*
J. f. O. 1875, p. 325 (Cuba).—CORY, List Bds. W. I. p. 27 (1885).

Accidental in Cuba.

GENUS Numenius LINN.

Numenius LINNÆUS, Syst. Nat. 1746.

Numenius hudsonicus LATH.

Numenius hudsonicus LATH. Ind. Orn. II. p. 712 (1790).—LAWR. Ann.
Lyc. N. Y. VIII. p. 100 (1867) (Sombrero) ; *ib.* Pr. U. S. Nat. Mus.
I. p. 238 (1878)(Antigua) ; *ib.* p. 242 (Barbuda) ; *ib.* p. 277 (Grenada).
—CORY. List Bds. W. I. p. 27 (1885).—WELLS, List Bds. Grenada,
p. 8 (1886).
Numenius hudsonius GUNDL. Anal. Soc. Esp. Hist. Nat. VII, p. 367 (1878)
(Porto Rico).

Winter visitant to the Antilles.

Numenius borealis (FORST.).

Scolopax borealis FORST. Phil. Trans. LXII, p. 411 (1772).
Numenius borealis GUNDL. Anal. Soc. Esp. Hist. Nat. VII, p. 367 (1878)
(Porto Rico).—CORY, List Bds. W. I. p. 27 (1885).—WELLS, List
Bds. Grenada, p. 8 (1886).

West Indies in winter ; reported from Porto Rico and Grenada.

Numenius longirostris WILS.

Numenius longirostris WILS. Am. Orn. VIII, p. 24 (1814).—DENNY,
P. Z. S. 1847 p. 39 (Jamaica).—LEMB. Aves Cuba, p. 88 (1850).—
BREWER, Pr. Bost. Soc. Nat. Hist. VII, p. 308 (1860) (Cuba).—
MARCH, Pr. Acad. Nat. Sci. Phila. 1864. p. 68 (Jamaica). — GUNDL.
Repert. Fisico-Nat. Cuba, I. p. 352 (1866) ; *ib.* J. f. O. 1875, p. 320
(Cuba).—LAWR. Pr. U. S. Nat. Mus. I, p. 197 (1878) (St. Vincent).
—A. & E. NEWTON, Handb. Jamaica, p. 116 (1881).—CORY. List
Bds. W. I. p. 27 (1885).

Greater Antilles in winter.

FAMILY CICONIIDÆ.

GENUS Tantalus LINN.

Tantalus LINNÆUS, Syst. Nat. 10th ed. 1758.

Tantalus loculator LINN.

Tantalus loculator LINN. Syst. Nat. I, p. 240 (1766).—D'ORB. in La
Sagra's Hist. Nat. Cuba. Ois. p. 219 (1840).—DENNY, P. Z. S.
1847. p. 39 (Jamaica).—GUNDL. J. f. O. 1856, p. 348 (Cuba).—
BREWER, Pr. Bost. Soc. Nat. Hist. VII, p. 308 (1860) (Cuba).—

GUNDL. J. f. O. 1862, p. 83 (Cuba): *ib.* Repert. Fisico-Nat. Cuba, I,
p. 351 (1866); *ib.* J. f. O. 1875. p. 313 (Cuba).—CORY, List Bds. W.
I. p. 27 (1885).

Recorded from Cuba and Jamaica.

FAMILY IBIDIDÆ.

GENUS **Guara** REICH.

Guara REICHENBACH, Syst. Avium, 1852. p. xiv.

Guara alba (LINN.).

Scolopax alba LINN. Syst. Nat. I, 10th ed. p. 145 (1758).
Ibis alba DENNY, P. Z. S. 1847, p. 39 (Jamaica).—LEMB. Aves Cuba, p.
86 (1850).
Endocimus albus BREWER, Pr. Bost. Soc. Nat. Hist. VII, p. 308 (1860)
(Cuba).—GUNDL. Repert. Fisico-Nat. Cuba, I, p. 352 (1866): *ib.*
J. f. O. 1875. p. 315 (Cuba); *ib.* Anal. Soc. Esp. Hist. Nat. VII, p.
364 (1878) (Porto Rico).—CORY, Bds. Haiti & San Domingo, p.
150 (1885); *ib.* List Bds. W. I. p. 27 (1885).

Common in most of the Greater Antilles.

Guara rubra (LINN.).

Tantalus ruber LINN. Syst Nat. I, p. 241 (1766).
Ibis rubra D'ORB. in La Sagra's Hist. Nat. Cuba, Ois. p. 228 (1840).—
GOSSE, Bds. Jam. p. 348 (1847).—DENNY, P. Z. S. 1847, p. 39
(Jamaica).—MARCH, Pr. Acad. Nat. Sci. Phila. 1864. p. 65 (Jamaica).
Endocimus ruber GUNDL. J. f. O. 1862, p. 83 (Cuba).—A. & E. NEWTON,
Handb. Jamaica, p. 112 (1881).—CORY, List Bds. W. I. p. 27 (1885).

Recorded from Cuba and Jamaica.

GENUS **Plegadis** KAUP.

Plegadis KAUP, Skizz. Entuv. Gesch. p. 82, 1829.

Plegadis antumnalis (HASSELQ.).

Tringa autumnalis HASSELQ. Reise nach Paläst Deutsch Ausg. 1762, p.
306.
Tantalus falcinellus LINN. Syst. Nat. I, p. 241 (1766).
Ibis falcinellus LEMB. Aves Cuba, p. 87 (1850).

Falcinellus erythrorhynchus CAB. J. f. O. 1856. p. 349 (Cuba) ; *ib.* GUNDL.
    1882, p. 84 (Cuba).
Falcinellus ordii BREWER. Pr. Bost. Soc. Nat. Hist. VII. p. 308 (1860)
    (Cuba).—GUNDL. Repert. Fisico-Nat. Cuba, I. p. 352 (1866) : *ib.*
    J. f. O. 1875. p. 318 (Cuba) ; *ib.* Anal. Soc. Esp. Hist. Nat. VII. p.
    366 (1878) (Porto Rico).
Plegadis falcinellus CORY, List Bds. W. I. p. 27 (1885).

Accidental in the Greater Antilles.

Plegadis guarauna (LINN.) is claimed to have occurred in
the West Indies, but I can find no satisfactory record of its
capture.

FAMILY **Plataleidæ**.

GENUS **Ajaja** REICH.

Ajaja REICHENBACH, Handb. xvi, 1851.

Ajaja ajaja (LINN.).

Platalea ajaja LINN. Syst. Nat. I, 10th ed. p. 140 (1758) ; *ib.* 12th ed. p.
    231 (1766).—D'ORB. in La Sagra's Hist. Nat. Cuba, Ois. p. 216
    (1840).—GOSSE, Bds. Jam. p. 346 (1847).—GUNDL. J. f. O. 1856,
    p. 347 (Cuba).—BRYANT, Pr. Bost Soc. Nat. Hist. VII, p. 121
    (1859) (Bahamas) ; *ib.* BREWER, p. 308 (1860) (Cuba).—ALBRECHT,
    J. f. O. 1862. p. 206 (Jamaica).—MARCH, Pr. Acad. Nat. Sci. Phila.
    1864, p. 65 (Jamaica).—GUNDL. Repert. Fisico-Nat. Cuba, I, p.
    351 (1866) ; *ib.* J. f. O. 1875. p. 311 (Cuba).—LAWR. Pr. U. S. Nat.
    Mus. I. p. 275 (1878) (Grenada) (?).—CORY, Bds. Bahama I. p.
    164 (1880).—A. & E. NEWTON. Handb. Jamaica, p. 112 (1881).—
    TRISTRAM. Ibis, 1884. p. 168 (San Domingo).
Platalea ajaja DENNY. P. Z. S. 1847. p. 39 (Jamaica).
Ajaja rosea CORY, List Bds. W. I. p. 28 (1885).

Resident in the Bahamas and Greater Antilles.

    .

FAMILY PHŒNICOPTERIDÆ.

GENUS **Phœnicopterus** LINN.

Phœnicopterus LINNÆUS, 1748 ; *ib.* Syst. Nat. I, p. 230, 1766.

Phœnicopterus ruber LINN.

Phœnicopterus ruber LINN. Syst. Nat. I, 10th ed. p. 139 (1758); *ib.* 12th
ed. p. 230 (1766).—GOSSE, Bds. Jam. p. 390 (1847).—DENNY, P. Z.
S. 1847, p. 39 (Jamaica).—GUNDL. J. f. O. 1856, p. 342 (Cuba).—
SALLÉ, P. Z. S. 1857, p. 236 (San Domingo).—BRYANT, Pr. Bost.
Soc. Nat. Hist. VII, p. 121 (1859) (Bahamas); *ib.* XI, p. 97 (1867)
(San Domingo); *ib.* BREWER, VII, p. 308 (1860) (Cuba).—
ALBRECHT, J. f. O. 1862, p. 206 (Jamaica).—MARCH, Pr. Acad. Nat.
Sci. Phila. 1864, p. 65 (Jamaica).—GUNDL. Repert. Fisico-Nat.
Cuba, I, p. 386 (1866); *ib.* J. f. O. 1874. p. 314 (Porto Rico); *ib.*
1875, p. 368 (Cuba); *ib.* Anal. Soc. Esp. Hist. Nat. VII, p. 398
(1878) (Porto Rico).—CORY, Bds. Bahama I. p. 180 (1880).—A. &
E. NEWTON, Handb. Jamaica, p. 112 (1881).—CORY, Bull. Nutt.
Orn. Club, VI, p. 155 (1881) (Haiti).—TRISTRAM, Ibis, 1884, p.
168 (San Domingo). — CORY, Bds. Haiti & San Domingo, p. 165
(1885); *ib.* List Bds. W. I. p. 28 (1885).

Resident and not uncommon in the Bahamas and Greater
Antilles.

FAMILY ARDEIDÆ.

GENUS **Ardea** LINN.

Ardea LINNÆUS, Syst. Nat. I, p. 233, 1766.

Ardea herodias LINN.

Ardea herodias LINN. Syst. Nat. I, p. 237 (1766).—D'ORB. in La Sagra's
Hist. Nat. Cuba, Ois. p. 199 (1840).—GOSSE, Bds. Jam. p. 346
(1847).—A. & E. NEWTON, Ibis, 1859, p. 263 (St. Croix) (?).—
BRYANT, Pr. Bost. Soc. Nat. Hist. VII, p. 120 (1859) (Bahamas);
ib. BREWER, p. 308 (1860) (Cuba).—SCL. P. Z. S. 1861, p. 81
(Jamaica).—GUNDL. J. f. O. 1862, p. 82 (Cuba).—MARCH, Pr.
Acad. Nat. Sci. Phila. 1864, p. 63 (Jamaica).—GUNDL. Repert.
Fisico-Nat. Cuba, I, p. 347 (1866).—LAWR. Ann. Lyc. N. Y. VIII,
p. 98 (1867) (Sombrero).—SUNDEV. Oefv. K. Vet. Akad. For.
1869, p. 589 (St. Bartholomew); *ib.* p. 602 (Porto Rico).—GUNDL.
J. f. O. 1875, p. 296 (Cuba).—LAWR. Pr. U. S. Nat. Mus. I, p. 196
(1878) (St. Vincent); *ib.* p. 236 (Antigua); *ib.* p. 240 (Barbuda);
ib p. 274 (Grenada); *ib.* p. 359 (Martinique).—GUNDL. Anal. Soc.
Esp. Hist. Nat. VII, p. 352 (1878) (Porto Rico).—SCL. P. Z. S.
1879. p. 765 (Montserrat).—CORY. Bds. Bahama I. p. 166 (1880).—
A. & E. NEWTON, Handb. Jamaica, p. 111 (1881).—CORY, List Bds.
W. I. p. 28 (1885).—WELLS, List Bds. Grenada, p. 9 (1886).
Ardea herodias ALBRECHT, J. f. O. 1862, p. 206 (Jamaica).

This species ranges throughout the West Indies.

Ardea occidentalis AUD.

Ardea occidentalis AUD. Orn. Biog. III, p. 542 (1835).—LEMB. Aves
    Cuba, p. 82 (1850).—MARCH, Pr. Acad. Nat. Sci. Phila. 1864, p.
    63 (Jamaica).—A. & E. NEWTON, Handb. Jamaica, p. 111 (1881).—
    CORY, List Bds. W. I. p. 28 (1885).
Herodias occidentalis BREWER, Pr. Bost. Soc. Nat. Hist. VIII, p. 308
    (1860) (Cuba).
? Ardea wurdemannii MARCH, Pr. Acad. Nat. Sci. Phila. 1864, p. 64
    (Jamaica).
Audubonia occidentalis GUNDL. Repert. Fisico-Nat. Cuba, I, p. 348
    (1866); *ib.* J. f. O. 1874, p. 313 (Porto Rico); *ib.* 1875, p. 298
    (Cuba); *ib.* Anal. Soc. Esp. Hist. Nat. VII, p. 354 (1878) (Porto
    Rico).

Recorded from Porto Rico, Cuba, and Jamaica.

Ardea egretta GMEL.

Ardea egretta GMEL. Syst. Nat. I, p. 629 (1788).—BRYANT, Pr. Bost.
    Soc. Nat. Hist. VIII, p. 120 (1859) (Bahamas); *ib.* X, p. 257 (1866)
    (Porto Rico).—SUND. Oefv. K. Vet. Akad. For. 1869, p. 602
    (Porto Rico).—CORY, Bds. Bahama I. p. 167 (1880).—A. & E
    NEWTON, Handb. Jamaica, p. 111 (1881).—CORY, List Bds. W. I.
    p. 28 (1885).
Ardea alba D'ORB. in La Sagra's Hist. Nat. Cuba, Ois. p. 191 (1840).
Egretta luce GOSSE, Bds. Jam. p. 346 (1847) (?).—ALBRECHT, J. f. O.
    1862, p. 206 (Jamaica).
Ardea abba DENNY, P. Z. S. 1847, p. 39 (Jamaica).
Herodias egretta GUNDL. J. f. O. 1856, p. 341 (Cuba).—BREWER, Pr.
    Bost. Soc. Nat. Hist. VII, p. 308 (1860) (Cuba).—SCL. P. Z. S.
    1861, pp. 70, 80 (Jamaica).—ALBRECHT, J. f. O. 1862, p. 205 (Ja-
    maica).—MARCH, Pr. Acad. Nat. Sci. Phila. 1864, p. 63 (Jamaica).
    —GUNDL. Repert. Fisico-Nat. Cuba, I, p. 348 (1866); *ib.* J. f. O.
    1875, p. 299 (Cuba).—LAWR. Pr. U. S. Nat. Mus. I, p. 241 (1878)
    (Barbuda) (?).—GUNDL. Anal. Soc. Esp. Hist. Nat. VIII, p. 355
    (1878) (Porto Rico).
Herodius luce SALLÉ, P. Z. S. 1857, p. 236 (San Domingo).
Ardea lence BRYANT, Pr. Bost. Soc. Nat. Hist. XI, p. 97 (1867) (San
    Domingo).

Bahamas and Greater Antilles.

Ardea candidissima GMEL.

Ardea candidissima GMEL. Syst. Nat. II, p. 633 (1788).—D'ORB. in La
    Sagra's Hist. Nat. Cuba, Ois. p. 196 (1840).—BRYANT, Pr. Bost.

Soc. Nat. Hist. VII, p. 120 (1859) (Bahamas); *ib.* XI, p. 97 (1867) (San Domingo).—SCL. P. Z. S. 1879. p. 765 (Montserrat).—CORY, Bds. Bahama I. p. 167 (1880).—A. & E. NEWTON, Handb. Jamaica, p. 111 (1881).—CORY. Bds. Haiti & San Domingo, p. 153 (1885); *ib.* List Bds. W. I. p. 28 (1885).—TRISTRAM, Ibis, 1884, p. 168 (San Domingo).

Egretta candidissima GOSSE. Bds. Jam. p. 336 (1847).—MARCH, Pr. Acad. Nat. Sci. Phila. 1864, p. 63 (Jamaica).

Herodias candidissima SALLE, P. Z. S. 1857. p. 236 (San Domingo).— BREWER, Pr. Bost. Soc Nat. Hist. VII, p. 308 (1860) (Cuba).

Garzetta candidissima SCL. P. Z. S. 1861. p. 81 (Jamaica).—ALBRECHT, J. f. O. 1862, p. 205 (Jamaica).—GUNDL. Repert. Fisico-Nat. Cuba, I, p. 349 (1866); *ib.* J. f. O. 1875. p. 304 (Cuba); *ib.* Anal. Soc. Esp. Hist. Nat. VII, p. 357 (1878) (Porto Rico).—LAWR. Pr. U. S. Nat. Mus. I, p. 66 (1878) (Dominica); *ib.* p. 196 (St. Vincent); *ib.* p. 236 (Antigua); *ib.* p. 274 (Grenada).—WELLS, List Bds. Grenada, p. 9 (1886).

Bahamas and Antilles.

Ardea rufa BODD.

Ardea rufa BODD. Tabl. P. E. p. 54 (1783).—CORY, Bds. Bahama I. p. 170 (1880).—A. & E. NEWTON, Handb. Jamaica, p. 111 (1881).— CORY, Bull. Nutt. Orn. Club, VI. p. 154 (1881) (Haiti); *ib.* Bds. Haiti & San Domingo, p. 152 (1885); *ib.* List Bds. W. I. p. 28 (1885).

Ardea rufescens LEMB. Aves Cuba, p. 83 (1850).

Ardea cubensis LEMB. Aves Cuba, p. 84 (1850).

Herodias rufescens BREWER, Pr. Bost. Soc. Nat. Hist. VII, p. 308 (1860) (Cuba).—GUNDL. J. f. O. 1862. p. 82 (Cuba).

Herodias pealii BREWER, Pr. Bost. Soc. Nat. Hist. VII, p. 308 (1860) (Cuba).—GUNDL. J. f. O. 1862. p. 82 (Cuba).

Dimigretta rufa MARCH, Pr. Acad. Nat. Sci. Phila. 1864, p. 63 (Jamaica).

Demiegretta rufa GUNDL. Repert. Fisico-Nat. Cuba, I, p. 348 (1866); *ib.* J. f. O. 1875, p. 302 (Cuba).

Demiegretta pealii GUNDL. Repert. Fisico-Nat. Cuba, I, p. 348 (1866); *ib.* J. f. O. 1875, p. 301 (Cuba).

Bahamas and Greater Antilles.

Ardea cærulea LINN.

Ardea cærulea LINN. Syst. Nat. I. 10th ed. p. 143 (1758); *ib.* 12th ed. p. 238 (1766). — D'ORB. in La Sagra's Hist. Nat. Cuba, Ois. p. 201 (1840). — DENNY. P. Z. S. 1847. p. 36 (Jamaica). — BRYANT, Pr. Bost. Soc. Nat. Hist. VII. p. 120 (1859) (Bahamas); *ib.* X. p. 257 (1866) (Porto Rico). — TAYLOR. Ibis, 1864, p. 171 (Porto Rico).—

SUNDEV. Oefv. K. Vet. Akad. For. 1869. p. 588 (St. Bartholomew) ;
ib. p. 602 (Porto Rico). — ALLEN. Bull. Nutt. Orn. Club, V, p. 169
(1880) (Santa Lucia). — CORY. Bds. Bahama I. p. 171 (1880). — A. &
E. NEWTON, Handb. Jamaica. p. 111 (1881). — CORY. Bull. Nutt.
Orn. Club, VI. p. 155 (1881) (Haiti) ; *ib.* Bds. Haiti & San Domingo, p. 154 (1885); *ib.* List. Bds. W. I. p. 28 (1885). — TRISTRAM,
Ibis, 1884. p. 168 (San Domingo).

"*Egretta nivea* GOSSE. Bds. Jam. p. 334 (1847)"?

Egretta cærulea GOSSE, Bds. Jam. p. 337 (1847).

Herodias cærulea BREWER Pr. Bost. Soc. Nat. Hist. VII, p. 308 (1860)
(Cuba). — GUNDL. J f O. 1862, p. 83 (Cuba).

Florida cærulea SCL. P. Z. S. 1861, p. 81 (Jamaica). — ALBRECHT, J. f. O.
1862, p. 205 (Jamaica). — MARCH, Pr. Acad. Nat. Sci. Phila. 1864,
p. 62 (Jamaica). — GUNDL. Repert. Fisico-Nat. Cuba, I, p. 349
(1866); *ib.* J. f. O. 1875, p. 305 (Cuba); *ib.* Anal. Soc. Esp. Hist.
Nat. VII, p. 357 (1878) (Porto Rico). — LAWR. Pr. U. S. Nat. Mus.
I, p. 66 (1878) (Dominica) ; *ib.* p. 196 (St. Vincent) ; *ib.* p. 236
(Antigua) ; *ib.* p. 241 (Barbuda) ; *ib.* p. 274 (Grenada) ; *ib.* p. 359
(Martinique). — WELLS, List Bds. Grenada. p. 9 (1886).

Bahamas and Antilles.

Ardea virescens LINN.

Ardea virescens LINN. Syst. Nat. I, 10th ed. p. 144 (1758) ; *ib.* 12th ed.
p. 238 (1766). — D'ORB. in La Sagra's Hist. Nat. Cuba, Ois. p. 203
(1840). — DENNY. P. Z. S. 1847 p. 39 (Jamaica). — BRYANT, Pr. Bost.
Soc. Nat. Hist. VII, p. 120 (1859) (Bahamas). — SUNDEV. Oefv. K.
Vet. Akad. For. 1869. p. 589 (St. Bartholomew) ; *ib.* p. 602 (Porto
Rico). — CORY, Bds. Bahama I, p. 171 (1880) ; *ib.* Bull. Nutt. Orn.
Club. VI, p. 155 (1881) (Haiti) ; *ib.* Bds. Haiti & San Domingo, p.
155 (1885) ; *ib.* List Bds. W. I. p. 28 (1885). — TRISTRAM. Ibis, 1884,
p. 168 (San Domingo). — CORY. Ibis, 1886, pp. 472, 474, 475 (Bar-
badoes, La Desirade and Grand Terre) ; *ib.* Auk, III, p. 502 (1886)
(Grand Cayman).

Herodias virescens GOSSE. Bds. Jam. p. 340 (1847). — ALBRECHT, J. f. O.
1862, p. 205 (Jamaica).

Butorides virescens SALLÉ, P. Z. S. 1857. p. 236 (San Domingo). — A. &
E. NEWTON, Ibis, 1859, p. 261 (St. Croix). — CASSIN. Pr. Acad.
Nat. Sci. Phila. 1860, p. 378 (St. Thomas) ; *ib.* MARCH. 1864. p. 64
(Jamaica). — SCL. P. Z. S. 1871. p. 273 (Santa Lucia). — LAWR. Pr.
U. S. Nat. Mus I, p. 66 (1878) (Dominica): *ib.* p. 196 (St. Vin-
cent); *ib.* p. 236 (Antigua); *ib.* p. 241 (Barbuda); *ib.* p. 275
(Grenada); *ib.* p. 359 (Martinique); *ib.* p. 460 (Guadeloupe). —
SCL. P. Z. S. 1879. p. 765 (Montserrat). — ALLEN. Bull. Nutt. Orn.
Club, V, p. 169 (1880) (Santa Lucia). — LISTER, Ibis, 1880. p. 44
(St. Vincent). — A. & E. NEWTON. Handb. Jamaica. p. 111 (1881). —
WELLS, List Bds. Grenada. p. 9 (1886).

Ocniscus virescens BREWER, Pr. Bost. Soc. Nat. Hist. VII. p. 308 (1860)
(Cuba). — GUNDL. Repert. Fisico-Nat. Cuba. I, p. 349 (1866); *ib.*
J. f. O. 1875, p. 307 (Cuba); *ib.* Anal. Soc. Esp. Hist. Nat. VII, p.
359 (1878) (Porto Rico).
Butorides brunnescens? MARCH, Pr. Acad. Nat. Sci. Phila. 1864, p. 64
(Jamaica).
Ardea (Butorides) virescens BRYANT, Pr. Bost. Soc. Nat. Hist. XI, p. 97
(1867) (San Domingo'.

Common throughout the Bahamas and Antilles.

Ardea brunnescens "GUNDL."

Ardea brunnescens "GUNDL. Mss." — LEMB. Aves Cuba, p. 84 (1850).—
REICH. J. f. O. 1877. p. 255.—CORY, List Bds. W. I. p. 28 (1885).
Ocniscus brunnescens CAB. J. f. O. 1856, p. 344. — BREWER, Pr. Bost. Soc.
Nat. Hist. VII, p. 308 (1860). — GUNDL. Repert. Fisico-Nat. Cuba,
I, p. 350 (1866,; *ib.* J. f. O. 1875, p. 308.
Butorides brunnescens BAIRD, Bds. N. Am. p. 677 (1858); *ib.* Cat. Am.
Bds. No. 494 (1859).—GUNDL. Ann. N. Y. Lyc. Nat. Hist. 1862. p.
271.—GRAY, Handl. Bds. III, p. 32 (1871). — BD. BWR. & RIDGW.
Hist. N. Am. W. Bds. I, p. 49 (1884).

SP. CHAR. — Pilium and occipital crest greenish black, showing a green
gloss in the light; whole throat and neck rich rufous brown, showing a tinge of orange brown on the chin; back feathers slaty gray;
wing-coverts not margined with white, but showing slight brownish edgings; otherwise resembling *A. virescens.*
  Length, 19.00; wing, 6.50; tail, 2.75; tarsus, 2.10; bill, 2.60.

HABITAT.   Cuba.

Ardea tricolor ruficollis (GOSSE).

Ardea leucogastra D'ORB. in La Sagra's Hist. Nat. Cuba, Ois. p. 200
(1840).
Egretta ruficollis GOSSE, Bds. Jam. p. 338 (1847). — ALBRECHT, J. f. O.
1862, p. 205 (Jamaica).
Herodias ludoviciana BREWER, Pr. Bost. Soc. Nat. Hist. VII, p. 308 (1860)
(Cuba).
Herodias ruficollis GUNDL. J. f. O. 1862, p. 83? (Cuba).
Demiegretta ludoviciana MARCH, Pr. Acad. Nat. Sci. Phila. 1864. p. 63
(Jamaica).
Demiegretta ruficollis GUNDL. Repert. Fisico-Nat. Cuba, I, p. 348 (1866);
ib. J. f. O. 1875, p. 303 (Cuba); *ib.* Anal. Soc. Esp. Hist. Nat. VII,
p. 356 (1878) (Porto Rico).
Ardea leucogastra var. *leucoprymna* CORY, Bds. Bahama I. p. 168 (1880).
Ardea ludoviciana A. & E. NEWTON, Handb. Jamaica. p. 111 (1881).
Ardea tricolor CORY, List Bds. W. I. p. 28 (1885).

Ardea cyanirostris * CORY, Bds. Bahama I. p. 168 (1880). — ALLEN, Bull. Nutt. Orn. VII, p. 21 (1881). — CORY, List Bds. W. I. p. 28 (1885).
Ardea tricolor ruficollis CORY, Auk, III, p. 502 (1886).

Common in the Bahama Islands and Greater Antilles (breeds).

GENUS **Nycticorax** STEPH.

Nycticorax STEPHENS, Gen. Zool. XI, p. 608, 1819.

Nycticorax violaceus (LINN.).

Ardea violacea LINN. Syst. Nat. I, p. 238 (1766); SUNDEV. Oefv. K. Vet. Akad. For. 1869, p. 589 (St. Bartholomew); *ib.* p. 603 (Porto Rico).

Nycticorax violacea D'ORB. in La Sagra's Hist. Nat. Cuba, Ois. p. 213 (1840).

Nycticorax violaceus BRYANT, Pr. Bost. Soc. Nat. Hist. VII, p. 120 (1859) (Bahamas); *ib.* BREWER, p. 308 (1860) (Cuba).—SCL. P. Z. S. 1861, pp. 70, 81 (Jamaica).—ALBRECHT. J. f. O. 1862, p. 206 (Jamaica).— SCL. P. Z. S. 1871. p. 273 (Santa Lucia).—A. & E. NEWTON, Handb. Jamaica, p. 111 (1881). — CORY, Auk, III, p. 502 (1886) (Grand Cayman).

Nyctherodius violaceus A. & E. NEWTON, Ibis, 1859, p. 262 (St. Croix).— CASSIN, Pr. Acad. Nat. Sci. Phila. 1860, p. 379 (St. Thomas); *ib.* MARCH, 1864, p. 65 (Jamaica).—GUNDL. Repert. Fisico-Nat. Cuba, I, p. 351 (1866).—LAWR. Ann. Lyc. N. Y. VIII, p. 98 (1867) (Sombrero).—GUNDL. J. f. O. 1875, p. 311 (Cuba); *ib.* Anal. Soc. Esp. Hist. Nat. VII, p. 363 (1878) (Porto Rico).—WELLS, List Bds. Grenada, p. 9 (1886).

Nyctiardea violacea LAWR. Pr. U. S. Nat. Mus. I, p. 275 (1878) (Grenada); *ib.* p. 460 (Guadeloupe).—ALLEN, Bull. Nutt. Orn. Club, V, p. 169 (1880) (Santa Lucia).—CORY, Bds. Bahama I. p. 173 (1880); *ib.* List Bds. W. I. p. 28 (1885).

Common in many portions of the West Indies. It has been recorded from the Bahamas, Cuba, Jamaica, Porto Rico, Grenada, Guadeloupe, Santa Lucia, St. Thomas, St. Bartholomew, Sombrero, St. Croix, and Grand Cayman.

---

* In originally naming *A. cyanirostris* I considered it distinct from *A. ruficollis*, from the totally different coloration of the bill and legs supposed to occur only in the breeding season. Since that time specimens have been taken in the winter months representing this same state of plumage. *A. tricolor* undoubtedly assumes a yellow bill during most of the year, even if Audubon was wrong in his statement to the effect that it had a yellow bill in the breeding season. It is probable that the two birds are identical, but it is also possible that the Bahama bird may be distinct, and colonies occasionally wander to Florida, where it has been found breeding. This would account for the blue-billed specimens being taken in Florida, and would not prove its specific identity with *A. tricolor*.

Nycticorax nycticorax nævius (Bodd.).

Ardea nævia Bodd. Tabl. Pl. Enl. 1783, p. 56.
Nycticorax vulgaris D'Orb. in La Sagra's Hist. Nat. Cuba, Ois. p. 208
(1840).
Nycticorax americanus Gosse, Bds. Jam. p. 344 (1847). — Albrecht,
     J. f. O. 1862, p. 206 (Jamaica).
Nycticorax gardeni Brewer, Pr. Bost. Soc. Nat. Hist. VII, p. 308 (1860)
     (Cuba).—Gundl. J. f. O. 1862. p. 83 (Cuba).—A. & E. Newton,
     Handb. Jamaica, p. 111 (1881).
Nyctiardea gardeni March, Pr. Acad. Nat. Sci. Phila. 1864, p. 65 (Ja-
     maica).—Gundl. Repert. Fisico-Nat. Cuba, I, p. 350 (1866); *ib.*
     J. f. O. 1875, p. 310 (Cuba); *ib.* Anal. Soc. Esp. Hist. Nat. VII, p.
     362 (1878) (Porto Rico).
Ardea nycticorax Sundev. Oefv. K. Vet. Akad. For. 1869, p. 602 (Porto
     Rico).
Nyctiardea grisea nævia Cory, List Bds. W. I. p. 28 (1885).
Nycticorax nycticorax nævius Zeledon, Pr. U. S. Nat. Mus. VIII, p. 113
     (1885).

Recorded from Greater Antilles.

Genus Botaurus Steph.

Botaurus Stephens, Shaw's Gen. Zool. XI, p. 592, 1819.

Botaurus lentiginosus (Mont.).

Ardea lentiginosa Mont. Orn. Dict. Suppl. 1813.—Lemb. Aves Cuba,
     p. 82 (1850).
Botaurus minor Gosse, Bds. Jam. p. 346 (1847).—Albrecht, J. f. O.
     1862, p. 206 (Jamaica).—Gundl. J. f. O. 1862, p. 83 (Cuba).
Botaurus lentiginosus Brewer, Pr. Bost. Soc. Nat. Hist. VII, p. 308
     (1860) (Cuba).—March, Pr. Acad. Nat. Sci. Phila. 1864, p. 65 (Ja-
     maica).—Gundl. Repert. Fisico-Nat. Cuba, I, p. 350 (1866); *ib.*
     J. f. O. 1874, p. 313 (Porto Rico); *ib.* 1875, p. 309 (Cuba); *ib.* Anal.
     Soc. Esp. Hist. Nat. VII, p. 361 (1878) (Porto Rico).—A. & E.
     Newton. Handb. amaica, p. 111 (1881).—Cory, List Bds. W. I.
     p. 28 (1885).

Accidental in Cuba, Jamaica, and Porto Rico.

Genus Ardetta Gray.

Ardetta Gray, List of Gen. App. p. 13, 1842.

Ardetta exilis (GMEL.).

Ardea exilis GMEL. Syst. Nat. I, p. 648 (1788).—D'ORB. in La Sagra's Hist. Nat. Cuba, Ois. p. 205 (1840).—BRYANT. Pr. Bost. Soc. Nat. Hist. X, p. 257 (1866) (Porto Rico).—SUNDEV. Oefv. K. Vet. Akad. For. 1869, p. 602 (Porto Rico).
Ardeola exilis GOSSE, Bds. Jam. p. 343 (1847).
Ardetta exilis BREWER, Pr. Bost. Soc. Nat. Hist. VII, p. 308 (1860) (Cuba).—SCL. P. Z. S. 1861, p. 81 (Jamaica).—ALBRECHT, J. f. O. 1862, p. 206 (Jamaica).—MARCH, Pr. Acad. Nat. Sci. Phila. 1864, p. 64 (Jamaica).—GUNDL. Repert. Fisico-Nat. Cuba, I, p. 350 (1866); *ib.* J. f. O. 1875. p. 308 (Cuba); *ib.* Anal. Soc. Esp. Hist. Nat. VII, p. 360 (1878) (Porto Rico).—CORY, Bds. Bahama I. p. 174 (1880).—A. & E. NEWTON, Handb. Jamaica, p. 111 (1881).—CORY, List Bds. W. I. p. 28 (1885).

Recorded from the Bahamas, Cuba, Jamaica, and Porto Rico.

FAMILY GRUIDÆ.

GENUS Grus LINN.

Grus LINNÆUS, Syst. Nat. 1735.

Grus mexicana MULL.

Grus mexicana MÜLL. Syst. Nat. Suppl. p. 110 (1776).—CORY, List Bds. W. I. p. 29 (1885).
Grus poliophaea LEMB. Aves Cuba, p. 80 (1850).—BREWER, Pr. Bost. Soc. Nat. Hist. VII, p. 308 (1860) (Cuba).
Grus canadensis GUNDL. J. f. O. 1856, p. 339 (Cuba); *ib.* 1862, p. 81 (Cuba).—BREWER, Pr. Bost. Soc. Nat. Hist. VII, p. 308 (1860) (Cuba)?—GUNDL. Repert. Fisico-Nat. Cuba, I, p. 347 (1866); *ib.* J. f. O. 1875. p. 293 (Cuba).

Accidental in Cuba.

FAMILY ARAMIDÆ.

GENUS Aramus VIEILL.

Aramus VIEILLOT, Analyse, 1816.

Aramus giganteus (BONAP.).

Rallus giganteus BR. Journ. Acad. Nat. Sci. Phila. V, p. 31 (1825).
Aramus guarauna D'ORB. in La Sagra's Hist. Nat. Cuba, Ois. p. 256 (1840).—BREWER, Pr. Bost. Soc. Nat. Hist. VII, p. 308 (1860) (Cuba).

Aramus scolopaceus GOSSE, Bds. Jam. p. 355 (1847).—SALLÉ, P. Z. S.
    1857 p. 236 (San Domingo).
Aramus giganteus SCL. P. Z. S 1861, p. 81 (Jamaica).—GUNDL. Repert.
    Fisico-Nat. Cuba, I, p. 360 (1866).—BRYANT, Pr. Bost. Soc. Nat.
    Hist. X, p. 257 (1866) (Porto Rico) ; *ib.* XI, p. 97 (1867) (San Do-
    mingo).—SUNDEV. Oefv. K. Vet. Akad. For. 1869, p. 601 (Porto
    Rico).—GUNDL. J. f. O. 1875, p. 353 (Cuba); *ib.* Anal. Soc. Esp.
    Hist. Nat. VII, p. 387 (1878) (Porto Rico).—A. & E. NEWTON,
    Handb. Jamaica, p 115 (1881).—CORY, List Bds. W. I. p. 29 (1885).
Notherodius scolopaceus GUNDL. J. f. O. 1862, p. 89 (Cuba).
Aramus scolopaceus giganteus CORY, Bull. Nutt. Orn. Club, VI, p. 155
    (1881) (Haiti).
Aramus pictus CORY, Bds. Haiti and San Domingo, p. 157 (1885).

Common in the Greater Antilles.

FAMILY JACANIDÆ.

GENUS **Jacana** BRISS.

Jacana BRISSON, Orn. V, p. 121 (1760).

Jacana spinosa (LINN).

Fulica spinosa LINN. Syst. Nat. 1758, p. 152.
Parra jacana D'ORB. in La Sagra's Hist. Nat. Cuba, Ois. p. 249 (1840).—
    GUNDL. J. f. O. 1856, p. 425 (Cuba). — BREWER, Pr. Bost. Soc. Nat. Hist.
    VII, p. 308 (1860) (Cuba). — GUNDL. J. f. O. 1862, p. 89 (Cuba); *ib.*
    Repert. Fisico-Nat. Cuba, I, p. 360 (1866); *ib.* J. f. O. 1875, p. 338 (Cuba);
    ib. Anal. Soc. Esp. Hist. Nat. VII, p. 385 (1878) (Porto Rico).
Parra violacea CORY, Bull. Nutt. Orn. Club, VI, pp. 130, 155 (1881) (Haiti); *ib.*
    List Bds. W. I. p. 29 (1885).
Parra gymnostoma CORY, Bds. Haiti & San Domingo, p. 159 (1885).
Jacana spinosa ELLIOT, Auk, p. 297 (1888).

SP. CHAR. *Male:* — Bill and comb pale orange; bare skin at the base of the lower
    mandible pale bluish white; head, neck, and upper breast dark lustrous
    green; back and wing-coverts purple, shading into rich golden brown near
    the rump; rump and tail-coverts purple; underpart, dark purple, showing
    a tinge of dark rufous on the crissum; most of the primaries and second-
    aries bright yellow, edged with brown; tail rufous brown; carpal spur pale
    orange; legs and feet dull olive; iris brown.
        Length, 9; wing, 5.50; tail, 2.25; tarsus, 2.25; bill, 1.40.

HABITAT. Cuba, Haiti, and San Domingo.

Since the publication of the ' Birds of Haiti and San Domingo,'
I have examined several specimens of *Jacana* from Cuba, and
they agree exactly with the San Domingo bird, but all are con-
siderably larger and brighter than specimens from Mexico and
Central America; the coloration of the wattles in life is, I believe,
also different.

FAMILY RALLIDÆ.

GENUS Rallus LINN.

Rallus LINNÆUS, Syst. Nat. I, p. 261, 1766.

Rallus maculatus BODD.

Rallus maculatus BODD. ex Buff. Pl. Enl. p. 775 (1783). — SCHLEG. Mus.
   Pays-Bas, Ralli, p. 13 (1865). — SCL. & SAL. P. Z. S. 1868, p.
   444; *ib.* Nom. Avium Neotr. p. 139 (1873). — CORY, List Bds. W.
   I. p. 29, (1885).
Rallus variegatus GMEL. Syst. Nat I p. 718 (1788). — BURM. Syst. Ueb.
   III. p. 382. — D ORB in La Sagra's Hist Nat. Cuba, Ois. p. 261
   (1840). — BREWER, Pr. Bost. Soc. Nat. Hist. VII, p. 308 (1860).
Aramides maculatus HARTL. Ind. Az. p. 23. — GRAY. Gen. Bds. p. 594
   (1844-49).
Pardirallus variegatus Br. Compt. Rend. XLIII, p. 599. — GRAY, Handl.
   Bds. III, p. 56 (1871).
Limnopardalus variegatus GUNDL. J f. O 1856, p. 428; *ib.* Repert. Fisico-
   Nat. Cuba, I, p. 361 (1866): *ib.* J. f. O. 1875, p. 357.

SP. CHAR. — General plumage dull black; feathers on the back heavily
   marked with brown; whole body heavily blotched and mottled with
   white; rump brown; thighs smoky brown, marked with white above;
   chin nearly white; legs (in skin) pale yellow; bill green; a spot of
   scarlet on the base of the lower mandible.
      Length (skin), 12; wing, 4.75; tail, 3; tarsus, 1.50; bill, 1.95.

Recorded from Cuba.

Rallus elegans AUD.

Rallus elegans AUD. Orn. Biog. III, p. 27 (1835). — GUNDL. J. f. O. 1856,
   p. 427 (Cuba). — BREWER, Pr. Bost. Soc. Nat. Hist. VII, p. 308
   (1860) (Cuba). — GUNDL. Repert. Fisico-Nat. Cuba, I, p. 360 (1866):
   ib. J. f. O 1875, p. 355 (Cuba). — CORY, List Bds. W. I. p. 29 (1885).

Accidental in Cuba.

Rallus virginianus LINN.

Rallus virginianus LINN. Syst. Nat. I, p. 263 (1766).—BREWER, Pr. Bost.
Soc. Nat. Hist. VII. p. 308 (1860) (Cuba).—GUNDL. Repert. Fisico-
Nat. Cuba, I, p. 361 (1866) ; *ib.* J. f. O. 1875. p. 357 (Cuba).—CORY,
List Bds. W. I. p. 29 (1885).
Accidental in Cuba.

Rallus longirostris crepitans (GMEL.).

Rallus crepitans GMEL. Syst. Nat. I, p. 713 (1788). — BRYANT, Pr. Bost.
Soc. Nat. Hist. VII, p. 122 (1859) (Bahamas).
Rallus longirostris CORY Bds. Bahama I. p. 176 (1880).
Rallus longirostris crepitans RIDGW. Bull. Nutt. Orn. Club, V, p. 140
(1880).—CORY, List Bds. W. I. p. 29 (1885).
Bahamas in winter.

Rallus longirostris caribæus RIDGW.

Rallus longirostris D'ORB. in La Sagra's Hist. Nat. Cuba, Ois, p. 260
(1840). — GOSSE, Bds. Jam. p. 364 (1847).—A. & E. NEWTON, Ibis,
1859. p. 260.—CASSIN, Pr. Acad. Nat. Sci. Phila. 1860, p. 378. — A.
& E. NEWTON, Handb. Jamaica, p. 114 (1881).
Rallus crepitans BREWER, Pr. Bost. Soc. Nat. Hist. VII, p. 308 (1860).—
SCL. P. Z. S. 1861, p. 81.—ALBRECHT, J. f. O. 1862, p. 206.—GUNDL.
Repert. Fisico-Nat. Cuba, I, p. 361 (1866) ; *ib.* J. f. O. 1875, p. 356;
ib. Anal. Soc. Exp. Hist. Nat. VII, p. 388 (1878).—LAWR. Pr. U. S.
Nat. Mus. I, pp. 461, 487 (1878).
"*Rallus elegans* MARCH, Pr. Acad. Nat. Sci. Phila. 1864, p. 69"?
Rallus longirostris var. *caribæus* RIDGW Bull. Nutt. Orn. Club, V,
p. 140 (1880). — BD. BWR. & RIDGW. Hist. N. Am. W. Bds. I,
p. 339 (1884).
Rallus longirostris caribæus CORY, List Bds. W. I. p. 29 (1885).

This form is very closely allied to the North American bird ;
the principal difference being that those from the West Indies
usually show olivaceous striping on the back, more or less distinct.

HABITAT.   Antilles.

Rallus coryi MAYNARD.

Rallus coryi MAYNARD American Exchange and Mart, Boston, Jan 15
(1887) : *ib.* Feb. 5 (1887)

SP. CHAR. — "Above pale yellowish brown, streaked with pale ashy;
wings light reddish, becoming paler on the outer edges ; beneath pale

ashy tinged with reddish across the breast, becoming white on the
throat and abdomen, banded faintly on sides and flanks with white
and pale ashy." (Maynard, orig. descr., l. c.)

Length, 11.50; wing, 6; tail, 2.10; tarsus, 1.75; bill, 2.15.

HABITAT. Andros Island, Bahamas.

GENUS Porzana VIEILL.

Porzana VIEILLOT, Analyse, p. 61, 1816.

Porzana concolor (GOSSE).

Rallus concolor GOSSE, Bds. Jam. p. 369 (1847).—ALBRECHT, J. f. O. 1862,
    p. 206.—MARCH, Pr. Acad. Nat. Sci. Phila. 1864, p. 69.
Corethrura cayennensis MOORE, P. Z. S. 1859, p. 64.—SCL. & SALV. Ibis,
    1859, p. 230.
Corethrura guatemalensis LAWR. Pr. Acad. Nat. Sci. Phila. 1863, p. 106.
Rallina castanea SCHLEG. Mus. Pays-Bas, Ralli, p. 17 (1865).
Porzana concolor SCL. & SALV. P. Z. S. 1868, p. 452; *ib.* Nom. Avium
    Neotr. p. 140 (1873).—A. & E. NEWTON, Handb. Jamaica, p. 114
    (1881).—CORY, List Bds. W. I p. 30 (1885).
Rufirallus concolor GRAY, Handl. Bds. III, p. 61 (1871).

SP. CHAR. *Male:*—Head dark olive, showing a tinge of rufous on the
forehead; back olive, shading into rufous brown on the wing-
coverts; sides of the head pale reddish brown, brightest on the
cheeks; chin white, shading into clear reddish brown on the lower
throat and breast, rest of underparts reddish brown; under surface
of wings dull slate color.

The sexes are similar.

Length (skin), 9.25; wing, 5.25; tarsus, 1.75; bill, 1.10.

HABITAT. Jamaica.

Porzana flaviventris (BODD.).

Rallus flaviventer "BODD. Pl. Enl. (1783)."
Rallus minutus GMEL. Syst. Nat. I. p. 719 (1788).—LEMB. Aves Cuba,
    p. 109 (1850).—BREWER, Pr Bost. Soc. Nat. Hist. VII. p. 308 (1860).
Ortygometra minuta BURM. Syst. Ueb. III, p. 388.—GOSSE, Bds. Jam.
    p. 372 (1847).
Rallus superciliaris VIEILL. Nouv. Dict. XXVIII, p. 565.
Ortygometra flaviventris GRAY, Gen. Bds III, p. 593 (1844-49).—HARTL.
    Ind. Az. p. 24.
Lateriallus gossei BP. Compt. Rend. XLIII, p. 599 (1856).
Erythra minuta BP. Compt. Rend. XLIII, p. 600 (1856).
Crybastus gossii CAB. J. f. O. 1856, p. 428.—GUNDL. Repert. Fisico-Nat.
    Cuba, 1, p. 36; (1866); *ib.*, J. f. O. 1875, p. 358.
Crex minuta SCL. P. Z. S. 1861, p. 81.

Porzana minuta MARCH, Pr. Acad. Nat. Sci. Phila. 1864, p. 69.
Porzana flaviventer SCHLEG. Mus. Pays-Bas, I, p. 31 (1865).—A. & E.
    NEWTON, Handb. Jamaica, p. 114 (1881).
Porzana flaviventris SCL. & SALV. P. Z. S. 1868, p. 455; *ib.* Nom. Avium
    Neotr. p. 140 (1873).—CORY, List Bds. W. I. p. 30 (1885).
Crybastus minutus GRAY, Handl. Bds. III, p. 61 (1871).

SP. CHAR.—Top of head and a line through the eye, from the bill, dark
brown, darkest on the crown, and shading into light brown on the
back of the neck; middle back dark brown and black, streaked with
white; rump and upper tail-coverts chestnut brown, sometimes
touched with white; tail-feathers black, edged with brown, and
dotted with white; wing-coverts light cinnamon-brown; scapularies
marked with black and white; quills pale brown; outer web of first
primaries dull white · underparts white; throat white, tinged with
very pale yellowish brown; flanks barred with white and black;
bill dark.

    Length (skin), 5.75; wing, 2.75; tail, 1.15; tarsus, .85; bill, .60.
HABITAT. Cuba and Jamaica.

Porzana jamaicensis (GMEL.).

Rallus jamaicensis GMEL. Syst. Nat. I, p. 718 (1788).
Ortygometra jamaicensis GOSSE, Bds. Jam. p. 375 (1847).—BREWER, Pr.
    Bost. Soc. Nat. Hist. VII, p. 308 (1860) (Cuba).
Creciscus jamaicensis GUNDL. J. f. O. 1856, p. 428 (Cuba); *ib.* 1875, p. 360
    (Cuba); *ib.* 1881, p. 401 (Cuba).
Porzana jamaicensis SCL. P. Z. S. 1861, p. 81 (Jamaica).—ALBRECHT,
    J. f. O. 1862, p. 206 (Jamaica).—MARCH, Pr. Acad. Nat. Sci. Phila.
    1864, p. 69 (Jamaica).—A. & E. NEWTON, Handb. Jamaica, p. 114
    (1881).—CORY, List Bds. W. I. p. 30 (1885).

Recorded from Cuba and Jamaica.

Porzana carolina (LINN.).

Rallus carolinus LINN. Syst. Nat. I, p. 363 (1766).—D'ORB. in La Sagra's
    Hist. Nat. Cuba, Ois. p. 262 (1840).—BRYANT, Pr. Bost. Soc. Nat.
    Hist. X, p. 257 (1866) (Porto Rico).—SUNDEV. Oefv. K. Vet. Akad.
    For. 1869, p. 587 (St. Bartholomew); *ib.* p. 601 (Porto Rico).
Ortygometra carolina GOSSE, Bds. Jam. p. 371 (1847).—BREWER, Pr.
    Bost. Soc. Nat. Hist. VII, p. 308 (1860) (Cuba).
Porzana carolina GUNDL. J. f. O. 1856, p. 428 (Cuba).—A. & E. NEWTON,
    Ibis, 1859, p. 260 (St. Croix).—SCL. P. Z. S. 1861, p. 81 (Jamaica).
    —MARCH Pr. Acad. Nat. Sci. Phila. 1864, p. 69 (Jamaica).—GUNDL.
    Repert. Fisico-Nat. Cuba, I, p. 361 (1866).—BRACE, Pr. Bost. Soc.
    Nat. Hist. XIX, p. 241 (1877) (Bahamas).—GUNDL. J. f. O. 1875, p.
    358 (Cuba); *ib.* Anal. Soc. Esp. Hist. Nat. VII, p 390 (1878) (Porto

Rico).—CORY, Bds. Bahama I. p. 176 (1880).—A. & E. NEWTON,
Handb. Jamaica, p. 114 (1881).—CORY, List Bds. W. I. p. 30
(1885).—WELLS, List Bds. Grenada, p. 9 (1886).

This species is found throughout the West Indies ; numerous
references from the Bahamas and Antilles.

Porzana noveboracensis (GMEL.).

Fulica noveboracensis GMEL. Syst. Nat. I. p. 701 (1788).
Porzana noveboracensis BD. BWR. & RIDGW. Hist. N. Am. W. Bds. I, p.
375 (1884) (Cuba).—CORY, List Bds. W. I. p. 30 (1885).

Accidental in Cuba.

GENUS Gallinula BRISS.

Gallinula BRISSON, Orn. VI, p. 3, 1760.

Gallinula galeata (LICHT.).

Crex galeata LICHT. Verz. Doubl. p. 826 (1823).
Gallinula chloropus D'ORB. in La Sagra's Hist. Nat. Cuba, Ois. p. 268
(1840).
Gallinula galeata GOSSE, Bds. Jam. p. 381 (1847).—SALLÉ, P. Z. S. 1857,
p. 237 (San Domingo).—A. & E. NEWTON, Ibis, 1859, p. 260 (St.
Croix). — BRYANT, Pr. Bost. Soc. Nat. Hist. VII, p. 122 (1859)
(Bahamas); *ib.* BREWER, p. 307 (1860) (Cuba). — CASSIN, Pr.
Acad. Nat. Sci. Phila. 1860, p. 378 (St. Thomas). — SCL. P. Z. S.
1861, p. 81 (Jamaica). — MARCH, Pr. Acad. Nat. Sci. Phila. 1864,
p. 69 (Jamaica).—BRYANT, Pr. Bost. Soc. Nat. Hist. X, p. 257 (1866)
(Porto Rico). — SUNDEV. Oefv. K. Vet. Akad. For. 1869, p. 601
(Porto Rico).—GUNDL. J. f. O. 1875, p. 360 (Cuba); *ib.* Anal. Soc.
Esp. Hist. Nat. VII, p. 391 (1878) (Porto Rico).—LAWR. Pr U. S.
Nat. Mus. I, p. 276 (1878) (Grenada); *ib.* p. 461 (Guadeloupe).—
ALLEN, Bull. Nutt. Orn. Club, V, p. 169 (1880) (Santa Lucia). —
CORY, Bds. Bahama I. p. 177 (1880). — A. & E. NEWTON, Handb.
Jamaica, p. 115 (1881). — CORY. Bull. Nutt. Orn. Club, VI, p. 155
(1881) (Haiti); *ib.* Bds. Haiti & San Domingo, p. 161 (1885); *ib.*
List. Bds. W. I. p. 30 (1885). — WELLS, List Bds. Grenada, p. 9
(1886).—CORY, Auk, III, p. 502 (1886) (Grand Cayman); *ib.* Ibis,
1886. p. 474 (Marie Galante).
Gallinula galeata GUNDL. Repert. Fisico-Nat. Cuba, I, p. 362 (1866).

Common in the Bahamas and Antilles.

GENUS Ionornis REICH.

Ionornis REICHENBACH, Syst. Av. p. 21, 1853.

Ionornis martinica (Linn.).

Fulica martinica Linn. Syst. Nat. I, p. 259 (1766).
Porphyrio martinica D'Orb. in La Sagra's Hist. Nat. Cuba, Ois. p. 265
    (1840).—Gosse, Bds. Jam. p. 377 (1847).—Brewer, Pr. Bost. Soc.
    Nat. Hist. VII, p. 307 (1860) (Cuba).—Cory, Bds. Bahama I. p.
    178 (1880); *ib.* Bull. Nutt. Orn. Club, VI, p. 155 (1881) (Haiti).
Gallinula martinica Bryant, Pr. Bost. Soc. Nat. Hist. VII, p. 122 (1859)
    (Bahamas); *ib.* X, p. 257 (1866) (Porto Rico).—March. Pr. Acad.
    Nat. Sci. Phila. 1864, p. 69 (Jamaica). — Sundev. Oefv. K. Vet.
    Akad. For. 1869, p. 601 (Porto Rico).
Porphyrio martinicus Scl. P. Z. S. 1861. p. 81 (Jamaica); *ib.* 1872. p. 653
    (Santa Lucia).—Lawr. Pr. U. S. Nat. Mus. I, p. 197 (1878) (St.
    Vincent); *ib.* p. 487 (Dominica). — Allen. Bull. Nutt. Orn. Club,
    V, p. 169 (1880) (Santa Lucia).—A. & E. Newton, Handb. Jamaica,
    p. 115 (1881).
Porphyrula martinica Gundl. Repert. Fisico-Nat. Cuba, I, p. 362 (1866);
    ib. J. f. O. 1874, p. 314 (Porto Rico); *ib.* 1875, p. 361 (Cuba); *ib.*
    Anal. Soc. Esp. Hist. Nat. VII, p. 392 (1878) (Porto Rico).
Ionornis martinica Cory, Bds. Haiti & San Domingo. p. 162 (1885); *ib.*
    List Bds. W. I. p. 30 (1885). — Wells, List Bds. Grenada, p. 9
    (1886).

Common in the Bahamas and Antilles.

Genus Fulica Linn.

Fulica Linnæus, Syst. Nat. 1735; *ib.* I, p. 152, 1758.

Fulica americana Gmel.

Fulica americana Gmel. Syst. Nat. I. p. 704 (1788). — Gosse, Bds. Jam.
    p. 384 (1847).—A. & E. Newton, Ibis, 1859. p. 260 (St. Croix)?—
    Bryant, Pr. Bost. Soc. Nat. Hist. VII, p. 122 (1859) (Bahamas);
    ib. Bds. p. 307 (1860) (Cuba).—Scl. P. Z. S. 1861. p. 81 (Ja-
    maica).—March. Pr. Acad. Nat. Sci. Phila. 1864, p. 69 (Jamaica).
    —Gundl. Repert. Fisico-Nat. Cuba, I. p. 363 (1866).—Sundev. Oefv.
    K. Vet. Akad. For. 1869, p. 587 (St. Bartholomew); *ib.* p. 601
    (Porto Rico).—Gundl. J. f. O. 1875, p. 363 (Cuba); *ib.* Anal. Soc.
    Esp. Hist. Nat. VII, p. 394 (1878) (Porto Rico). — Cory. Bds.
    Bahama I. p. 178 (1880). — A. & E. Newton, Handb. Jamaica. p.
    115 (1881).—Cory. Bull. Nutt. Orn. Club, VI, p. 155 (1881) (Haiti);
    ib. Bds. Haiti & San Domingo, p. 163 (1885); *ib.* List Bds. W. I.
    p. 30 (1885).—Wells, List Bds. Grenada, p. 9 (1886).
Fulica atra D'Orb. in La Sagra's Hist. Nat. Cuba, Ois. p. 211 (1840).

Common in the Bahamas and Antilles.

Fulica caribæa RIDGW.

Fulica caribæa RIDGW. Pr. U. S. Nat. Mus. VII, p. 358 (1884). — CORY,
List Bds. W. I. p. 30 (1885).

"SP. CHAR.—Similar to *F. americana*, but differing in the slenderer bill
and in the form and color of the frontal shield. Frontal shield oval
or elliptical, much wrinkled, .70-.90 inch long, and .35-.50 wide,
in the breeding season; its color, pale brownish (whitish in life?)
instead of chestnut or liver brown, as in *F. americana*." (Ridgw.
l. c. orig. descr.).

HABITAT. Guadeloupe and St. John.

FAMILY ANATIDÆ.

GENUS Anser BRISS.

Anser BRISSON, Orn. 1760.

Anser albifrons gambeli.

Anser albifrons LEMB. Aves Cuba, p. 112 (1850).—CORY, List Bds. W. I.
p. 30 (1885).
Anser gambeli HARTL. Rev. Mag. Zool. 1852, p. 7.—CAB. J. f. O. 1857, p.
226 (Cuba).—BREWER, Pr. Bost. Soc. Nat. Hist. VII, p. 308 (1860)
(Cuba).
Anser gambelii GUNDL. Repert. Fisico-Nat. Cuba, I. p. 387 (1866); *ib.*
J. f. O. 1875, p. 375 (Cuba).—CORY, List Bds. W. I. p. 30 (1885).

Accidental in Cuba in winter.

GENUS Chen BOIE.

Chen BOIE, Isis, 1822, p. 563.

Chen hyperborea (PALL.).

Anser hyperboreus PALL. Spic. Zool. VI, pp. 80, 25 (1769).—LEMB. Aves
Cuba, p. 111 (1850).— BREWER, Pr. Bost. Soc. Nat. Hist, VII, p.
308 (1860) (Cuba).—MARCH, Pr. Acad. Nat. Sci. Phila. 1864, p. 70
(Jamaica).—CORY, Bds. Bahama I. p. 182 (1880).
Chen hyperboreus GOSSE, Bds. Jam. p. 408 (1847).—ALBRECHT, J. f. O.
1862, p 207 (Jamaica). — GUNDL. Repert. Fisico-Nat. Cuba, I. p.
387 (1866); *ib.* J. f. O. 1875, p. 371 (Cuba); *ib.* Anal. Soc. Esp.
Hist. Nat. VII, p. 399 (1878) (Porto Rico). — A. & E. NEWTON,
Handb. Jamaica, p. 112 (1881).—CORY, List Bds. W. I. p. 30 (1885).

Accidental in Bahamas, Cuba, Jamaica and Porto Rico.

Chen cærulescens (LINN.).

Anas cærulescens LINN. Syst. Nat. I, 10th ed. p. 124 (1758); *ib.* 12th ed.
p. 198 (1766).
Chen cærulescens GUNDL. Repert. Fisico-Nat. Cuba, I. p. 387 (1866); *ib.*
J. f. O. 1875. p. 374 (Cuba).—CORY, List. Bds. W. I. p. 30 (1885).
Anser cærulescens BRYANT, Pr. Bost. Soc. Nat. Hist. XI, p. 70 (1867)
(Bahamas).

Recorded from the Bahamas and Cuba. Possibly not separable from the preceding species, of which it may prove to be a
race.

GENUS Branta SCOPOLI.

Branta SCOPOLI. Ann. i Hist. Nat. p. 67, 1769.

Branta canadensis (LINN.).

Anas canadensis LINN. Syst. Nat. I, p. 198 (1766).
Branta canadensis BANNISTER. Pr. Acad. Nat. Sci. Phila. 1870, p. 131.
Berniela canadensis A. & E. NEWTON, Handb. Jamaica, p. 112 (1881).—
CORY, Revised List Bds. W. I. p. 30 (1886).

Recorded from Jamaica.

GENUS Dendrocygna SWAINS.

Dendrocygna SWAINSON. Classif. Birds, II, p. 365, 1837.

Dendrocygna arborea (LINN.).

Anas arborea LINN. Syst. Nat. I, p. 207 (1766).—GMEL. Syst. Nat. I,
p. 540 (1788).—VIEILL. Enc. Méth. p. 141 (1823).—D'ORB. in La
Sagra's Hist. Nat. Cuba, I, p. 291 (1840).—SUNDEV. Oefv. K. Vet.
Akad. For. 1869. p. 603.
Anas jacquini GMEL. Syst. Nat. I, p. 536 (1788).
Dendrocygna arborea EYTON, Mon. Anat. p. 110 (1838).—GOSSE, Bds.
Jam. p. 395 (1847).—CAB. J. f. O. 1857, p. 227.—A. & E. NEWTON,
Ibis, 1859. p. 366.—BRYANT, Pr. Bost. Soc. Nat. Hist. VII, p. 122
(1859); *ib.* XI, p. 70 (1866).—ALBRECHT, J. f. O. 1862, p. 206.—
MARCH, Pr. Acad. Nat. Sci. Phila. 1864, p. 70.—GUNDL. Repert.
Fisico-Nat. Cuba, I, p. 387 (1866); *ib.* Anal. Soc. Esp. Hist. Nat.
VII, p. 400 (1878).—SCL. & SALV. P. Z. S. 1876, p. 375.—CORY,
Bds. Bahama I. p. 183 (1880).—A. &. E. NEWTON, Handb. Jamaica,
p. 112 (1881).—CORY, Bds. Haiti & San Domingo, p. 166 (1885);
ib. List Bds. W. I. p. 30 (1885).—BD. BWR. & RIDGW. Hist. N. Am.
W. Bds. I, p. 480 (1884).

Dendrocygnus arborea BREWER, Pr. Bost. Soc. Nat. Hist. VII, p. 308
(1860).
Dendrocygna autumnalis? TAYLOR, Ibis, 1864. p. 172.

SP. CHAR. *Male:*—Head with black band on the crown, continuing in
narrow stripes to the nape; forehead and over the eye reddish
brown, shading into dull white on the throat, and mottled brown
and white on the sides of the head and neck; breast and upper parts
brown, the feathers broadly edged with tawny; rump and tail black;
underparts brownish white, heavily spotted and banded upon the
sides, the spots becoming very small and faint upon the abdomen;
most of the primaries slate-color, becoming brownish at the tips;
legs and bill black.

Length, 21.00; wing, 11.25; tarsus, 2.60; bill 2.00.

HABITAT. Bahamas and Antilles.

Dendrocygna autumnalis (LINN.).

Anas autumnalis LINN. Syst. Nat. I, p. 205 (1766).
Dendrocygna autumnalis GOSSE, Bds. Jam. p. 398 (1847).—ALBRECHT.
J. f. O. 1862, p. 206 (Jamaica).—MARCH, Pr. Acad. Nat. Sci. Phila.
1864. p. 70 (Jamaica).—A. & E. NEWTON, Handb. Jamaica, p. 112
(1881).—CORY, List Bds. W. I. p. 30 (1885).

Accidental in Jamaica.

Dendrocygna viduata (LINN.).

Anas viduata LINN. Syst. Nat. I, p. 205 (1766).—GMEL. Syst. Nat. I,
p. 536 (1788).—VIEILL. Enc. Méth. p. 132 (1823).
Dendrocygna viduata EYTON, Mon. Anat. p. 110 (1838).—ALBRECHT,
J. f. O. 1861, p. 214.—GUNDL. Repert. Fisico-Nat. Cuba, I, p. 388
(1866); *ib.* J. f. O. 1875, p. 377.—SCL. & SALV. P. Z. S. 1876, p.
376.—BD. BWR. & RIDGW. Hist. N. Am. W. Bds. I, p. 481 (1884).—
CORY, List Bds. W. I. p. 30 (1885).
Dendrocygnus viduata BREWER, Pr. Bost. Soc. Nat. Hist. VII, p. 308
(1860).

SP. CHAR. *Male:*—Entire front of head, including eye, cheeks and chin,
white, tinged with brown; a patch of white on the middle of the
throat, connecting with the white upper throat and chin by a narrow
white line; rest of head and neck black; breast and upper back
rufous brown; sides of the body thickly banded with narrow black
and white lines; centre of the belly and lower breast black; feathers on the back edged with tawny; wings black; carpus and
shoulder chestnut brown; wing-coverts showing an olive tinge; bill
black; feet black.

The sexes are similar.

Length, 19.00; wing, 9.00; tail, 4.00; tarsus, 2.00; bill, 2.00.

Given by authors as occurring in Cuba ; by some, claimed to
have been introduced.

Genus Anas Linn.

Anas Linnæus, Syst. Nat. I, 10th ed. p. 122. 1758; *ib.* 12th ed. p. 194,
1766.

Anas strepera Linn.

Anas strepera Linn. Syst. Nat. I, 10th ed. p. 125 (1758); *ib.* 12th ed.
p. 200 (1766).—March, Pr. Acad. Nat. Sci. Phila. 1864. p. 72 (Jamai-
ca).—A. & E. Newton. Handb. Jamaica, p. 113 (1881).
Chaulelasmus streperus Gosse, Bds. Jam. p. 408 (1847).—Gundl. Repert.
Fisico-Nat. Cuba, I, p 389 (1866); *ib.* J. f. O. 1875, p. 381 (Cuba).
Anas streperus Cory, List Bds. W. I. p. 30 (1885).

Cuba and Jamaica in winter.

Anas boschas Linn.

Anas boschas Linn. Syst. Nat. I, p. 205 (1766).—Gosse, Bds. Jam. p.
408 (1847).—Bryant, Pr. Bost. Soc. Nat. Hist. VII. p. 122 (1859)
(Bahamas); *ib.* Brewer. p. 308 (1860) (Cuba).—March, Pr. Acad.
Nat. Sci. Phila. 1864, p. 72 (Jamaica).—Gundl. Repert. Fisico-Nat.
Cuba. I. p. 388 (1866); *ib.* J. f. O. 1875, p. 378 (Cuba).—Cory. Bds.
Bahama I. p. 184 (1880); *ib.* List Bds. W. I. p. 30 (1885).—Wells,
List Bds. Grenada, p. 10 (1886).
Anas boscas A. & E. Newton, Handb. Jamaica, p. 113 (1881).

Accidental in Cuba, the Bahamas, Jamaica, and Grenada.

Anas obscura Gmel.

Anas obscura Gmel. Syst. Nat. I, p. 541 (1788).—Gosse, Bds. Jam. p. 408
(1847).—March, Pr. Acad. Nat. Sci. Phila. 1864, p. 72 (Jamaica).
—A. & E. Newton, Handb. Jamaica. p. 113 (1881).—Bd. Bwr. &
Ridgw. Hist. N. Am. W. Bds. I, p. 499 (1884) (Cuba?).—Cory,
Revised List Bds. W. I. p. 30 (1886).
Anas fulvigula? Cory, List Bds. W. I. p. 30 (1885).

Cuba? Jamaica.
It is uncertain whether the Dusky Duck which, it is claimed.
occurs in Jamaica, is *Anas fulvigula* Ridgw. or this species.
Both occur in Florida.
Anas maxima. described by Gosse (Bds. Jam. p. 399, 1847),
is supposed to be a hybrid.

GENUS Dafila STEPH.

Dafila STEPHENS, Shaw's Gen. Zool. XII, p. 126, 1824.

Dafila bahamensis (LINN.).

Anas bahamensis LINN. Syst. Nat. I, p. 199 (1766).—GMEL. Syst. Nat. I,
p. 516 (1788).—MAX. Beitr. p. 925 (1831).—SUNDEV. Oeiv. K Vet.
Akad. For. 1869, p. 591.
Anas rubirostris VIEILL. Nouv. Dict. V, p. 108 (1816).
Anas ilathera VIEILL. Enc. Méth. p. 152 (1823).
Anas urophasianus VIG. Zool. Journ. IV, p. 357 (1829).
Phasianurus vigorsii WAGL. Isis, 1832, p. 1235.
Anas fimbriata MERREM, Ersch u. Gruber's Ency. sect. 1, XXXV, p. 35.
Dafila urophasianus EYTON, Mon. Anat. p. 112 (1838).
Pœcilonetta bahamensis EYTON, Mon. Anat. p. 116 (1838).—GOSSE, Bds.
Jam. p. 408 (1847).—SCL. P. Z. S. 1860, p. 389.—ABBOTT, Ibis,
1861, p. 160.—ALBRECHT, J. f. O. 1862, p. 207.—MARCH, Pr. Acad.
Nat. Sci. Phila. 1864. p. 71.—GUNDL. J. f. O. 1874, p. 314; *ib.* Anal.
Soc. Esp Hist. Nat. VII, p. 403 (1878); *ib.* J. f. O. 1881, p. 400.
Dafila bahamensis HARTL. Ind. Az. p. 27 (1847).—CAB. in Schomb.
Guian III, p. 763 (1848).—SCL. & SALV. P. Z. S. 1876, p. 393.—
LAWR Pr. U. S. Nat. Mus. I, p. 487 (1878).—A. & E. NEWTON,
Handb. Jamaica, p 113 (1881).—CORY, Bds. Bahama I. p. 185
(1880); *ib.* Bds. Haiti & San Domingo, p. 167 (1885); *ib.* List Bds.
W. I p. 31 (1885)

SP. CHAR. *Male:*—General plumage tawny, mottled and streaked with
brown; wings banded with lustrous green, black and tawny, in the
order given; top of head and nape brown, finely mottled with dark
brown; rest of head and throat white; a triangular patch on each
side of the upper mandible lake red; tail tawny, becoming pale at
the tip; legs black.

Length, 19.00; wing, 8.00; tail, 4.75; tarsus, 1.25; bill, 1.95.

HABITAT. Bahamas and Antilles.

Dafila acuta (LINN.).

Anas acuta LINN. Syst. Nat. I, p. 202 (1766).—LEMB. Aves Cuba, p. 113
(1850).
Dafila acuta GOSSE, Bds. Jam. p. 408 (1847).—CAB. J. f. O. 1857, p. 227
(Cuba).—BREWER, Pr. Bost. Soc. Nat. Hist. VII, p. 308 (1860)
(Cuba).—ALBRECHT, J. f. O. 1862, p. 207 (Jamaica).—MARCH. Pr.
Acad. Nat. Sci. Phila. 1864, p. 71 (Jamaica).—GUNDL. Report.
Fisico-Nat. Cuba, I, p. 388 (1866); *ib.* J. f. O. 1875, p. 378 (Cuba);
ib. Anal. Soc. Esp. Hist. Nat. VII, p. 402 (1878) (Porto Rico).—
A. & E. NEWTON, Handb. Jamaica, p. 113 (1881).—CORY, List Bds.
W. I. p. 31 (1885).

Recorded from Cuba, Jamaica, and Porto Rico.

GENUS Mareca STEPH.

Mareca STEPHENS, Shaw's Gen. Zool. XII, pt. II. p. 130, 1824.

Mareca americana (GMEL.).

Anas americana GMEL. Syst. Nat. II. p. 526 (1788).—D'ORB. in La Sagra's
     Hist. Nat. Cuba, Ois. p. 293 (1840).—SUNDEV. Oefv. K. Vet. Akad.
     For. 1869, p. 603 (Porto Rico).
Mareca americana GOSSE. Bds Jam. p. 408 (1847).—CAB. J. f. O. 1857,
     p. 227 (Cuba). — BREWER, Pr. Bost. Soc. Nat. Hist. VII. p. 308
     (1860)(Cuba).—NEWTON, Ibis, 1860. p. 308 (St. Thomas).—MARCH,
     Pr. Acad. Nat. Sci. Phila. 1864. p. 71 (Jamaica).—GUNDL. Repert.
     Fisico-Nat. Cuba, I, p. 388 (1866); *ib.* J. f. O. 1875, p. 378 (Cuba);
     ib. Anal. Soc. Esp. Hist. Nat. VII. p. 402 (1878) (Porto Rico).—
     A. & E. NEWTON, Handb. Jamaica, p. 113 (1881).—CORY, List Bds.
     W. I. p. 31 (1885).

Accidental in winter in the West Indies ; records from Cuba,
Jamaica, Porto Rico, and St. Thomas.

Cairina moschata is given by numerous writers from Cuba
and Jamaica. It is claimed to have been introduced.

GENUS Querquedula STEPH.

Querquedula STEPHENS, Shaw's Gen. Zool. XII, p. 149, 1824.

Querquedula discors (LINN.).

Anas discors LINN. Syst. Nat. I, p. 205 (1766).—D'ORB. in La Sagra's
     Hist. Nat. Cuba, Ois. p. 294 (1840).—SUNDEV. Oefv. K. Vet Akad.
     For. 1869, p. 591 (St. Bartholomew); *ib.* p. 603 (Porto Rico).
Cyanopterus discors GOSSE, Bds. Jam. p. 401 (1847).
Cyanopterus inornatus GOSSE, Bds. Jam. p. 402 (1847).—ALBRECHT,
     J. f. O. 1862, p. 206 (Jamaica).
Querquedula discors SALLÉ, P. Z. S. 1857, p. 237 (San Domingo).—BRY-
     ANT, Pr. Bost. Soc. Nat. Hist. VII, b. 122 (1859) (Bahamas).—SCL.
     P. Z. S. 1861, p. 82 (Jamaica).—MARCH, Pr. Acad. Nat. Sci. Phila.
     1864, p 71 (Jamaica).—GUNDL. Repert. Fisico-Nat. Cuba. I. p. 389
     (1866).—LAWR. Ann. Lyc. N. Y. VIII, p. 101 (1867) (Sombrero).—
     GUNDL. J. f. O. 1874 , p. 314 (Porto Rico) : *ib.* 1875. p. 380 (Cuba);
     ib. Anal. Soc. Esp. Hist. Nat. VII. p. 404 (1878) (Porto Rico).—
     CORY, Bds. Bahama I. p. 186 (1880).—A. & E. NEWTON, Handb.
     Jamaica, p. 113 (1881).—CORY. Bds. Haiti & San Domingo. p. 168
     (1883).— CORY, List Bds. W. I. p. 31 (1885).—WELLS, List Bds.
     Grenada. p. 10 (1886).
Pterocyanea discors BREWER, Pr. Bost. Soc. Nat. Hist. VII, p. 308 (1860)
     (Cuba).

Querquedula inornata March. Pr. Acad. Nat. Sci. Phila. 1864, p. 71 (Jamaica).—A. & E. Newton. Handb. Jamaica, p. 113 1881).
Anas (Querquedula) discors Bryant Pr. Bost. Soc. Nat. Hist. XI, p. 97 (1867) (San Domingo).

Abundant throughout the Bahamas and Antilles.

Querquedula carolinensis (Gmel.).

Anas carolinensis Gmel. Syst. Nat. I, p. 533 (1788).—Lemb. Aves Cuba, p. 114 (1850).
Querquedula carolinensis Gosse. Bds. Jam. p. 408 (1847).—Brewer, Pr. Bost Soc. Nat. Hist. VII. p. 308 (1860) (Cuba).—Gundl. J. f. O. 1862, p 92 (Cuba).—March. Pr. Acad. Nat. Sci. Phila. 1864. p. 72 (Jamaica).—Cory, Bds. Bahama I. p. 187 (1880).—A. & E. Newton, Handb. Jamaica. p. 113 (1881).—Cory. List Bds. W. I. p. 31 (1885).—Wells. List Bds. Grenada. p. 10 (1886).
Nettion carolinensis Bryant, Pr. Bost. Soc. Nat. Hist. VII. p. 122 (1859) (Bahamas).—Gundl. Repert. Fisico-Nat. Cuba, I, p. 389 (1866); *ib.* J. f. O. 1875 p. 381 (Cuba).

This species has been taken in winter in the Bahama Islands, Cuba, Jamaica, and Grenada.

Querquedula cyanoptera is given by Brewer as occurring in Cuba (Pr. Bost. Soc. Nat. Hist. VII, p. 308, 1860). If the species in question was correctly identified, it is of rare occurrence in the West Indies.

Nyroca ferruginea is given by W. T. March, as occurring in Jamaica (Pr. Acad. Nat. Sci. Phila. 1864. p. 72). This record is undoubtedly incorrect. The bird in question was probably some other species wrongly identified, possibly *Querquedula cyanoptera*.

Genus Spatula Boie.

Spatula Boie. Isis, 1822, p. 564.

Spatula clypeata (Linn.).

Anas clypeata Linn. Syst. Nat. I, 10th ed. p. 124 (1758): *ib.* 12th ed. p. 200 (1766).—Lemb. Aves Cuba, p. 115 (1836).—Sundev. Oefv. K. Vet. Akad. For. 1869. p. 603 (Porto Rico).
Anas mexicana D'Orb. in La Sagra's Hist. Nat. Cuba, Ois. p. 299 (1840).
Rhyncaspis clypeata Gosse. Bds. Jam. p. 408 (1847).
Rhynchaspis clypeata Brewer, Pr. Bost. Soc. Nat. Hist. VII. p. 308 (1860) (Cuba).

Spatula clypeata NEWTON, Ibis, 1860, p. 308 (St. Thomas).—SCL. P. Z. S.
    1861, p. 82 (Jamaica).—MARCH. Pr. Acad. Nat. Sci. Phila. 1864, p.
    71 (Jamaica).—GUNDL. Repert. Fisico-Nat. Cuba, I, p. 389 (1866);
    ib. J. f. O. 1874. p. 314 (Porto Rico); *ib.* 1875, p. 379 (Cuba); *ib.*
    Anal. Soc. Esp. Hist. Nat. VII, p. 404 (1878) (Porto Rico).—A. & E.
    NEWTON, Handb. Jamaica, p. 113 (1881).—CORY, List Bds. W. I. p.
    31 (1885).
Querquedula clypeata ALBRECHT, J. f. O. 1862, p. 207 (Jamaica).

West Indies in winter; records from Cuba, Jamaica, Porto
Rico, and St. Thomas.

GENUS Aix BOIE.

Aix BOIE. Isis, 1828, p. 329.

Aix sponsa (LINN.).

Anas sponsa LINN. Syst. Nat. I, p. 207 (1766).—D'ORB. in La Sagra's
    Hist. Nat. Cuba, Ois. p. 288 (1840).
Aix sponsa GOSSE, Bds. Jam. p. 408 (1847).—CAB. J. f. O. 1857, p. 226
    (Cuba).—MARCH, Pr. Acad. Nat. Sci. Phila. 1864, p. 72 (Jamaica).
    —GUNDL. Repert. Fisico-Nat. Cuba. I, p. 389 (1866); *ib.* J. f. O.
    1875, p. 381 (Cuba).—A. & E. NEWTON. Handb. Jamaica, p. 113
    (1881).—CORY, List Bds. W. I. p. 31 (1885).
Dendrocygans sponsa BREWER, Pr. Bost. Soc. Nat. Hist. VII, p. 308 (1860)
    (Cuba).

Cuba and Jamaica in winter.

GENUS Aythya BOIE.

Aythya BOIE. Isis, 1822, p. 564.

Aythya affinis (EYTON).

Fuligula affinis EYTON, Mon. Anat. p. 157 (1838). — GOSSE, Bds. Jam.
    p. 408 (1847).—ALBRECHT, J. f. O. 1862, p. 207 (Jamaica).—CORY,
    Bds. Bahama I, p. 187 (1880); *ib.* List Bds. W. I. p. 31 (1885).
Anas marila D'ORB. in La Sagra's Hist. Nat. Cuba, Ois. p. 295 (1840)(?)
Fulix affinis A. & E. NEWTON. Ibis, 1859, p. 366 (St. Croix) (?)—MARCH,
    Pr. Acad. Nat. Sci. Phila. 1864, p. 71 (Jamaica).—GUNDL. Repert.
    Fisico-Nat. Cuba, I, p. 390 (1866); *ib.* J. f. O. 1874, p. 314 (Porto
    Rico); *ib.* 1875, p. 382 (Cuba); *ib.* Anal. Soc. Esp. Hist. Nat. VII,
    p. 405 (1878) (Porto Rico).
Fuligula marila BREWER, Pr. Bost. Soc. Nat. Hist. VII, p. 308 (1860)
    (Cuba).

Fuligula mariloides GUNDL. J. f. O. 1862, p. 92 (Cuba).
Nyroca affinis A. & E. NEWTON. Handb. Jamaica, p. 113 (1881).

Recorded from Porto Rico, Cuba, Bahamas, Jamaica, and St. Croix.

Aythya collaris (DONOV.).

Anas collaris DONOV. Brit. Birds. VI. pl. 47 (1809).
Fuligula rufitorques GOSSE, Bds. Jam. p. 408 (1847).—LEMB. Aves Cuba, p. 117 (1850).—BREWER. Pr. Bost. Soc. Nat. Hist. VII, p. 308 (1860) (Cuba).—ALBRECHT, J. f. O. 1862, p. 207 (Jamaica).
Fulix collaris BRYANT. Pr. Bost. Soc. Nat. Hist. VII. p. 122 (1859) (Bahamas).—MARCH. Pr. Acad. Nat. Sci. Phila. 1864. p. 72 (Jamaica).—GUNDL. Repert. Fisico-Nat. Cuba. I, p. 390 (1866); *ib.* J. f. O. 1875, p. 383 (Cuba); *ib.* Anal. Soc. Esp. Hist. Nat. VII, p. 406 (1878) (Porto Rico).
Anas (Fuligula) rufitorques GUNDL. J. f. O. 1871, p. 283 (Cuba).
Fuligula collaris CORY, Bds. Bahama I. p. 188 (1880); *ib.* List Bds. W. I. p. 31 (1885).
Nyroca collaris A. & E. NEWTON, Handb. Jamaica, p. 113 (1881).

Bahamas, Cuba, Jamaica, and Porto Rico, in winter.

Aythya vallisneria (WILS.).

Anas vallisneria WILS. Am. Orn. VIII. p. 103 (1814).—CORY, List Bds. W. I. p. 30 (1885).
Nyroca vallisneria BREWER, Pr. Bost. Soc. Nat. Hist. VII, p. 308 (1860) (Cuba).
Aythya vallisneria MARCH. Pr. Acad. Nat. Sci. Phila. 1864, p. 72 (Jamaica)
Aythya vallisneria GUNDL. Repert. Fisico-Nat. Cuba, I, p. 390 (1866)); *ib.* J. f. O. 1875, p. 382 (Cuba).
Nyroca vallisneria A. & E. NEWTON, Handb. Jamaica. p. 113 (1881).

Recorded from Cuba and Jamaica.

Aythya americana (EYTON).

Fuligula americana EYTON, Mon. Anat. p. 155 (1838).—GOSSE, Bds. Jam. p. 408 (1847).—CORY, List Bds. W. I. p. 31 (1885).
Aythya americana BRYANT, Pr. Bost. Soc. Nat. Hist. VII, p. 122 (1859) (Bahamas). — MARCH, Pr. Acad. Nat. Sci. Phila. 1864, p. 72 (Jamaica).
Fuligula ferina var. americana CORY, Bds. Bahama I. p. 189 (1880).
Nyroca americana A. & E. NEWTON. Handb. Jamaica, p. 113 (1881).

A winter visitant; records from the Bahamas, Cuba, and Jamaica.

GENUS **Charitonetta** STEJN.

Charitonetta STEJNEGER, Orn. Expl. Kamtsch. p. 163, 1885.

Charitonetta albeola (LINN.).

Anas albeola LINN. Syst. Nat. I, p. 199 (1766).
Clangula albeola BREWER, Pr. Bost. Soc. Nat. Hist. VII, p. 308 (1860)
    (Cuba).—CORY, List Bds. W. I. p. 31 (1885).
Bucephala albeola GUNDL. Repert. Fisico-Nat. Cuba, I, p. 390 (1866); *ib.*
    J. f. O. 1875, p. 383 (Cuba).

Accidental in Cuba in winter.

GENUS **Glaucionetta** STEJN.

Glaucionetta STEJNEGER, Pr. U. S. Nat. Mus. VIII, p. 409, 1885.

Glaucionetta clangula americana (BONAP.).

Clangula americana BP. Comp. List, 1838, p. 58.
Clangula glaucion LAWR. Pr. U. S. Nat. Mus. I. p. 241 (1878) (Barbuda).
Clangula glaucion americana BD. BWR. & RIDGW. Hist. N. Am W. Bds.
    II, p. 44 (1884) (Cuba) (?).
Clangula glaucium CORY, Revised List Bds. W. I. p. 31 (1886).

Recorded from Cuba and Barbuda.

GENUS **Œdemia** FLEMING.

Œdemia FLEMING, Philos. Zool. II, p. 260, 1822.

Œdemia perspicillata (LINN.).

Anas perspicillata LINN. Syst. Nat. I, p. 201 (1766).
Œdemia perspicillata GOSSE, Bds. Jam. p. 408 (1847).—ALBRECHT, J. f. O.
    1862, p. 207 (Jamaica).—MARCH, Pr. Acad. Nat. Sci. Phila. 1864,
    p. 72 (Jamaica).—A. & E. NEWTON, Handb. Jamaica, p. 113 (1881).
    —CORY, List Bds. W. I. p. 31 (1885).

Claimed to have occurred in Jamaica.

GENUS **Erismatura** BONAP.

Erismatura BONAPARTE, Saggio Distr. Met. p. 143, 1832.

Erismatura rubida (WILS.).

Anas rubida WILS. Am. Orn. VIII, pp. 128-130 (1814).
Erismatura spinosa GOSSE. Bds. Jam. p. 404 (1847).—ALBRECHT, J. f. O.
    1862 p 207 (Jamaica).
Fuligula rubida LEMB. Aves Cuba, p. 118 (1850).
Erismatura rubida BRYANT, Pr. Bost. Soc. Nat. Hist. VII, p. 122 (1859)
    (Bahamas).—GUNDL. Repert. Fisico-Nat. Cuba, I, p. 390 (1866);
    ib. J f. O. 1874, p. 314 (Porto Rico); *ib.* 1875, p. 384 (Cuba); *ib.*
    Anal Soc. Esp. Hist. Nat. VII, p. 407 (1878) (Porto Rico).—CORY,
    Bds Bahama I. p 189 (1880).—A. & E. NEWTON, Handb. Jamaica,
    p. 113 (1881).—CORY, List Bds. W. I. p. 31 (1885).—WELLS, List
    Bds. Grenada, p. 10 (1886).

Occurs in winter in the West Indies ; records from Porto Rico,
Cuba, Jamaica, Grenada, and the Bahamas.

GENUS Nomonyx RIDGW.

Nomonyx RIDGWAY, Pr. U. S. Nat. Mus. II, p. 15, March 27, 1880.

Nomonyx dominicus (LINN.).

Anas dominica LINN Syst. Nat. I, p. 201 (1766).—SUNDEV. Oefv. K. Vet.
    Akad. For 1869 p 603 (Porto Rico).
Anas spinosa? D'ORB. in La Sagra's Hist. Nat. Cuba. Ois. p. 297 (1840)?
Erismatura dominica A. & E. NEWTON, Ibis, 1859, p. 367 (St. Croix) (?).
    —GUNDL. Repert. Fisico-Nat. Cuba, I, p. 391 (1866); *ib.* J. f. O.
    1874, p. 314 (Porto Rico); *ib.* 1875. p. 314 (Cuba); *ib.* Anal. Soc.
    Esp Hist Nat. VII, p. 408 (1878) (Porto Rico).
Dendrocygnus spinosa BREWER, Pr. Bost. Soc. Nat. Hist. VII. p. 308 (1860)
    (Cuba)
Nomonyx dominicus CORY. List Bds. W. I. p. 31 (1885).

SP. CHAR. *Male:*—Top of head brownish black ; a stripe of brown through
    the eye, and a parallel stripe of the same color below, separated by
    a narrow stripe of tawny : a narrow tawny superciliary stripe ;
    throat tawny brown, the feathers marked with chestnut, heaviest on
    the lower part ; underparts dull white, marked with yellowish
    brown ; feathers of the back having the centres black, and heavily
    edged with chestnut ; quills and tail dark brown ; secondaries white,
    tipped with brown, forming a large white patch on the wing. In
    some plumages the male is described as having the entire head
    black.

    The female differs from the male in lacking the chestnut marking
on the upper parts which is replaced by pale brown, bill dark
brown. almost black

    Length, 12.00 wing, 5.30 ; tail 3.10 ; tarsus, 90 ; bill, 1.30.

HABITAT Antilles.

GENUS Lophodytes REICH.

Lophodytes REICHENBACH, Syst. Av. p. IX, 1852.

Lophodytes cucullatus (LINN.).

Mergus cucullatus LINN. Syst. Nat. I, 10th ed. p. 129 (1758); *ib.* 12th ed.
   p. 207 (1766).—BREWER. Pr. Bost. Soc. Nat. Hist. VII, p. 308 (1860)
   (Cuba).—GUNDL. J. f. O. 1862. p. 93 (Cuba).—CORY, List Bds.
   W. I. p. 31 (1885).
Lophodytes cucullatus GUNDL. Repert. Fisico-Nat. Cuba, I, p. 391 (1866);
   ib. J. f. O. 1875, p. 385 (Cuba).

Accidental in Cuba.

FAMILY FREGATIDÆ.

GENUS Fregata CUV.

Fregata CUVIER, Lec. d'Anat. Comp. I, tabl. II, 1799-1800.

Fregata aquila (LINN.).

Pelecanus aquilus LINN. Syst. Nat. I, 10th ed. p. 133 (1758); *ib.* 12th ed.
   p. 216 (1766).
Fregata aquila D'ORB. in La Sagra's Hist. Nat. Cuba, Ois. p. 309 (1840).
   —BREWER. Pr. Bost. Soc. Nat. Hist. VII, p. 308 (1860) (Cuba).—
   LAWR. Pr. U. S. Nat. Mus. I, p. 65 (1878) (Dominica)(?) *ib.* p. 195
   St. Vincent); *ib.* p. 236 (Antigua); *ib.* p. 240 (Barbuda); *ib.* p. 274
   (Grenada); *ib* p 359 (Martinique).—ALLEN Bull. Nutt. Orn.
   Club, V, p. 169 (1880) (Santa Lucia).—A. & E. NEWTON Handb.
   Jamaica, p. 112 (1881).—CORY. List Bds. W. I p. 31 (1885).
Fregata aquilus GOSSE, Bds. Jam. p. 422 (1847).
Tachypetes aquilus A. & E. NEWTON Ibis, 1859. p. 369 (St. Croix).—
   BRYANT, Pr. Bost. Soc. Nat. Hist. VII, p. 126 (1859) (Bahamas).—
   GUNDL. Repert. Fisico-Nat. Cuba, I, p. 396 (1866); *ib.* J. f. O. 1874,
   p. 315 (Porto Rico); *ib.* Anal. Soc. Esp. Hist. Nat. VII, p. 421
   (1878) (Porto Rico).—BRYANT, Pr. Bost. Soc. Nat. Hist. XI, p. 98
   (1867) (San Domingo).—CORY, Bds. Bahama I. p, 200 (1880); *ib.*
   Bull. Nutt. Orn. Club, VI, p. 155 (1881) (Haiti); *ib.* Bds. Haiti &
   San Domingo, p. 173 (1885).
Tachypetes aquila WELLS, List Bds. Grenada, p. 10 (1886).

Abundant in the Bahamas and throughout the Antilles.

Family PELECANIDÆ.

Genus Pelecanus Linn.

Pelecanus Linnæus, Syst. Nat. 1735; *ib.* 10th ed. I, p. 132, 1758.

Pelecanus fuscus Linn.

Pelecanus fuscus Linn. Syst. Nat. I, p. 215 (1766).—D'Orb. in La Sagra's
Hist. Nat. Cuba, Ois. p. 300 (1840).—Gosse, Bds. Jam. p. 409
(1847).—A. & E. Newton. Ibis, 1859, p. 368 (St. Croix).—Bryant,
Pr. Bost. Soc. Nat. Hist. VII, p. 122 (1859) (Bahamas); *ib.* Brewer,
p. 308 (1860) (Cuba).—Gundl. Repert. Fisico-Nat. Cuba, I, p. 394
(1866).—Lawr. Ann. Lyc. N. Y. VIII, p. 101 (1867) (Sombrero).—
Sundev. Oefv. K. Vet. Akad. For. 1869. p. 603 (Porto Rico).—Lawr.
Pr. U. S. Nat. Mus. I, p. 66 (1878) (Dominica); *ib.* p. 196 (St.
Vincent); *ib.* p. 236 (Antigua); *ib.* p. 240 (Barbuda); *ib.* p. 274
(Grenada); *ib.* p 359 (Martinique).—Gundl. Anal. Soc. Esp. Hist.
Nat. VII, p. 416 (1878) (Porto Rico).—Cory, Bds. Bahama I. p.
196 (1880).—A. & E. Newton, Handb. Jamaica. p. 112 (1881).—
Cory, Bull. Natt. Orn. Club, VI, p. 155 (1881) (Haiti); *ib.* Bds.
Haiti & San Domingo, p. 172 (1885); *ib.* List Bds. W. I. p. 32
(1885).—Wells, List Bds. Grenada, p. 10 (1886).

The Brown Pelican is common in the Bahama Islands, the
Greater Antilles, and in many of the Lesser Antilles.

Family PHALACROCORACIDÆ.

Genus Phalacrocorax Briss.

Phalacrocorax Brisson. Orn. VI, p. 511, 1760.

Phalacrocorax dilophus floridanus (Aud.).

Phalacrocorax floridanus Aud. Orn. Biog. III, p. 387 (1835).—Lemb.
Aves Cuba, p. 119 (1850)—Brewer, Pr. Bost. Soc. Nat. Hist. VII,
p. 308 (1860) (Cuba).—Gundl. J. f. O. 1862, p. 95 (Cuba).
Graculus floridanus Bryant, Pr. Bost. Soc. Nat. Hist. VII, p. 128 (1859)
(Bahamas).—Gundl. Repert. Fisico-Nat. Cuba, I, p. 394 (1866);
ib. J. f. O. 1875. p. 400 (Cuba).
Graculus dilophus var. *floridanus* Cory, Bds. Bahama I. p. 198 (1880).
Phalacrocorax dilophus floridanus Ridgw. Nom. N. A. Bds. No. 643 *a*
(1881).—Cory, List. Bds. W. I. p. 32 (1885).

Accidental in the Bahamas and Cuba.

Phalacrocorax mexicanus (Brandt).

Carbo mexicanus Brandt, Bull. Sc. Ac. Imp. St. Pet. III, p. 56 (1837).
Phalacrocorax resplendens Lemb. Aves Cuba, p. 119 (1850) —Gundl.
　　J. f. O. 1862. p. 95 (Cuba).
Phalacrocorax townsendi Lemb. Aves Cuba, p. 120 (1850).—Gundl.
　　J. f. O. 1862. p. 95 (Cuba).
Phalacrocorax mexicanus Brewer, Pr. Bost. Soc. Nat. Hist. VII, p. 308
　　(1860) (Cuba).—Cory, List Bds. W. I. p. 32 (1885).
Graculus mexicanus Gundl. Repert. Fisico-Nat. Cuba, I, p. 395 (1866);
　　ib. J. f. O. 1875, p. 401 (Cuba).

Recorded from Cuba.

Family ANHINGIDÆ.

Genus Anhinga Briss.

Anhinga Briss. Orn. VI, p. 476, 1760.

Anhinga anhinga (Linn.).

Plotus anhinga Linn. Syst. Nat. I. p. 218 (1766).—Lemb. Aves Cuba, p.
　　120 (1850).—Brewer, Pr. Bost. Soc. Nat. Hist. VII, p. 308 (1860)
　　(Cuba).—Gundl. J. f. O. 1862, p. 96 (Cuba); *ib.* Repert. Fisico-
　　Nat. Cuba, I, p. 395 (1866); *ib.* J. f. O. 1875, p. 405 (Cuba).—Cory,
　　List Bds. W. I. p. 32 (1885).

This species is stated to be common in many parts of Cuba.

Family SULIDÆ.

Genus Sula Briss.

Sula Brisson, Orn. VI, p. 495. 1760.

Sula cyanops (Sundev.).

Dysporus cyanops Sundev. Phys. Tidskr. Lund. pt. 5 (1837).
Sula dactylatra? Bryant, Pr. Bost. Soc. Nat. Hist. VII, p. 125 (1859)
　　(Bahamas); XI, p 97 (1867) (San Domingo).—Cory, Bds. Ba-
　　hama I. p. 194 (1880).
Sula cyanops Cory, Bds. Haiti & San Domingo, p. 170 (1885); *ib.* List
　　Bds. W. I. p. 32 (1885).

Sp. Char.—Large. General color white; remiges and greater wing-coverts
dark brown; middle rectrices hoary white, tipped with brown; rest
of tail dark brown, white at the base; feet reddish? gular sac bluish.
Wing, 16.00; tail, 7.70; bill, 3.90; tarsus, 1.85.

Habitat. West Indies, Bahamas, breeding (*Bryant*).

Sula sula (Linn.).

Pelecanus sula Linn. Syst. Nat. ed. 12, I, p. 218 (1766).
Pelecanus leucogastra Bodd. Tabl. Pl. Enl. p. 57 (1783).
Sula fusca D'Orb. in La Sagra's Hist. Nat. Cuba, Ois. p. 306 (1840).—
Gosse, Bds. Jam. p. 417 (1847).—Sallé, P. Z. S. 1857, p. 237 (San
Domingo).—Brewer, Pr. Bost. Soc. Nat. Hist. VII, p. 308 (1860)
(Cuba).—Albrecht, J. f. O. 1862, p. 207 (Jamaica); *ib.* Gundl.
p. 95 (Cuba).—Bryant, Pr. Bost. Soc. Nat. Hist. XI, p. 97 (1867)
(San Domingo).
Sula fiber Gosse, Bds. Jam. p. 417 (1847).—Bryant, Pr. Bost. Soc. Nat.
Hist. VII, p. 123 (1859) (Bahamas). — Albrecht, J. f. O. 1862, p.
207 (Jamaica).—Lawr. Ann. Lyc. N. Y. VIII, p. 101 (1867) (Som-
brero); Pr. U. S. Nat. Mus. I, p. 196 (1878) (St. Vincent); *ib.*
p. 274 (Grenada).—Allen, Bull. Nutt. Orn. Club, V. p. 169 (1880)
(Santa Lucia).—Cory, Bds. Bahama I. p. 191 (1880).—Wells,
List. Bds. Grenada, p. 11 (1886).
Dysporus sula A. & E. Newton, Ibis, 1859, p. 369 (St. Croix).
Dysporus fiber Gundl. Repert. Fisico-Nat. Cuba, I, p. 395 (1866); *ib.*
J. f. O. 1874, p. 314 (Porto Rico); *ib.* 1875, p. 402 (Cuba); *ib.* Anal.
Soc Esp. Hist. Nat. VII, p. 418 (1878) (Porto Rico).
Dysporus leucogaster Sundev. Oefv. K. Vet. Akad. For. 1869, p. 591 (St
Bartholomew).
Sula leucogastra A. & E. Newton. Handb. Jamaica, p. 112 (1881).—
Cory, Bds. Haiti & San Domingo, p. 171 (1885); *ib.* List Bds. W. I.
p. 32 (1885).
Sula sula Ridgw. Pr. U. S. Nat. Mus. VIII, p. 356 (1885).

Sp. Char. *Adult:*—Head, throat, upper part of breast, and entire upper
plumage dark olive brown; underparts white; gular sac pale yellow; upper mandible greenish; feet pale yellowish green; iris yellowish.
Length, 27.00; wing, 15.50; tail, 8.00; tarsus, 1.60; bill, 4.00.

Habitat. Antilles.

Sula piscator (Linn.).

Pelecanus piscator Linn. Syst. Nat. I, 10th ed. p. 134 (1758); *ib.* 12th ed.
p. 217 (1766).
Sula parva? Gosse, Bds Jam. p. 219 (1847).—Wells, List Bds. Grenada, p. 11 (1886).

Sula piscator GOSSE, Bds. Jam. p. 418 (1847).—A. & E. NEWTON, Handb
Jamaica, p. 112 (1881).—CORY, List Bds. W. I. p. 32 (1885).—
WELLS, List Bds. Grenada, p. 11 (1886).
Dysporus hernandezi GUNDL. J. f. O. 1878, p. 298 (Cuba).
Dysporus piscator GUNDL. J. f. O. 1881, p. 401 (Cuba).

SP. CHAR. *Adult Male:*—General plumage white, showing a buff tinge on
the head and neck; shafts of the tail-feathers pale yellow; remiges
and most of the wing-coverts slaty gray, showing an ash tinge; feet
reddish.
Young in first plumage:—General plumage grayish brown above; dull
gray beneath, sometimes whitish; plumage very variable.
Length, 28.00; wing, 14.50; tail, 8.00; tarsus, 2.10; bill, 3.30.

HABITAT.   West Indies.

FAMILY PHAËTHONTIDÆ.

GENUS Phaëthon LINN.

Phaëthon LINNÆUS, Syst. Nat. 1736; *ib.* I, p. 134, 1758.

Phaëthon flavirostris BRANDT.

Phaëthon flavirostris BRANDT, Bull. Soc. Acad. St. Petersb. II, p. 349
(1837).—LAWR. Pr. U. S. Nat. Mus. I, p. 65 (1878) (Dominica).—
ib. p. 240 (Barbuda); *ib.* p. 359 (Martinique).—CORY, Bds. Baha-
ma I. p. 204 (1880); *ib.* Bds. Haiti & San Domingo, p. 175 (1885);
ib. List Bds. W. I. p. 33 (1885).
Phaëton flavirostris? BRYANT, Pr. Bost. Soc. Nat. Hist. VII, p. 128
(1859) (Bahamas).—*ib.* BREWER, p. 308 (1860) (Cuba).—GUNDL.
Repert. Fisico-Nat. Cuba, I. p. 395 (1866).—LAWR. Ann. Lyc. N.
Y. VIII, p. 103 (1867) (Sombrero).—GUNDL. J. f. O. 1874, p. 314
(Porto Rico); *ib.* 1875, p. 403 (Cuba); *ib.* 1878, p. 163 (Porto
Rico); *ib.* Anal. Soc. Esp. Hist. Nat. VII, p. 419 (1878) Porto
Rico).

Adult Male:—Bill pale orange yellow; general plumage white, sometimes slightly rosy-tinted; most of primaries showing much black;
a streak passing through the eye; some of the wing-coverts and
shafts of tail-feathers black; tail extended into two very long feathers which are reddened; tarsus bluish; iris black; webs and toes
black.
Length, including tail-feathers, 31.30; wing, 11.00; tail, 21.00;
tarsus, .90; bill, 2.00.

HABITAT.   Bahamas and Antilles.

Phaëthon æthereus LINN.

Phaëthon æthereus LINN. Syst. Nat. I, 10th ed. p. 134 (1758); *ib.* 12th
ed. p. 219 (1766).—GOSSE, Bds. Jam. p. 430 (1847).—SUNDEV. Oefv.
K. Vet. Akad. For 1869, p. 590 (St. Bartholomew); *ib.* p. 603
(Porto Rico).—LAWR. Pr. U. S. Nat. Mus. I, p. 195 (1878) (St.
Vincent); *ib.* p. 274 (Grenada); *ib.* p. 460 (Guadeloupe).—A. & E.
NEWTON, Handb. Jamaica, p. 112 (1881).—CORY, List Bds. W. I.
p. 33 (1885); *ib.* Ibis, 1886, p. 474 (La Desirade).—WELLS, List
Bds. Grenada. p 11 (1886).
Phaëton æthereus D'ORB. in La Sagra's Hist. Nat. Cuba, Ois, p. 312
(1840).—GUNDL. J. f. O. 1862, p. 96 (Cuba); *ib.* ALBRECHT, p. 207
(Jamaica).

SP. CHAR.—Bill red; General plumage white; a black crescent in front
of the eye; a stripe extending from the eye to the occiput; outer
webs of outer primaries, and most of the primary coverts, black;
rest of upper surface irregularly barred with dull black; flanks
striped; elongated central tail-feathers white, basal portion of the
shafts black; tarsus yellowish orange this color reaching to the
first joint of the toes, including the web between the inner and hind
toes; rest of feet black

Length, 31.00 wing, 12.00 bill, 2.45.

Recorded from Cuba, Jamaica, Porto Rico, St. Vincent,
Grenada, Gaudeloupe, St. Bartholomew, and La Desirade.

FAMILY RYNCHOPIDÆ.

GENUS Rynchops LINN.

Rynchops LINNÆUS. Syst. Nat. I, 10th ed. p. 228, 1758; *ib.* 12th ed. p.
228, 1776.

Rynchops nigra LINN.

Rynchops nigra LINN. Syst. Nat. I, 10th ed. p. 228 (1758); *ib.* 12th ed. p.
228 (1766).—A. & E. NEWTON, Ibis, 1859, p. 371 (St. Croix)?—CORY.
List Bds. W. I. p. 33 (1885).
Rhyncops nigra GUNDL. Repert. Fisico-Nat. Cuba, I, p. 393 (1866); *ib.*
J. f. O. 1875, p. 395 (Cuba).

Accidental in Cuba and St. Croix.

FAMILY LARIDÆ.

GENUS Larus LINN.

Larus LINNÆUS, Syst. Nat. I. p. 136, 1758.

Larus atricilla LINN.

Larus atricilla LINN. Syst. Nat. I, 10th ed. p. 136 (1758); *ib.* 12th ed. p.
225 (1766).—D'ORB. in La Sagra's Hist. Nat. Cuba, Ois. p. 315
(1840).—BRYANT. Pr. Bost. Soc. Nat. Hist. VII, p. 134 (1859) (Bahamas); *ib.* BREWER, p. 308 (1860) (Cuba).—SUNDEV. Oefv. K.
Vet. Akad. For. 1869, p. 590 (St. Bartholomew); *ib.* p. 603 (Porto
Rico).—LAWR. Pr. U. S. Nat. Mus. I, p. 238 (1878) (Antigua); *ib.*
p 142 (Barbuda); *ib.* p. 277 (Grenada); *ib.* p. 462 (Gaudeloupe).—
CORY, Bds. Bahama I. p. 208 (1880).—A. & E. NEWTON, Handb.
Jamaica, p. 117 (1881).—CORY, Bds. Haiti & San Domingo, p. 177
. (1885); *ib.* List Bds. W. I. p. 33 (1885).—WELLS, List Bds. Gren-
ada, p. 11 (1886).
Nema atricilla GOSSE. Bds. Jam. p. 437 (1847).—ALBRECHT, J. f. O. 1862,
p. 207 (Jamaica).
Chrœcocephalus atricilla A. & E. NEWTON. Ibis, 1859 p. 371 (St. Croix).
Chroicocephalus atricilla GUNDL. Repert. Fisico-Nat. Cuba, I, p. 391
(1866); *ib.* J. f. O. 1874, p. 314 (Porto Rico); *ib.* 1875. p. 385
(Cuba); *ib.* Anal. Soc. Esp. Hist. Nat. VII, p. 408 (1878) (Porto
Rico).

Common throughout the West Indies.

Larus argentatus BRÜNN.

Larus argentatus BRÜNN. Orn. Bor. p. 44 (1764).—BREWER, Pr. Bost. Soc.
Nat. Hist. VII. p. 308 (1860) (Cuba).—ALBRECHT. J. f. O. 1861, p.
215 (Cuba).—CORY, List Bds. W. I. p. 33 (1885).
Larus marinus LEMB. Aves Cuba, p. 122 (1850).- BREWER, Pr. Bost. Soc.
Nat. Hist. VII, p. 308 (1860) (Cuba).—GUNDL. J. f. O. 1862, p. 95
(Cuba)
Larus zonorhynchus GUNDL. J. f. O. 1862. p. 94 (?) (Cuba).
Larus smithsonianus GUNDL. Repert. Fisico-Nat. Cuba, I, p. 391 (1866);
ib. J. f. O. 1875. p. 387 (Cuba).

Cuba and Bahamas in winter.

Larus franklinii SWAINS.

Larus franklinii SWAINS. & RICH. F. B. A. II. p. 424, pl. 71 (1831).
Larus franklini SUNDEV. Oefv. K. Vet. Akad. For. 1869, p. 390 (St. Bartholomew).—CORY, List Bds. W. I. p. 33 (1885).

Recorded from St. Bartholomew.

Larus philadelphia is claimed to have been *seen* at Long Island, Bahamas. There is no actual record of the capture of this species in the West Indies.

GENUS Gelochelidon BREHM.

Gelochelidon BREHM, Naturg. Vög. Deutschl. 1831, p. 774.

Gelochelidon nilotica (HASSELQ.).

Sterna nilotica HASSELQ. Reise nach Pal. Deutschl. Ausg. 1762, p. 325.
Sterna anglica MONT. Orn. Dict. Suppl. 1813. — D'ORB. in La Sagra's Hist. Nat. Cuba, Ois. p. 321 (1840).—MOORE, Pr. Bost. Soc. Nat. Hist. XIX, p. 141 (1877) (Bahamas).—CORY, Bds. Bahama I. p. 209 (1880); *ib.* List Bds. W. I. p. 33 (1885).
Gelochelidon aranea BREWER, Pr. Bost. Soc. Nat. Hist. VII, p. 308 (1860) (Cuba).—GUNDL. J. f. O. 1862, p. 91 (Cuba).
Gelochelidon anglica GUNDL. Repert. Fisico-Nat. Cuba, I, p. 392 (1866); *ib.* J. f. O. 1875, p. 388 (Cuba).
Gelochelidon nilotica STEJN. Auk, I, p. 366 (1884).

Bahamas and Antilles.

GENUS Sterna LINN.

Sterna LINNÆUS. Syst. Nat. I, ed. 10, p. 137 (1758); *ib.* ed. 12, p. 227 (1766).

Sterna maxima BODD.

Sterna maxima BODD. Tabl. Pl. Enl. p. 58 (1783). — SAUNDERS, P. Z. S. 1876, p. 655 (W. I.).—LAWR. Pr. U. S. Nat. Mus. I, p. 198 (1878) (St. Vincent): *ib.* p. 488 (Antigua); *ib.* p. 242 (Barbuda); *ib.* p. 277 (Grenada); *ib.* p. 462 (Gaudeloupe).—A. & E. NEWTON, Handb. Jamaica, p. 117 (1881).—CORY, Bds. Haiti & San Domingo, p. 178 (1885); *ib.* List. Bds. W. I. p. 33 (1885).
Sterna cayennensis D'ORB. in La Sagra's Hist. Nat. Cuba, Ois. p. 319 (1840).
Thalassens cayanus GOSSE, Bds. Jam. p. 431 (1847).
Thalasseus regius A. & E. NEWTON, Ibis, 1859, p. 371 (St. Croix).— GUNDL. Repert. Fisico-Nat. Cuba, I, p. 392 (1866).—LAWR. Ann. Lyc. N. Y. VIII, p. 103 (1867) (Sombrero).— GUNDL. J. f. O. 1874, p. 314 (Porto Rico): *ib.* 1875. p. 388 (Cuba); *ib.* Anal. Soc. Esp. Hist. Nat. VII, p. 410 (1878) (Porto Rico).
Sterna regia BRYANT, Pr. Bost. Soc. Nat. Hist. VII, p. 134 (1859) (Bahamas); *ib.* XI, p. 98 (1867) (San Domingo).—SCL. P. Z. S. 1861,

p. 82 (Jamaica).—CORY. Bds. Bahama I. p. 210 (1880).—WELLS,
List Bds. Grenada. p. 11 (1886).
Gelochelidon cayennensis BREWER, Pr. Bost. Soc. Nat. Hist. VII, p. 368
(1860) (Cuba).

Bahamas and Antilles.

Sterna sandvicensis acuflavida.

Sterna cantiaca GMEL. Syst. Nat. I, p. 606 (1788)?—CORY, Bds. Bahama
I, p. 211 (1880) ; *ib.* List Bds. W. I. p. 33 (1885).
Sterna acuflavida CABOT, Pr. Bost. Soc. Nat. Hist. II, p. 257 (1847).—
BRYANT, *ib.* VII, p 134 (1859) (Bahamas).
Thalasseus acuflavidus BREWER, Pr. Bost. Soc. Nat. Hist. VII, p. 308
(1860) (Cuba).—GUNDL. Repert. Fisico-Nat. Cuba, I. p. 392 (1856) ;
ib. J. f. O. 1874, p. 311 (Porto Rico) ; *ib* 1875. p. 390 (Cuba) ; *ib.*
Anal. Soc. Esp. Hist. Nat. VII. p. 411 (1878) (Porto Rico).
Thalasseus acuflavida GUNDL. J. f. O. 1862. p. 94 (Cuba).
Sterna sandvicensis acuflavida RIDGW. Water Bds. N. Am. II, p. 288
(1884).

Bahamas and Antilles.

Sterna hirundo LINN.

Sterna hirundo LINN. Syst. Nat. I, ed. 10. p. 137 (1758) ; *ib.* ed. 12, p. 227
(1766).—CORY, Bds. Bahama I. p. 211 (1880) ; *ib.* List Bds. W. I.
p. 33 (1885).
Sterna wilsoni BRYANT, Pr. Bost. Soc. Nat. Hist. VII, p. 134 (1859) (Bahamas.)

Accidental in the Bahama Islands.

Sterna anosthætus SCOP.

Sterna anosthætus SCOP. Del. Faun. et Flor. Ins. II, No. 72 (1786).
Haliplana discolor COUES, Ibis, 1864, p. 392.—LAWR. Ann. Lyc. N. Y.
VIII, p. 104 (1867) (Sombrero).
Sterna anosthæta CORY, Bds. Bahama I. p. 215 (1880) ; *ib.* List Bds.
W. I. p. 33 (1885).—WELLS, List Bds. Grenada, p. 11 (1886).
Haliplana anæstheta GUNDL. J. f. O. 1881, p 400 (Cuba).

SP. CHAR. — Bill black ; cap black: forehead white, *extending like two
horns over each eye and reaching behind them ;* upper back grayish,
shading into the white on the sides of the neck ; upper plumage
grayish brown ; underparts white ; primaries dark brown, the first

and second showing a clear band of white on the inner webs, not reaching within an inch of the tips, and gradually fading on the others; upper tail-coverts slaty gray; outer tail-feathers *almost entirely white*, showing a tinge of brownish near the tip; legs and feet black; iris brown.

Length, 14.25; wing, 10.00; tail, 6.25; tarsus, .70; bill, 1.50.

Common in the Bahama Islands; breeds. Cuba, Sombrero; probably occurs throughout the West Indies.

Sterna fuliginosa Gmel.

Sterna fuliginosa Gmel. Syst. Nat. I, p. 605 (1788). — D'Orb. in La Sagra's Hist. Nat. Cuba, Ois. p. 319 (1840). — Bryant, Pr. Bost. Soc. Nat. Hist. VII, p. 134 (1859) (Bahamas); *ib*. XI, p. 98 (1867) (San Domingo). — Sundev. Oefv. K. Vet. Akad. For. 1869, p. 589 (St. Bartholomew); *ib*. p. 603 (Porto Rico).—Lawr. Pr. U. S. Nat. Mus. I, p. 68 (1878) (Dominica); *ib*. p. 277 (Grenada); *ib*. p. 462 (Guadeloupe). — Cory, Bds. Bahama I. p. 214 (1880).— A. & E. Newton, Handb. Jamaica, p. 117 (1881). — Cory, Bds. Haiti & San Domingo, p. 181 (1885); *ib*. List Bds. W. I. p. 33 (1885).— Wells, List Bds. Grenada, p. 11 (1886).
Hydrochelidon fuliginosa Gosse, Bds. Jam. p. 433 (1847). — Brewer, Pr. Bost. Soc. Nat. Hist. VII, p. 308 (1860) (Cuba).—Albrecht. J. f. O. 1862, p. 207 (Jamaica).
Onychoprion fuliginosus A. & E. Newton, Ibis, 1859. p. 371 (St. Croix)? —Cassin, Pr. Acad. Nat. Sci. Phila, 1860, p. 379 (St. Thomas).
Haliplana fuliginosa Gundl. Repert. Fisico-Nat. Cuba, I, p. 393 (1866); *ib*. J. f. O. 1875. p. 393 (Cuba); *ib*. 1878, p. 163 (Porto Rico); *ib*. Anal. Soc. Esp. Hist. Nat. VII, p. 414 (1878) (Porto Rico).

Bahamas and Antilles.

Sterna dougalli Mont.

Sterna dougalli Mont. Orn. Dict. Suppl. (1813). — Sundev. Oefv. Vet. Akad. For. 1869, p. 589 (St. Bartholomew). — Lawr. Pr. U. S. Nat. Mus. I. p. 488 (1878) (Dominica); *ib*. p. 238 (Antigua) *ib*. p. 277 (Grenada); *ib*. p. 360 (Martinique); *ib*. p. 462 (Guadeloupe). —Cory, List Bds. W. I. p. 33 (1885).—Wells. List Bds. Grenada, p. 11 (1886).
Sterna paradisea Gundl. Repert. Fisico-Nat. Cuba, I, p. 392 (1866); *ib*. J. f. O. 1875, p. 391 (Cuba); *ib*. 1878, p. 163 (Porto Rico); *ib*. Anal. Soc. Esp. Hist. Nat. VII, p. 411 (1878) (Porto Rico).—Cory, Bds. Bahama I. p. 212 (1880).

Common throughout the West Indies.

Sterna antillarum (LESS.).

Sternula antillarum LESS. Descr. Mam. et Ois. p. 256 (1847).
Sterna argentea GOSSE, Bds. Jam. p. 437 (2847). — ALBRECHT, J. f. O.
    1862, p. 207 (Jamaica).
Sterna minuta LEMB. Aves Cuba, p. 123 (1850).
Sternula frenata BREWER, Pr. Bost. Soc. Nat. Hist. VII, p. 308 (1860)
    (Cuba).—GUNDL. J. f. O. 1862, p. 93 (Cuba).
Sterna antillarum GUNDL. Repert. Fisico-Nat. Cuba, I, p. 393 (1866).—
    BRYANT, Pr. Bost. Soc. Nat. Hist. XI. p. 98 (1867) (San Domingo).
    —GUNDL. J. f. O. 1874, p. 314 (Porto Rico): *ib.* 1875, p. 391 (Cuba);
    ib. 1878, p. 163 (Porto Rico); *ib.* Anal. Soc. Esp. Hist. Nat. VII,
    p. 412 (1878) (Porto Rico). — SAUNDERS, P. Z. S. 1876, p. 661
    (Antilles).—LAWR. Pr. U. S. Nat. Mus. I, p. 68 (1878) (Dominica).
    — A. & E. NEWTON, Handb. Jamaica, p. 117 (1881). — CORY, Bds.
    Haiti & San Domingo, p. 179 (1885); *ib.* List Bds. W. I. p. 33
    (1885).
Sterna minuta americana SUNDEV. Oefv. K. Vet. Akad. For. 1869, p. 589
    (St. Bartholomew).
Sterna superciliaris CORY, Bds. Bahama I. p. 213 (1880).

Bahamas and Antilles.

GENUS Hydrochelidon BOIE.

Hydrochelidon BOIE, Isis, 1822, p. 563.

Hydrochelidon nigra surinamensis (GMEL.).

Rallus lariformis LINN. Syst. Nat. I, 10th ed. p. 153 (1758)?
Sterna surinamensis GMEL. Syst. Nat. I, 2nd part. p. 604 (1788).
Hydrochelidon nigra GOSSE, Bds. Jam. p. 437 (1847).—ALBRECHT. J. f. O.
    1862. p. 207 (Jamaica).—SAUNDERS, P. Z. S. 1876, p. 642 (W. I.).—
    A. & E. NEWTON, Handb. Jamaica. p. 117 (1881).
Sterna nigra LEMB. Aves Cuba, p. 124 (1850).
Hydrochelidon surinam BREWER, Pr. Bost. Soc. Nat. Hist. VII, p. 308
    (1860) (Cuba).
Hydrochelidon plumbea GUNDL. J. f. O. 1862, p. 93 (Cuba).
Hydrochelidon fissipes GUNDL. Repert. Fisico-Nat. Cuba. I, p. 393 (1866);
    ib. J. f. O. 1875. p. 393 (Cuba); *ib.* Anal. Soc. Esp. Hist. Nat. VII,
    p. 413 (1878) (Porto Rico).
Hydrochelidon lariformis COUES, Bds. N. W. p. 704 (1874).—CORY, List
    Bds. W. I. p. 34 (1885).
Hydrochelidon nigra surinamensis STEJN. Pr. U. S. Nat. Mus. V, p. 40
    (1882).

Antilles in winter.

Hydrochelidon leucoparcia (Natt.) is given by Mr. Ridgway as
accidental in the West Indies. (Ridgw. Man. N. A. Bds. p. 47, 1887.)

GENUS **Anous** LEACH.

Anous LEACH, Shaw's Gen. Zool. XIII, p. 139, 1826.

Anous stolidus (LINN.).

Sterna stolida LINN. Syst. Nat. I. 10th ed. p. 137 (1758); *ib.* 12th ed. p. 227 (1766).—D'ORB. in La Sagra's Hist. Nat. Cuba, Ois. p. 317 (1840).

Megalopterus stolidus GOSSE, Bds. Jam. p. 434 (1847).

Anous stolidus BRYANT, Pr. Bost. Soc. Nat. Hist. VII, p. 134 (1859)(Bahamas); *ib.* BREWER. p. 308 (1860) (Cuba).—GUNDL. Repert. Fisico-Nat. Cuba, I, p. 393 (1866).—LAWR. Ann. Lyc. N. Y. VIII, p. 105 (1867)(Sombrero).—SUNDEV. Oefv. K. Vet. Akad. For. 1869, p. 590 (St. Bartholomew).—GUNDL. J. f. O. 1875, p. 395 (Cuba) *ib.* Anal. Soc. Esp. Hist. Nat. VII, p. 415 (1878) (Porto Rico).—LAWR. Pr. U. S. Nat. Mus. I, p. 488 (1878) (Dominica): *ib.* p. 277 (Grenada). —CORY. Bds. Bahama I. p. 216 (1880).—GRISDALE, Ibis, 1882, p. 486 (Montserrat).—CORY, Bds. Haiti & San Domingo, p. 182 (1885); *ib.* List Bds. W. I. p. 34 (1885).—WELLS, List Bds. Grenada, p. 12 (1886).

Sterna (Anous) stolida BRYANT, Pr. Bost. Soc. Nat. Hist. XI, p. 97 (1867) (San Domingo).

Abundant in the Bahamas and Antilles.

FAMILY PROCELLARIIDÆ.

GENUS **Oceanites** KEYS. & BLAS.

Oceanites KEYSERLING & BLASIUS, Wirb Eur. I, p. xciii, 1840.

Oceanites oceanicus (KUHL).

Procellaria oceanica KUHL, Beitr. Zool. 1820, Mon. Proc. p. 136, pl. 10, fig. 1.

Thalassidroma wilsonii BRYANT, Pr. Bost. Soc. Nat. Hist., VII, p. 131 (1859) (Bahamas).

Oceanites wilsoni GUNDL. Repert. Fisico-Nat. Cuba, I, p. 394 (1866); *ib.* J. f. O. 1875, p. 396 (Cuba).

Oceanites oceanica CORY, Bds. Bahama I. p. 218 (1880).

Oceanites oceanicus CORY, List Bds. W. I. p. 34 (1886).—WELLS, List Bds. Grenada, p. 12 (1886).

Bahamas and Antilles.

Genus Æstrelata Bp.

Æstrelata Bonaparte, Consp. Avium, II, p. 188, 1856.

Æstrelata jamaicensis (Bancr.).

Procellaria jamaicensis Bancr. Zool. Journ. V, p. 81 (1828).
"*Œstrelata caribæa* Auct."
Blue Mountain Duck Gosse. Bds. Jam. p. 437 (1847).
Pterodroma caribbæa Carte, P. Z. S. 1866, p. 93, pl. X.
Petrodroma caribbæus Gray, Handl. Bds. III, p. 107 (1871).
Œstrelata jamaicensis A. & E. Newton. Handb. Jamaica, p. 117 (1881).—
    Bd. Bwr. & Ridgw. Hist. N. Am. W. Bds. II, p. 394 (1884).—Cory,
    List. Bds. W. I. p. 34 (1885).

Sp. Char.   *Male:*—General plumage dark sooty brown, paling slightly
    on the chin, forehead and upper part of the back, joining the neck;
    rump brownish black; upper tail-coverts dull white; quills and tail
    brownish black; bill and feet black.
      The sexes are similar.
      Length, 14.00; wing, 11.00; tail, 4.75; tarsus, 1.40; bill, 1.25 to
    1.50.

Habitat.   Jamaica.

Æstrelata hasitata (Kuhl).

Procellaria hasitata "Kuhl, Mon. Proc. Beitr. Zool. p. 142. No. 11 (1820)."
? *Procellaria diabolica* L'Herminier, MSS.—Lawr. Pr. U. S. Nat.
    Mus. 1879, p. 450 (Gaudeloupe).
Œstrelata hesitata Bd. Bwr. & Ridgw, Hist. N. Am. W. Bds. II, pp.
    394-395 (1884) (Haiti).—Cory, List Bds. W. I. p. 34 (1885).

It is probable that this bird is occasionally to be found in the
West Indies. One specimen, claimed to have been taken near
Haiti, is now in the British Museum.

Genus Puffinus Briss.

Puffinus Brisson, Orn. VI, p. 131, 1760.

Puffinus major Faber.

Puffinus major Faber. Prodr. Isl. Orn. p. 56 (1822).—Cory, Bds. Bahama
    I. p. 218 (1880).—*ib.* List Bds. W. I. p. 34 (1885).

Recorded from the Bahama Islands.

Puffinus auduboni Finsch.

Puffinus obscurus Bryant, Pr. Bost. Soc. Nat. Hist. VII, p. 132 (1859)
    (Bahamas).—Scl. P Z. S. 1879, p. 765 (Montserrat).—Cory, Bds.

Bahama I. p. 219 (1880); *ib.* Bds. Haiti & San Domingo, p. 184 (1885).

Procellaria obscura BRYANT, Pr. Bost. Soc. Nat. Hist. XI, p. 98 (1867) (San Domingo).

Puffinus auduboni FINSCH, P. Z. S. 1872, p. 111 (Bahamas). — GUNDL. J. f. O. 1881, p. 400 (Cuba). — CORY, List Bds. W. I. p. 34 (1885).

SP. CHAR. — Above glossy brown, shading into grayish upon the sides of the breast; below white; crissum brown and white; tail brown, the feathers faintly tipped with ashy; bill lead-color.

Length, 12.50; wing, 8.00; tail, 4.25; tarsus, 1.60; bill, 1.30.

Common among the Bahama Islands and the Greater Antilles, but shy and somewhat difficult to procure. They breed in holes in the rocks. Incubation commences about March 15.

FAMILY PODICIPIDÆ.

GENUS **Podiceps** LATH.

Podiceps LATHAM, Ind. Orn. II, p. 780, 1790.

Podiceps dominicus (LINN.).

Colymbus dominicus LINN. Syst. Nat. I, p. 223 (1766). — GUNDL. J. f. O. 1856 p. 430 (Cuba).

Colymbus dominicensis D'ORB. in La Sagra's Hist. Nat. Cuba, Ois. p. 282 (1840).

Podiceps dominicus GOSSE, Bds. Jam. p. 440 (1847). — SALLÉ, P. Z. S. 1857, p.237 (San Domingo). — SCL. P. Z. S. 1861, p. 82 (Jamaica). — MARCH, Pr. Acad. Nat. Sci. Phila. 1864, p. 70 (Jamaica). — GUNDL. Report. Fisico-Nat. Cuba, I, p. 386 (1866). — BRYANT, Pr. Bost. Soc. Nat. Hist. XI, p. 97 (1867) (San Domingo). — GUNDL. J. f. O. 1875, p. 365 (Cuba); *ib.* Anal. Soc. Esp. Hist. Nat. VII, p. 395 (1878) (Porto Rico). — CORY, Bds. Bahama I. p. 222 (1880). — A. & E. NEWTON, Handb. Jamaica, p. 117 (1881). — CORY, Bds. Haiti & San Domingo, p. 185 (1885); *ib.* List Bds. W. I. p. 34 (1885).

Sylbeocyclus dominicus BREWER, Pr. Bost. Soc. Nat. Hist. VII, p. 308 (1860) (Cuba).

Podilymbus dominicus TAYLOR, Ibis, 1864, p. 172 (Porto Rico).

WINTER PLUMAGE. *Male:* — Above dark brown with slight greenish reflections; sides of the head and throat ashy gray, continuous in a broad band around the neck; underparts silky white, mottled with dusky; outer primaries showing chocolate-brown, the others and secondaries white.

Length, 9.35; wing, 3.00; tarsus, 1.24; bill, .85.

HABITAT.　Bahamas and Greater Antilles.

Colymbus holboellii (Reinh.) is included by Mr. Wells in his list of the birds of Grenada (*Podiceps holbolli?* Wells, List of the Birds of Grenada, p. 12, 1886), but is probably some other species wrongly identified.

GENUS Podilymbus LESS.

Podilymbus LESSON, Traité d'Orn. I, p. 595 (1831).

Podilymbus podiceps (LINN.).

Colymbus podiceps LINN. Syst. Nat. I, p. 223 (1766). — SUNDEV. Oefv. K. Vet. Acad. For. 1869, p. 603 (Porto Rico).

Colymbus carolinensis D'ORB. in La Sagra's Hist. Nat. Cuba, Ois. p. 285 (1840).

Podilymbus carolinensis GOSSE, Bds. Jam. p. 438 (1847).

Sylbeocyclus carolinensis GUNDL. J. f. O. 1856, p. 431 (Cuba). — BREWER, Pr. Bost. Soc. Nat. Hist. VII, p. 308 (1860) (Cuba).

Podilymbus podiceps SCL. P. Z. S. 1861, p. 82 (Jamaica). — MARCH, Pr. Acad. Nat. Sci. Phila. 1864, p. 70 (Jamaica). — GUNDL. Repert. Fisico-Nat. Cuba, I, p. 386 (1866); *ib.* J. f. O. 1875, p. 307 (Cuba). — LAWR. Pr. U. S. Nat. Mus. I, p. 488 (1878) (St. Vincent and Grenada); *ib.* p. 242 (Barbuda). — GUNDL. Anal. Soc. Esp. Hist. Nat. VII, p. 397 (1878) (Porto Rico). — SCL. P. Z. S. 1870, p. 765 (Montserrat). — A. & E. NEWTON, Handb. Jamaica, p. 117 (1881). — CORY, List Bds. W. I. p. 34 (1885). — WELLS, List Bds. Grenada, p. 12 (1886).

Numerous West **Indian records.**

APPENDIX.

Mimocichla ravida CORY.

Mimocichla ravida CORY, Auk, III, pp. 499, 501 (1886).

SP. CHAR. — Bill large; general plumage dull, ashy, or brownish plumbeous; no stripes on the throat, which is the same color as the breast; a patch of dull white on the vent and under tail-coverts; three outer tail-feathers tipped with white on the inner webs; bill, bare space around the eye, and legs orange red; iris dull red.

Length, 9.50; wing, 5.25; tail, 4.40; tarsus, 1 50; bill, .90.

HABITAT. Grand Cayman.

Margarops montanus albiventris (LAWR.).

Margarops albiventris LAWRENCE, Ann. New York Acad. of Sciences, vol. IV, p. 23 (1887).

This form is described as somewhat smaller than the type, and darker above, with more white, and less brown, spots on the under parts.

It is very closely allied to the true *montanus*, from which the propriety of separating it is doubtful.

HABITAT. Grenada.

Margarops montanus rufus CORY.

Margarops montanus rufus CORY, Auk, vol. V, pt. 1 (1888).

SP. CHAR. — Slightly larger than the type, and much lighter brown, being light reddish brown instead of dark brown. Apparently a light-colored northern race of *M. montanus*.

HABITAT. Dominica.

Rhamphocinclus sanctæ-luciæ CORY.

Rhamphocinclus sanctæ-luciæ CORY, Auk, p. 94 (1887).
Rhamphocinclus brachyurus SCLATER, P. Z. S. (1871) p. 268. — SCLATER & SEMBER, P. Z. S. (1872) p. 648. — ALLEN, Bull. Nutt. Orn. Club, vol. V, p. 166 (1880).

(285)

Sp. Char. ♂. — Top of the head dark brown, showing a dull rufous tinge; back and rump rufous brown; lores and below the eye black, shading into brown on the ear-coverts; throat and breast pure white; belly white; sides of the body chocolate brown; wing and tail dark brown; bill very dark, nearly black; legs olive brown.

Length, 8; wing, 3.10; tail, 3.60; tarsus, 1.15; bill, .85.

Habitat.   St. Lucia, West Indies.

The St. Lucia bird differs from that found in Martinique, in having the upper parts brown, instead of dark slate-color.   The brown marking on the sides of the body is of a different shade, the black on the lores is more extended, the tail is broader, and the bird, generally, somewhat larger.

Thryothorus guadeloupensis Cory.

Thryothorus guadeloupensis Cory, Auk, III, p. 31 (1886); *ib.* Ibis (1886), pp. 471, 474.

Sp. Char. — Upper parts dark brown, showing darkest on the head; wing-coverts tipped with rufous brown; primaries and secondaries dark brown, the outer webs mottled with reddish brown, showing pale, indistinct bands on some of the inner secondaries; entire underparts tawny brown; under tail-coverts tawny, heavily marked with dark brown; bill yellowish brown, under mandible quite pale; legs and feet pale; iris yellow.

The tail of the specimen above described is lacking.

Length, —; wing, 1.95; tarsus, .78; bill, .75.

Habitat.   Grand Terre, Guadeloupe.

Dendroica vitellina Cory.

Dendroica vitellina Cory, Auk, III, pp. 497, 501 (1886). — Ridgw. Proc. U. S. Nat. Mus. p. 574 (1887).

Sp. Char. — Somewhat resembling *D. discolor* in general appearance, but larger, and having the entire underparts bright yellow, with no black streaks on the sides, but showing traces of olive on the sides and flanks; upper parts dull green, pale yellowish green on the rump; quills dark brown, edged with pale yellowish green on outer webs, inner webs edged with dull white; a distinct wing band of yellow; tail-feathers edged with yellowish green on the outer webs; two outer feathers heavily marked with white on the terminal portion of the inner webs, narrowly showing on the third feather; a superciliary line of bright yellow nearly, if not quite, reaching the occiput; bill horn color; feet dull black.

Length, 4.30; wing, 2.50; tail, 2.10; tarsus, .70; bill, .50.

Habitat.   Grand Cayman.

Dendroica aurocapilla Ridgw.

Dendroica aurocapilla Ridgw. Proc. U. S. Nat. Mus. p. 572 (1887).
Dendroica petechia gundlachi Cory, Auk, p. 501 (1886).

Sp. Char. — " Similar to *D. ruficertex* Ridgw. from Cozumel, but crown much paler (orange instead of rufous), and rufous chestnut streaks of breast, etc., much narrower." (Ridgw. l. c.)

Habitat. Island of Grand Cayman.

Geothlypis coryi Ridgw.

Geothlypis coryi Ridgw. Auk, III, pp. 334, 335 (1886).

Sp. Char. — " In plumage much resembling *G. beldingi nobis* (from lower California), but yellow of lower parts with less of an orange tint, the sides and upper parts without any olive brown tinge; the flanks bright greenish-yellow, and the yellow posterior border to the black 'mask' much narrower and less purely yellow. Form very different, the bill about twice as large and of different shape. Female very different from that of any other known species, being bright olive green above and entirely pure gamboge yellow below; with ashy auriculars and yellowish forehead and superciliary stripe.

" *Adult male* (type No. 107,876, U. S. Nat. Mus. Eleuthera I., Bahamas, March 12, 1886, Chas. H. Townsend) : — Above, bright olive green, very slightly tinged with ashy on top of head; lower parts, including flanks, entirely rich gamboge yellow; forehead (back to about .35 from nostril), lores, orbital region, malar region, and auriculars, uniform deep black, bordered posteriorly by gamboge yellow (less distinct across crown); bill blackish, paler along tomia and at base of lower mandible; legs and feet light brown; wing, 2.60; tail, 2.50; culmen, .75; bill from nostril, .45; depth of bill at base, .20; width, .20; tarsus, .90.

" *Adult female* (No. 107,875, U. S. Nat. Mus. same locality and date, J. E. Benedict) : — Similar to the male, except in color of the head, which lacks entirely any black, the forehead, cheeks, and superciliary region being olive yellowish, lores grayish, and auriculars ashy; flanks and under tail-coverts rather paler and more olivaceous yellow than in the male. Wing, 2.45; tail, 2.50; culmen, .75; bill from nostril, .45; depth at base, .20; width, .18; tarsus, .87." (Ridgw. orig. descr. l. c.)

Habitat. Eleuthera, Bahama Islands.

Geothlypis tanneri Ridgw.

Geothlypis tanneri Ridgw. Auk, III, p. 335 (1886).

Sp. Char. — " Similar to *G. coryi*, but bill more robust and straighter, black of forehead more extended, yellow posterior border to 'mask' paler, and changing to yellowish gray across crown, olive green of upper parts much duller, and yellow of lower parts less intense

"*Adult male* (type No. 108,402, U. S. Nat. Mus. Abaco I., Bahamas, April 3, 1886, Chas. H. Townsend) : — Wing, 2.65 ; tail, 2.50 ; culmen, .75 ; bill from nostril, .43 ; depth at base, .22 ; width, .22 ; tarsus, .88.

" *Adult female* (No. 108,496, same locality and date, Willard Nye) : — In plumage, nearly intermediate between the same sex of *G. rostratus* and *G. coryi*, having more and brighter yellow on lower parts than the former, and less than the latter ; head, however, more as in *G. rostratus*, the distinct yellow superciliary stripe of *coryi* being absent, and the forepart of crown tinged with reddish brown. Wing, 2.35 ; tail, 2.50 ; culmen, .70 ; bill from nostril, .42 ; depth at base, .22 ; width, .22 ; tarsus, .85." (RIDGW. orig. descr. l. c.)

HABITAT.   Abaco, Bahama Islands.

Certhiola sharpei CORY.

Certhiola sharpei CORY, Auk, III, pp. 497, 501 (1886).— RIDGW. Proc. U. S. Nat. Mus. p. 574 (1887).

SP. CHAR. — Throat ash gray, darker than in *C. caboti* or *C. bahamensis*, but much lighter than *C. flaveola;* underparts yellow, brightest upon the breast, and dullest, with a slight olive tinge, on the belly and flanks ; top of the head and stripe through the eye dull black ; superciliary stripe white ; back dull black, showing an ashy tinge ; rump yellow ; quills dark brown, edged with white ; carpus edged with bright yellow ; tail tipped with white, heaviest on the outer feathers ; bill and feet dull black.

Length, 4.10 ; wing, 2.45 ; tail, 1.80 ; tarsus, .80 ; bill, .52.

HABITAT.   Grand Cayman.

Vireo caymanensis CORY.

Vireo caymanensis CORY, Auk (1887), p. 6. — RIDGW. Proc. U. S. Nat. Mus. p. 573 (1887).

SP. CHAR. — (♂ Coll. C. B. Cory, No. 6273.) Upper parts dull olive, brightest on the rump and upper tail-coverts ; crown darker than the back, showing a slight brownish tinge ; underparts dull yellowish white, faintly tinged with olive on the sides and flanks ; upper throat dull white ; a dull white superciliary stripe from the upper mandible ; a stripe of slaty brown from the upper mandible passing through and back of the eye ; quills dark brown, narrowly edged with dull green on the outer webs, most of the inner feathers showing a white edging on the basal portion of the inner webs ; tail dull olive brown, the feathers showing green on the edges ; upper mandible dark ; lower mandible pale ; feet slaty brown.

Length, 5.40 ; wing, 2.75 ; tail, 2.25 ; tarsus, .75 ; bill, .52.

HABITAT.   Island of Grand Cayman, West Indies.

This interesting species seems to be restricted to Grand Cayman. It frequents the large trees on the north side of the island.

Calliste cucullata (SWAIN.).

Aglaia cucullata " SWAINSON, Orn. Dr. pl. 7."
Calliste cucullata GRAY, Gen. B. II, p. 366; SCLATER, Contr. Orn. p. 63 (1851);
id. Syn. Av. Tan. p. 79; id. Cat. Bds. Brit. Mus. p. 113 (1886); WELLS,
List Bds. Grenad. p. 3 (1886).
Calliste versicolor LAWRENCE, Proc. U. S. Nat. Mus. vol. IX, p. 613.

SP. CHAR. — Top of the head dark chestnut; back distinctly green, with golden
reflections; ear-coverts black; underparts ochraceous, showing a faint
bluish tinge; cressum tinged with rufous brown; under wing-coverts white.
Length, 5.25; wing, 2.30; tail, 2.10; tarsus, .60; bill, .35.

HABITAT. Grenada.

It has been supposed that Swainson's type of this species came
from Venezuela; but it is probable that it is restricted to the island
of Grenada, and does not occur at all on the continent.

Spindalis salvini CORY.

Spindalis salvini CORY, Auk, III, pp. 499, 501 (1886).

SP. CHAR. — Top of the head and cheeks black; a superciliary and malar stripe
of white; a patch of yellowish orange on the upper throat, separated from
the white malar stripe by a line of black which reaches the breast; chin
dull white; lower throat and breast chestnut; back dark olive, separated
from the black of the head by a chestnut collar; lower back yellowish; a
patch of chestnut on the carpus; rump brownish orange; belly and under
tail-coverts dull white; tail black, the two outer feathers heavily marked
with white, the third feather tipped with white on the inner web, the two
central feathers narrowly edged with white on the inner webs; bill horn
color; feet slate brown. General appearance of *Spindalis pretrii*, but
having the bill heavier and throat marking unlike those of that species, besides other minor differences.
Length, 5.50; wing, 3.35; tail, 2.75; tarsus, .70; bill, .50.

HABITAT. Grand Cayman.

Spindalis zena townsendi RIDGW.

Spindalis zena townsendi RIDGW. Proc. U. S. Nat. Mus. X, p. 3 (1887).

Described as being " similar to *S. zena*, but with the back either
entirely olive or much mixed with this color, instead of being uniform deep black."

HABITAT. Abaco I., Bahamas.

The material before me tends to show that the same variation in
coloration occurs at times in *S. zena* from New Providence.

Loxigilla barbadensis Cory.

Loxigilla barbadensis Cory, Auk, III, p. 382 (1886); ib. Ibis (1886), p. 472.

Sr. Char.   Male : —General appearance of L. noctis. Upper parts dull olive brown; underparts ashy brown, palest on the throat; under tail-coverts pale rufous brown; a faint tinge of reddish brown is sometimes perceptible on the throat and in front of the eye, but is not constant, and is lacking in several specimens; quills brown, the outer webs edged with brownish white; wing-coverts edged with red brown; tail olive brown, showing numerous nearly obsolete bands, when held in the light; bill and feet dark brown, the latter nearly black.

Length (skin), .5; wing, 2.75; tail, .2; tarsus, .75; bill, .45.

The sexes are apparently similar.

Habitat.   Barbadoes.

Loxigilla richardsoni Cory.

Loxigilla richardsoni Cory, Auk, III, p. 382 (1886); ib. Ibis. (1886), pp. 472, 475.

Sr. Char.— Entire plumage dull black; no trace of rufous brown on the throat or above the eye; under wing-coverts dull black; inner web of outer tail-feather dark brown; legs and feet apparently pale.

Length (skin), 5.10; wing, 2.85; tail, 1.95; tarsus, .95; bill, .50.

Habitat.   St. Lucia.

Genus Volatinia Reich.

Volatinia " Reichenbach, Av. Syst. t. 79 (1850)."

Volatinia jacarina (Linn.).

Tanagra jacarina Linn. Syst. Nat. I, p. 314 (1766).
Passerina jacarini Vieill. Nouv. Dict. XXV, p. 14.
Spiza jacarina Tsch. Faun. Per. p. 220 (1844).
Volatinia jacarina Cab. Mus. Hein. p. 147.— Scl. P. Z. S. (1855), p. 160; ib. Cat. Am. Bds. p. 106 (1862).— Wells, List Bds. Grenada, p. 3 (1886).— Lawr. Proc. U. S. Nat. Mus. vol. IX, p. 615.

Sr. Char.   Male : — Entire plumage bluish black, distinct blue reflections; belly slightly tinged with brown; quills dark brown; a few white feathers at the carpus.

Sr. Char.   Female : — Above rusty brown; under plumage tawny brown, mottled with darker brown on the breast; quills and tail dull brown.

Length, 4.15; wing, 2; tail, 1.50; tarsus, .78; bill, .35.

The specimen above described was taken in Guiana, as I have never seen a West India example.

Recorded from Grenada.

GENUS Spermophila SWAINSON.

Spermophila "SWAINSON Zoöl. Journ. III, p. 348 (1827)."
Spermophila gutturalis (LICHT.).
" *Fringilla gutturalis* LICHT. Doubl. p. 26."
Loxia plebeia SPIX. Av. Bras. II, p. 46 (1825).
Fringilla melanocephala MAX. Beitr. III, p. 577 (1831).
Phonipara gutturalis Br. Consp. p. 494 (1850).
Spermophila gutturalis SCL. P. Z. S. (1855), p. 160; *ib.* Cat. Am. Bds. p. 105
    (1862). — WELLS, List Bds. Grenada, p. 3 (1886). — LAWR. Proc. U. S. Nat.
    Mus. vol. IX, p. 614.
Sporophila gutturalis CAB. Mus. Hein. p. 149.

SP. CHAR. *Male :* — Entire head dull black, tinged with green on the occiput;
    back and rump olive green; underparts pale yellow; quills and tail brown,
    tinged with olive on the edges; bill pale.

    A specimen in the Smithsonian Institution from Ecuador, labelled
    "female," has the entire plumage pale brownish olive, except on the
    belly, which is dull yellowish white.

    Length, 4.25; wing, .2; tail, 1.75; tarsus, .55; bill, .30.

Grenada.

Icterus bairdi CORY.

Icterus bairdi CORY, Auk, III, pp. 500, 502 (1886).

SP. CHAR. — Front of face and throat black; underparts bright yellow; back dull
    yellow, showing a faint tinge of olive on the upper back; tail and wings
    black; lesser wing-coverts bright yellow; greater secondary wing-coverts
    pure white, forming a broad white wing patch, some of the inner primaries
    delicately edged with white, showing more clearly on the inner secondaries;
    bill and feet black.

    Length, 7.25; wing, 3.75; tail, 3; tarsus, .85; bill, .85.

HABITAT. Grand Cayman.

Quiscalus caymanensis CORY.

Quiscalus caymanensis CORY, Auk, III, pp. 499, 502 (1886). — RIDGW. Proc. U.
    S. Nat. Mus. p. 574 (1887).

SP. CHAR. — General plumage purplish black, showing a greenish gloss on the
    back and rump; wing feathers showing a faint greenish gloss; quills and
    tail black; bill and feet black; iris yellow.

    Length, 9.75; wing, 5.30; tail, 4.50; tarsus, 1.25; bill, 1.10.

HABITAT. Grand Cayman.

Elainia barbadensis CORY.

Elainia martinica SCLATER, P. Z. S. p. 175 (1874). — CORY, Ibis, p. 472 (1886).
Elainia barbadensis CORY, Auk, vol. V, p. 47 (1888).

SP. CHAR. ♂ *ad.* — Upper parts dark olive, the basal portion of the feathers on
the crown white; outer webs of quills edged with dull brownish white; tail-
feathers olive brown, edged with olive on the outer webs; sides of the head
and cheeks dark olive; throat gray; breast and underparts olive gray faintly
tinged with yellow. It is larger than *E. martinica* and apparently darker.
Length, 6.25; wing, 3.55; tail, 3.30; tarsus, .95; bill, .45.

HABITAT. Barbadoes.

Elainea pagana (LICHT.).

Muscicapa pagana "LICHT. Doubl. p. 54."
Muscicapa brevirostris "MAX. Beitr. iii, p. 799."
Elainea pagana CAB. in Schomb. Guian. III, p. 701. — SCLATER, P. Z. S. 1859,
p. 46; *ib.* Cat. American Bds. p. 216 (1862).
Elainea martinica LAWR. Proc. U. S. Nat. Mus. vol. I, p. 270 (1878). — WELLS,
List Bds. Grenada, p. 4 (1886). — WELLS & LAWRENCE, Proc. U. S. Nat.
Mus. IX, p. 616 (1887).

SP. CHAR. ♂. — Above somewhat lighter in color than *E. martinica*, two distinct
yellowish white wing-bands; tertials heavily marked with white on the
outer webs; basal portion of the feathers on the crown dull white; throat
ash gray, shading into grayish olive on the breast; sides and flanks olive
gray, tinged with yellow; belly distinctly yellow; flanks olive, tinged with
yellow.
Length (skin), 6.40; wing, 3.30; tail, 3; tarsus, .90; bill, .38.

Grenada.

Grenada examples of this species apparently do not differ from
specimens from the continent and the island of Tobago.

Myiarchus oberi LAWR.

M. oberi is allied to *M. erythrocercus*, but differs from it in being
somewhat larger and darker. In a large series these characters,
though slight, seems constant. Neither *M. oberi* nor *M. erythrocercus* can be confounded with *M. tyrannulus* (Müll.), for *tyrannulus*,
as originally described by Brisson (*M. cayenensis* Brisson), has
the tail dark brown, the feathers not edged with rufous (see
p. 126).

Myiarchus denigratus CORY.

Myiarchus denigratus CORY, Auk, III, pp. 500, 502 (1886). — RIDGW. Proc. U. S. Nat. Mus. p. 574 (1887).

SP. CHAR. — Top of the head dark blackish brown; back dull olive brown; sides of the head, cheeks, and ear-coverts dark brown, slightly lighter than the crown; throat and breast ash gray, shading into dull yellowish white on the belly and crissum; quills very dark brown, some of the feathers edged with very pale rufous; tail-feathers dark brown, edged with pale rufous on the inner webs; bill and feet black.

Length, 6.25; wing, 3.15; tail, 3; tarsus, .80; bill, .60.

HABITAT. Grand Cayman I. The type is in my collection.

Myiarchus berlepschii CORY.

Myiarchus berlepschii CORY, Auk, V, p. 266 (1888).

SP. CHAR. — Top of the head dark brown; back and rump distinctly dark olive, shading into rufous on the tips of the upper tail-coverts; throat pale ashy, somewhat darker on the upper breast; rest of underparts pale yellow; primaries and secondaries dark brown, edged with dull rufous, the rufous color wanting on the terminal portion of the four outer primaries; the primaries show pale rufous on the basal portion of the inner webs; all the secondaries and tertials are broadly edged with pale rufous on the inner webs; tail dark brown; all the tail-feathers, excepting the two middle ones, have nearly the entire inner web bright rufous; a narrow line of brown separates the rufous from the shaft of the feather; bill dark brown; the feet are black.

Length, 7; wing, 3.50; tail, 3.25; tarsus, .95; bill, .65.

HABITAT. Island of St. Kitts, West Indies.

The underparts of this species are similar in marking and general coloration to *M. oberi;* but the throat is more ashy and lighter, and the yellow of the belly is paler. The back and head of *M. berlepschii* are darker, and the bird is smaller.

Blacicus flaviventris LAWR.

Blacicus flaviventris LAWRENCE, Proc. U. S. Nat. Mus. vol. IX, p. 617 (1887).

SP. CHAR. — "Upper plumage dark; hair brown, deeper in color on the crown; tail colored like the back; quills brownish black; wing-coverts dark brown, edged with clear pale rufous; throat of a dull pale fulvous; breast and abdomen pale yellow, intermixed with dusky; upper mandible black, the under clear light yellow; feet black. In size, about the same as

B. brunneicapillus. Wing, 2.50; tail, 2.50; tarsus, .50; bill, .50." (LAWR. orig. descr. l. c.)

HABITAT. Grenada.

Supposed to differ from *B. brunneicapillus* in being yellow on the abdomen instead of fulvous, and the tail not being tipped with fulvous.

Blacicus martinicensis CORY.

Blacicus martinicensis CORY, Auk, p. 96 (1886).

SP. CHAR.—Top of the head smoky black; back and upper tail-coverts dark olive; throat ashy, becoming tinged with tawny brown on the breast; belly dull rufous brown, extending upon the under tail-coverts; tail dark brown; upper mandible black; under mandible pale yellow; feet dark.

Length, 4.80; wing, 2.60; tail, 2.50; tarsus, .60; bill, .50.

HABITAT. Martinique.

Chætura cinereiventris SCL.

"*Cypselus acutus* MAX. Beitr. III, p. 351 (1831) (nec. Auct.").
Acanthylis spinicauda BURM. Syst. Ueb. II, p. 366.
Chætura cinereiventris SCL. Cat. Am. Bds. p. 283 (1862); *ib.* P. Z. S. (1863), p. 101; *ib.* (1865), p. 612.—WELLS, List Bds, Grenada, p. 4 (1886).

SP. CHAR.—Top of the head and back greenish black; rump gray; the shafts of the feathers dark, becoming pale at the tips; tail-feathers greenish black; spines extending about one quarter of an inch beyond the feathers; throat whitish, shading into gray on the breast and belly, faintly tinged with brown near the vent; under tail-coverts dull greenish black; quills dark brown.

Length, 4.20; wing, 4.30; tail, 1.50; tarsus, .40.

Grenada.

The specimen before me does not perfectly agree with another, labelled *C. cinereiventris*, from South America; but in the absence of more material for comparison, I refer it, provisionally, to that species. A bird in the Smithsonian collection, No. 84,841, is labelled *C. cinereiventris*. It was taken in Grenada in July, 1881, by Mr. J. G. W

Chætura brachyura (JARD.).

Acanthylis brachyura "JARDINE, Am. N. H. ser. I, XVIII, p. 120 (1846)."
Chætura poliura CORY, Ibis, p. 473 (1886), LAWR.

SP. CHAR. ♂.—Top of the head, back and wings black, showing a greenish gloss when held in the light; rump and upper tail-coverts ash gray, the tail-coverts extending to the tips of the tail-feathers; tail gray, darker than the coverts, and having the shafts black; underparts dull black, showing a slight greenish tinge when held in the light; chin brownish, showing lighter than the throat.

Length (skin), 4.25; wing, 5; tail, 1.25; tarsus, .38; bill, .15.

St. Vincent and Grenada.

This is apparently a good species, and is most certainly distinct from the *C. paliura* Temm. The upper tail-coverts are much elongated, reaching the tips of the tail-feathers. I have examined five specimens from St. Vincent and one from Grenada, in all of which the characters were constant.

Centurus caymanensis CORY.

Centurus caymanensis CORY, Auk, III, pp. 499, 502 (1886). — RIDGW. Proc. U. S. Nat. Mus. p. 574 (1887).

SP. CHAR. *Male:* — Forehead and sides of the head white; crown and nape bright crimson red; a tinge of red at the nostrils; throat dull white, shading into brownish white on the breast and belly; a patch of crimson red at the vent; back banded with dull white and black; rump dull white, marked irregularly with black; quills dark brown, heavily blotched with white on the basal portion of the inner webs; secondaries, and some of the inner primaries, heavily marked with white; tail brownish black, faintly tipped with tawny brown, the outer pair and two central feathers blotched with dull white; bill and feet black.

The female is similar to the male, but lacks the red crown, having only a nuclear patch of that color.

Length, 8.60; wing, 5; tail, 4; tarsus, .75; bill, 1.50.

This species lacks the black superciliary mark which is found in both the Cuban and Bahama species.

HABITAT. Grand Cayman.

Centurus nyeanus (RIDGW.).

Centurus nyeanus RIDGW. Auk, III, p. 336 (1886).

SP. CHAR. — " Similar to *C. superciliaris* (TEMM.) of Cuba, but much smaller; the white bars of upper parts and gray of lower parts almost entirely devoid of yellow tinge; red of belly and black superciliary spot more restricted, and outer webs of middle tail-feathers without spots.

" *Adult male* (type No. 107,996, U. S. Nat. Mus. Waitling's I., Bahamas, March 5, 1886, Willard Nye): — Frontlet bright scarlet, paler anteriorly and along lower edge; forehead (for about .30 of an inch back from base of culmen), lores, suborbital region, and auriculars white, the latter with a faint buffy grayish tinge; crown, occiput, and hind neck bright crimson scarlet, lighter posteriorly; back, scapulars, and rump barred with black and dull white, the two colors in about equal amount, the bars of each averaging about .08 of an inch in width; wing-coverts more broadly barred with black and pure white; alulæ and primary coverts uniform black, the exterior feather of the former with some white along edge; primaries black, irregularly spotted with white toward base, and more or less broadly tipped with white; upper tail-coverts white, rather distantly and irregularly barred with black; tail black, the inner webs of intermediæ marked with oblique quadrate spots of white, the outer webs with an irregularly wedge-shaped streak of white on basal half (chiefly concealed by coverts); exterior pair barred or transversely spotted with white on terminal portion; chin and upper part of throat grayish buffy white, gradually deepening into light buffy grayish on lower throat; foreneck, sides of neck and chest deeper grayish, this color assuming decidedly more of a buffy tinge on the breast, upper part of belly, and on flanks; central lower part of belly dull scarlet, the adjacent portions, including lower tail-coverts, dingy white, marked with V-shaped bars of blackish; bill entirely black; feet dusky. Wing, 5.20; tail, 4.00; culmen, 1.50; tarsus, .92." (RIDGW. orig. descr. l. c.)

HABITAT.　Wattling's Island, Bahamas.

Centurus blakei RIDGW.

Centurus blakei RIDGW. Auk, III, p. 337 (1886).

SP. CHAR. — "Similar to *C. nyeanus*, but much darker; the forehead pale drab, or light grayish buff, instead of pure white; auriculars deep light drab; foreneck and chest olivaceous drab, and lighter bars of back; scapulars and rump light dingy buff, instead of nearly pure white; frontlet dull orange red, instead of pure vermilion or scarlet.

"*Adult male* (type No. 108,618, Abaco I., Bahamas, April 2, 1886, Charles H. Townsend) : — Wing, 5.35; tail, 3.90; culmen, 1.45; tarsus, .90.

"*Adult female* (type No. 198,619, same locality, April 6, 1886, Willard Nye) : — Similar to the male, but crown and occiput ash gray, becoming gradually lighter anteriorly, the hinder portion (connecting superciliary spots) spotted with black; frontlet merely tinged with orange, and grayish of lower parts, as well as white bars of back, etc., much less strongly tinged with yellowish. Wing, 5.25; tail, 3.80; culmen, 1.35; tarsus, .85." (RIDGWAY, l. c.)

HABITAT.　Abaco, Bahamas.

Coccygus maynardi RIDGW.

Coccygus maynardi RIDGW. Man. N. A. Bds. p. 274 (1887).

SP. CHAR. — Smaller and paler than *C. minor*; underparts buff or ashy, not ochraceous, as in *minor*.

Length, 11.90; wing, 5.10; tail, 6.40; bill, 1.05.

HABITAT.　Bahama Islands.

Chrysotis caymanensis CORY.

Chrysotis caymanensis CORY, Auk, III, pp. 497, 502 (1886).

SP. CHAR. — Resembles *C. leucocephalus*, but differs from that species in the markings of the throat and head; forehead dull yellowish white, not reaching the eye; feathers of the head bluish green, tipped and edged with dull black; cheeks bright crimson red, the feathers mixed with green in some specimens; throat pale red, the feathers broadly edged with yellow; underparts bluish green, marked with dull red on the sides and belly; the feathers of the breast and belly edged with dull black; quills heavily edged with blue on the outer webs, inner webs dark brown; tail yellowish green, the outer web of outer feather pale blue, the basal portion of inner webs heavily marked with dark red.

Length, 11; wing, 8; tail, 5; tarsus, .70; bill, 1.05.

HABITAT. Grand Cayman.

Columbigallina passerina insularis RIDGW.

Columbigallina passerina insularis RIDGWAY, Proc. U. S. Nat. Mus. p. 574 (1887).

In the very large series which I have examined from the Bahamas, Grand Cayman, and elsewhere, I can find no differences which appear to be at all constant or restricted to any one locality. Mr. Maynard separated the Bahama bird under the name of *C. bahamensis* (Maynard, American Exchange and Mart, Vol. III, No. 4, p. 69, Feb. 5, 1887, Boston), describing it as " similar to the common ground dove, but with the bill constantly wholly black, and much smaller and paler."

Mr. Ridgway separates the Grand Cayman bird (l. c.) as *C. passerina insularis*, describing it as very similar to *C. bahamensis* Mayn, "but larger, and with the basal half (or more) of bill distinctly orange or yellowish." Part of the series examined is contained in my own collection, and comprises forty-one specimens from Grand Cayman and Cayman Brack, fifty-two from the Bahama Islands, and sixty-seven from Florida, Georgia, and the Greater and Lesser Antilles.

Vanellus vanellus (LINN.).

Tringa vanellus LINN. Syst. Nat. I, p. 148 (1758); *ib.* Syst. Nat. I, p. 248 (1766).
Vanellus cristatus WOLF & MYER, Hist. Nat. Ois. de. l'Allem, p. 110 (1805). — FEILDEN, Zoölogist, p. 301 (1888) (Barbadoes).

Vanellus vulgaris, BECHST. Orn. Taschenb, II, p. 313 (1803).— DRESSER, Hist.
　　Bds. Europe, VII, p. 28 (1871–1881).
Vanellus vanellus LICHT. Nom. Mus. Berol. p. 95 (1854). — A. O. U. Code and
　　Check List N. A. Bds. p. 160 (1886).

A single specimen of the European Lapwing has been recorded
from the West Indies, a bird of this species having been taken at
Barbadoes, Dec. 24, 1888. (Feilden, Zoölogist, p. 301, 1888.)

Ardea bahamensis BREWSTER.

Ardea bahamensis BREWSTER, Auk, V, p. 83 (1888).

SP. CHAR. — "Smaller than *A. virescens;* the general coloring much paler,
browner, or yellower, and more uniform; the forehead strongly tinged with
brownish; the light edging of the secondaries broader; the dorsal plumes
and rump only slightly, sometimes not at all, greenish; top of head dark
dull green, strongly tinged with brownish on the forehead; throat, jugulum,
and foreneck creamy white, with dusky spotting on the jugulum; remainder
of head and neck light chestnut, approaching cinnamon in places; forepart and sides of back rusty cinnamon; rump and most of upper tail-
coverts drab; dorsal plumes dull greenish, the central ones glaucous, with
a tinge of lilac; wings and tail dull green, the wing-coverts edged broadly
on both webs, the secondaries more narrowly on the outer webs only, with
rusty or whitish under wing-coverts; breast, abdomen, crissum, and sides
of the body light yellowish drab." (BREWSTER, orig. discr. l. c.)

HABITAT. Bahama Islands (Rum Kay, Watling's I., Abaco I.).

Hydrochelidon leucoptera (TEMM.).

This species is given by Mr. Ridgway as accidental in the West
Indies. (Man. N. A. Birds, p. 47, 1887.)

Stercorarius parasiticus (LINN.).

Larus parasiticus LINN. Syst. Nat. I, p. 136 (1758).
Stercorarius parasiticus SCHÄFF. Mus. Orn. p. 62 (1789). — A. O. U. Code and
　　Check List of N. A. Bds. p. 85 (1886).
Stercorarius crepidatus FEILDEN, Zoölogist, p. 350 (1888).

Col. H. W. Feilden records this species from Barbadoes, a single
specimen taken alive on July 10, 1888.

L ANTIC OCEA

20

VIRGIN ISLANDS

PORTO RICO
Culebra

Anegada

Tortola   Virgin Gorda

Sombrero

Desecheo
Mona

St John
Thomas

Dog I.

Anguilla
Eual I.
St Bartholomew

Vieque or
Crab I.

St Martins

Saba
St Eustatius
St Christopher
or St Kitts

Barbuda

Santa Cruz
or St Croix

Nevis

Antigua

I. Redond

Montserrat

Grande Terre I.
Desirade
Petite Terre I.
Marie-galante

Guadeloupe

L E S S E R

Ares I.

Les Saintes

Dominica

15

Martinique

St Lucia

St Vincent
Bequia
I. Canouan
I. Union
Carriacou

Barbadoes

A N T I L L E S

Grenada

Aves I.
Los Roques
Orchila

Blanquilla
Los Hermanos

L E E W A R D   I S L A N D S

Tobago

Tortuga

Margarita

0

TRINIDAD

10

65

60

C A R I B B E E   I S

W I N D W A R D   I S L A N D S

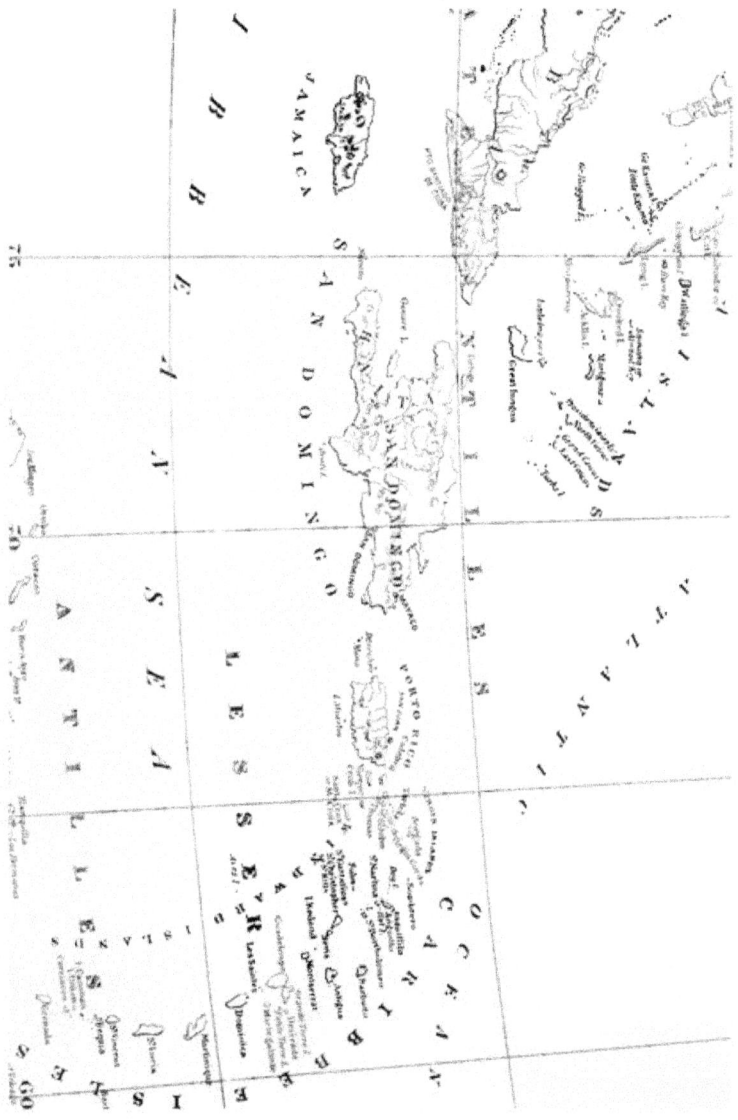